International Council of Archaeozoology (ICAZ)
Proceedings of the 9th conference of the ASWA (AA) Working Group
Archaeozoology of SouthWest Asia and Adjacent Areas

Archaeozoology of the Near East 9

In honour of Hans-Peter Uerpmann and François Poplin

edited by

Marjan Mashkour and Mark Beech

Volume 2

OXBOW | books
Oxford & Philadelphia

Published in the United Kingdom in 2017 by
OXBOW BOOKS
The Old Music Hall, 106–108 Cowley Road, Oxford, OX4 1JE

and in the United States by
OXBOW BOOKS
1950 Lawrence Road, Havertown, PA 19083

Hardcover Edition: ISBN 978-1-78297-844-2
Digital Edition: ISBN 978-1-78297-845-9 (epub)

A CIP record for this book is available from the British Library

Library of Congress Cataloging-in-Publication Data

Names: International Symposium on the Archaeozoology of Southwestern Asia and Adjacent Areas (9th : 2008 : Al Ain, United Arab Emirates) | Mashkour, M. (Marjan), editor. | Beech, Mark J., editor.
Title: Archaeozoology of the Near East 9 : proceedings of the 9th Conference of the ASWA (AA) Working Group : archaeozoology of Southwest Asia and adjacent areas / edited by Marjan Mashkour and Mark Beech.
Description: Oxford ; Philadelphia : Oxbow Books, 2016. | Includes bibliographical references. | Description based on print version record and CIP data provided by publisher; resource not viewed.
Identifiers: LCCN 2016040915 (print) | LCCN 2016032516 (ebook) | ISBN 9781782978459 (epub) | ISBN 9781782978473 (pdf) | ISBN 9781782978466 (mobi) | ISBN 9781782978442 (hardback) | ISBN 9781782948459 (digital edition)
Subjects: LCSH: Animal remains (Archaeology)--Middle East--Congresses. | Hunting and gathering societies--Middle East--Congresses. | Prehistoric peoples--Middle East--Congresses. | Middle East--Antiquities--Congresses.
Classification: LCC CC79.5.A5 (print) | LCC CC79.5.A5 I58 2008 (ebook) | DDC 930.1--dc23
LC record available at https://lccn.loc.gov/2016040915

Printed in the United Kingdom by Short Run Press, Exeter

For a complete list of Oxbow titles, please contact:

UNITED KINGDOM
Oxbow Books
Telephone (01865) 241249, Fax (01865) 794449
Email: oxbow@oxbowbooks.com
www.oxbowbooks.com

UNITED STATES OF AMERICA
Oxbow Books
Telephone (800) 791-9354, Fax (610) 853-9146
Email: queries@casemateacademic.com
www.casemateacademic.com/oxbow

Oxbow Books is part of the Casemate Group

Front cover: Baynunah Camel site – Abu Dhabi Tourism & Culture Authority (TCA Abu Dhabi) – United Arab Emirates

Contents

VOLUME 2

PART 4. PASTORALISM, NOMADISM AND MOBILITY

PART 5. EXPLOITATION OF ANIMALS IN THE ARABIAN PENINSULA

PART 6. RITUALS AND ANIMAL DEPOSITS

Part 4

Pastoralism, nomadism and mobility

17. Fish and mammal bones in the Abu Dhabi desert: evidence for Bedouin diet during the pre-oil era

Mark J. Beech, Hanae Sasaki, Tatsuo Sasaki, Walid Yasin Al-Tikriti and Mohammed Amer Al-Neyadi

Archaeological research in south-east Arabia has rarely concentrated on the recent historical period. This paper presents the results of an investigation of three Bedouin campsites located on the traditional migration route between Abu Dhabi and Liwa in the United Arab Emirates. Analysis of pottery from these sites dates them to between the latter half of the 18th century to the 20th century. Investigation of the animal bones from these sites provides a new insight into the animals accompanying the people on these regular migrations, as well as demonstrating the diet of pre-oil era populations in Abu Dhabi emirate.

Keywords Bedouin, diet, desert archaeology, dried fish, camels, caprines

Introduction

Archaeological research in south-east Arabia has rarely concentrated on the recent historical period. This is surprising considering the lack of written accounts documenting the life of traditional Bedouin communities inhabiting the deserts of south-east Arabia.

Our understanding of the life of traditional Bedouin society in this region is primarily dependent on photographs and written accounts made by foreign visitors who either lived or passed through the region. These included the German traveller Hermann Burckhardt who took some of the earliest recorded photographs in 1903 of Sheikh Zayed Bin Khalifa. The British explorer Wilfred Thesiger who crossed the Empty Quarter twice between 1945 and 1949 described in some detail the vanishing way of life of the Bedouin, vividly described in his classic travel book 'Arabian Sands' (Thesiger 1959). Thesiger took many photographs during his travels and donated his vast collection of negatives to the Pitt Rivers Museum in Oxford. A valuable series of photographs were also taken by the British oil company executive Ronald Codrai, who worked in the Trucial States from 1948 to 1955, and who then frequently visited the country until just before it became the United Arab Emirates in 1971 (Codrai 1990). Codrai managed to photograph all the rulers of the seven Emirates, as well as taking many important photos documenting aspects of daily life at that time.

Our knowledge of the historical past is supplemented by the more recent accounts of UAE-based historians (e.g. Heard-Bey 1996) and UAE nationals who have written memoirs reminiscing about their childhoods (e.g. Al Fahim 1998).

These photographs and accounts provide some clues as to the dress and general attire of these communities, as well as everyday activities such as tending camels, preparation and drinking of coffee, falconry, the collection and transportation of water, the general appearance of *'arish* (palm frond) type house structures, harvesting of dates, fishing-related activities, pearling, selling items in the souq, etc.

Some of these observers, like Thesiger, have glorified the nobility of Bedouin life, whilst more recently local cultural and media organisations have often used air-brushed images to create a romanticised nostalgic view of the past. In actual fact, older Emiratis who lived in the 1940s and 1950s talk about how hard life was, and mention the importance of small industries like firewood, charcoal, chickens, goats and vegetables, that allowed them to survive (Bristol-Rhys 2011). There have been some studies looking at travellers observations of Bedouin food (cf. e.g. Iddison 1996).

What do we actually mean when we use the generic term 'Bedouin'? It is unfair to characterise the entire population of the region as being desert nomads. The situation was of course much more complicated. Some communities were settled in economically viable locations, others were nomadic groups, combining herding, hunting and fishing. A key point is the seasonal nature of the climate in south-east Arabia. The extreme heat and humidity of coastal areas during the summer months led part of the population of Abu Dhabi to switch to the less humid interior oases, such as Al Ain and Liwa (Cordes & Scholz 1980; Heard-Bey 1996). This of course coincided with the beginning of the season for harvesting dates from the oasis date palm gardens. Pearling activities, however, required that part of the coastal communities remained on the coast during the summer months, when pearling was traditionally carried out. Trade and commerce also meant that some merchants remained in coastal ports. Another important phenomenon to consider in the more recent historical period is the growth of largely urbanised populations, who live with the trappings of urban life, but whom still annually migrated to the interior oases at the onset of the hot summer season.

Bedouin campsites in the Abu Dhabi desert

Since the mid-1990s both the Abu Dhabi Islands Archaeological Survey (ADIAS) team, as well as the former Al Ain Department of Antiquities and Tourism, were aware of the existence of a number of Bedouin campsites located between Abu Dhabi and Liwa, approximately 65km south-east of Abu Dhabi, the capital of the United Arab Emirates. These were located in two areas known as Tawi Beduwa Shwaiba and Muray (Fig. 17.1). Various local UAE nationals had brought to their attention the presence of complete pottery vessels collected in an area of sand dunes where hand-dug wells had been exploited, and where scatters of pottery, mammal and fish bone, and marine shells were observed. Some of these complete pottery vessels were curated by both aforementioned organisations, which are now integrated within the Abu Dhabi Tourism and Culture Authority (TCA Abu Dhabi).

In late April to early May 2008 a team of archaeologists from Japan's Kanazawa University and the former Abu Dhabi Authority for Culture and Heritage (ADACH) carried out a more detailed investigation of a number of these sites. This paper concerns an analysis of the fish and mammal bones retrieved from these remarkable historic desert campsites.

Methods

The survey and excavation team included the following members: Department of Archaeology, Kanazawa University, Japan: Professor Tatsuo Sasaki and Dr Hanae Sasaki; Historic Environment Department at the former Abu Dhabi Authority for Culture and Heritage (ADACH), now known as the Abu Dhabi Authority for Tourism and Culture (TCA Abu Dhabi): Mohammed Amer Al Neyadi (Head of Department), Dr Mark Beech (Head of Coastal Heritage and Palaeontology section), Dr Walid Yasin Al-Tikriti (Head of Archaeology section), Ali Al-Meqbali, Ahmed Al-Haj, Dia'eddin Tawalbeh, Hamdan Al-Rashedi, Mohammed Al-Dhaheri and Jaber Al-Merri (Archaeologists).

The three main sites investigated are difficult to access and required skilled off-road driving in the sand dunes to approach the sites. Nissan Patrol 4WD cars were used to approach the sites. It was necessary to deflate the tyres to provide better traction in the loose sands of the area, as well as to have safety and rescue equipment, including tow ropes, steel cable winches and shovels.

Local UAE nationals, who had been travelling through the area, first identified the sites investigated. Baqit Saeed Al Mansouri acted as a local guide helping us navigate to these sites. Initial work at the site in late April 2008 concentrated on GPS mapping of the precise location of these sites. Co-ordinates were taken using a Garmin handheld GPS in decimal degrees with WGS84 as the datum.

Samples of pottery, fish bones, mammal bones and marine shells were collected from the surface of each of the three sites under investigation. These were all random grab samples, with the exception of a small quantity of material collected during the excavation of the test trenches at Muray site 1. As the sample size is quite small all this material is simply treated on a site-by-site basis. The methodology adopted for studying the fish bones is that defined earlier by Beech (2004).

Location

The three principal sites investigated during this survey were Muray site 1 (also known as Bad Al Muray), Muray site 2, and Tawi Beduwa Shwaiba (Fig. 17.2).

Muray site 1 (Bad Al Muray) is located at 54.64250 E, 23.87444 N, approximately 70km south-east of Abu Dhabi, the capital city of the United Arab Emirates. The site was first discovered by a UAE national in December 2007. A team from the Historic Environment Department, including Ahmed El Haj, at the former Abu Dhabi Authority for Culture and Heritage visited the site during the same month, and collected some initial samples of pottery sherds from the site. The joint Japanese – ADACH team then made a survey on 28 April 2008, followed by trench excavations from 5–6 May 2008. A further pottery scatter covering an area of about 20×10m, consisting of earthenware potsherds and fish bones, was noted to the south-west of Muray site 1 on 6 May 2008. This was designated as Muray site 1 – area C.

Muray site 2 is located at 54.70105 E, 23.88772 N, just

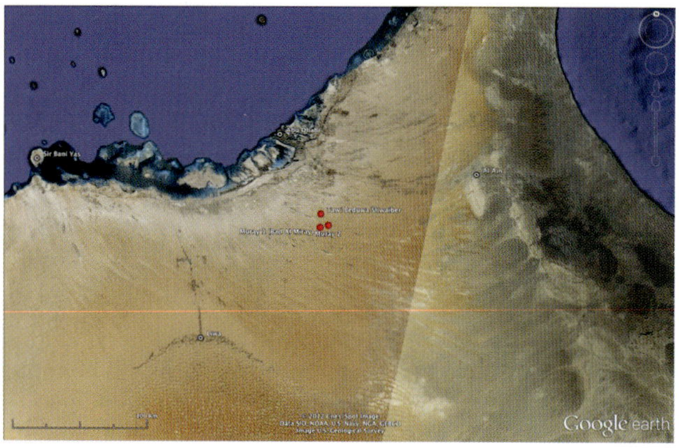

Fig. 17.1. Location of Tawi Beduwa Shwaiba, Muray site 1 (Bad Al Muray) and Muray site 2 in Abu Dhabi emirate. Scale bar is 100km.

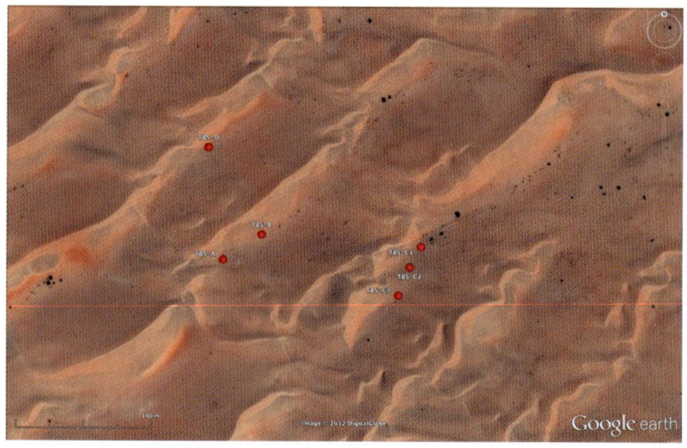

Fig. 17.3. Tawi Beduwa Shwaiba. Note the relative position of the sub-site areas: TBS-A, TBS-B, TBS-C1, TBS-C2, TBS-C3 and TBS-D. Scale bar is 100m.

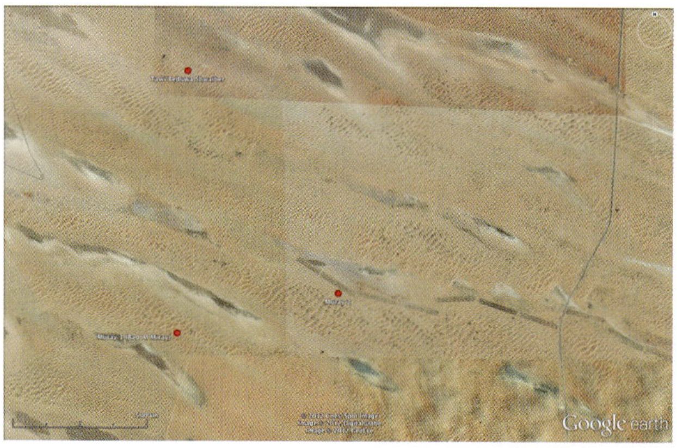

Fig. 17.2. Location of the sites studied. Note the locations of all three sites within the sand dunes. Scale bar is 5km.

Fig. 17.4. View of Muray site 1.

6km north-east of Muray site 1 (Bad Al Muray). The site was first discovered by a local UAE national in December 2007. The joint Japanese – ADACH team surveyed the site on 29 April and 6 May 2008.

Tawi Beduwa Shwaiba is located at 54.64781 E, 23.96403 N, approximately 60km south east of the city of Abu Dhabi. The site was first noted in 1992 in a brief visit by the former Al Ain Department of Antiquities and Tourism, who collected some ceramic finds from the site which are curated in the collections of Al Ain National Museum. In 1997 Mubarak Mohammed Al Mansouri, a UAE national who had known the area since his childhood, found a number of pottery vessels which had appeared due to recent dune movement. He collected the vessels and brought them to Sheikh Mohammed bin Butti Al Hamed, who at that time was the Governor of Al Gharbia. Escorted by Mubarak, a team from the Abu Dhabi Islands Archaeological Survey (ADIAS) then visited the site in September 1997. A number of complete vessels, as well as samples of potsherds, were collected and curated within the ADIAS archives. A preliminary assessment suggested that these were all of Late Islamic date, probably dating from the 17th century AD onwards. Mubarak Mohammed Al Mansouri described the site as having been 5 days camel ride from Liwa en route to Abu Dhabi. He remembered it as having been in use during his childhood, around 50 years ago, and that he had been informed that it had been in use as far back as the time of his grandfather's grandfather. A more detailed investigation and report were subsequently issued by the ADIAS team (ADIAS Newsletter 1997; Hellyer 1998). The joint Japanese–ADACH team re-surveyed Tawi Beduwa Shwaiba on 1–7 May 2008. More recent dune movements had exposed further traces of the archaeological remains. Four different areas were identified within the site, designated as areas A, B, C and D (Fig. 17.3).

Muray site 1 (Bad Al Muray)

Scattered ceramic sherds and animal bones (both mammal and fish bones) were discovered spread over an area measuring 54m north–south by 50m east–west. This was located on the lower southern slope of a sand dune inclined down gently to the north, and on the surface of an interdunal hollow besides a pronounced high sand dune to the north

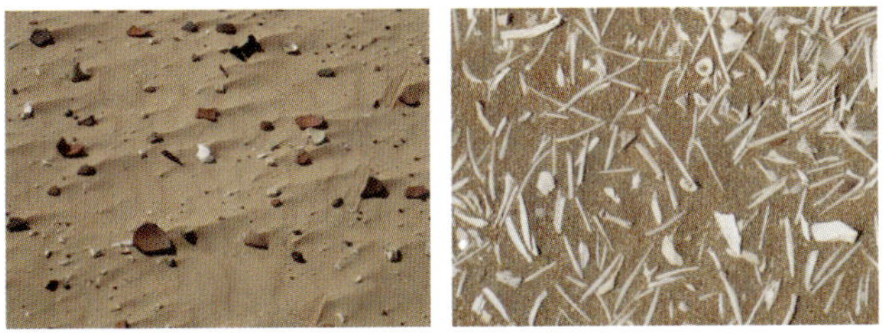

Fig. 17.5. Close-up view of the occupation deposits at Muray site 1.

Fig. 17.6. View of the two test trenches excavated at Muray site 1.

Fig. 17.7. View of Muray site 2.

Fig. 17.8. View of Tawi Beduwa Shwaiba, site TBS-B.

(Fig. 17.4). Sand dunes located to the north and south were about 10–15m in height from the base of the interdunal hollow. Vegetation was very sparse in this location and no trees were found around the site.

Many small scattered fragments of ceramics were noted on the surface, as well as small fish bones and scales (Fig. 17.5). A number of mammalian bones were also noted. These were numerous small sherds of unglazed reddish ware and small sherds of glazed ceramics. Unglazed sherds consisted of cooking pots and various jars used for preserving food items. One complete earthenware basin was found on the southern slope, almost at the southern limit of the pottery scatter. Large bowls of Iranian under glazed painted ware are dominant amongst the glazed wares. Brown glazed vases and large bowls from Bahla in Oman are the next most common type. Yellow glazed bowls and green glazed bowls are present in small quantities. Chinese blue-and-white bowls and dishes from Fujain province are present. These were typical export wares during the 19th century. Chinese small cups for coffee dating from the 20th century were also found. European ceramics such as bowls and plates from Holland and Britain were also recorded. Overall, the ceramic assemblage suggests that the site was occupied during the Late Islamic period, some time between the latter half of the 18th century right up until the mid-20th century.

Glass vessels were quite rare at the site, and other finds included only a few glass bangles and bronze rings. A single coin was also recovered, a Bahraini bronze coin dating to 1965.

The site appears to be an Islamic campsite, perhaps utilised by one family or several families. A local informant told us several interesting pieces of information. Namely, that 'water could once be found at a depth of only 2m below the interdunal hollow'; that date palms had once existed in this location; that the Manasir tribe had lived here approximately 100–130 years ago; and that the women had stayed here, whilst the men went to the coast during the winter time for fishing and pearling.

The test trenches excavated at Muray site 1 (Bad Al Muray) revealed that archaeological remains were only scattered directly on the surface, no stratified remains being recovered from buried layers. The profiles of the excavated trenches demonstrated a series of thin natural windblown accumulated layers beneath the surface finds (Fig. 17.6).

Table 17.1 Fishes represented at Muray and Tawi Beduwa Shwaiba

Taxon	*Common Name*	*Muray 1*	*Muray 2*	*Tawi Beduwa Shwaiba*
Chondrichthyes				
Indet	Shark/Ray/ Skate	4	0	2
Belonidae				
Indet	Needlefish	0	1	0
Serranidae				
Epinephelus sp.	Grouper	5	1	52
Lethrinidae				
Lethrinus sp.	Emperor	0	0	1
Sparidae				
Acanthopagrus sp.	Seabream	10	15	8
Argyrops Spinifer	King Soldier-bream	4	3	2
Rhabdosargus sp.	Haffara Seabream	12	3	5
Indet		11	5	6
Scombridae				
Thunnus sp.	Tuna	2	0	1
Osteoichthyes				
Indet		3	11	19
Total		51	39	96

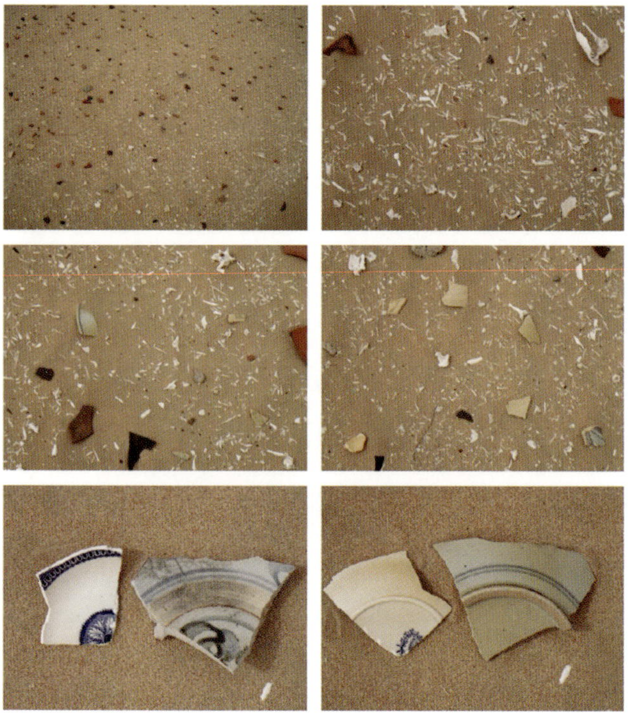

Fig. 17.9. Close-up view of the occupation deposits at Tawi Beduwa Shwaiba, site TBS-B.

Muray site 2

The site was first discovered by a local man in December 2007. The joint Japanese–ADACH team then surveyed the site on 29 April and 6 May 2008. Muray site 2 is located in similar terrain to Muray site 1. A scatter of ceramics, fish bones, camel and goat bones, approximately 26m north–south by 10m east–west, was identified on the surface of a narrow interdunal area amidst sand dunes (Fig. 17.7). The height of the surrounding sand dunes are approximately 10m above the level of the interdunal area.

There were small scattered fragments of earthenware cooking pots, basins and jars, as well as animal bones, visible on the surface. Small fish bones are scattered with grey ash on the surface of the sand. The scatter of earthenware ceramics is relatively small, as are the quantity of Chinese ceramic cups and dishes. Overall, the ceramic assemblage appears to be consistent with the occupation of the site being during the Late Islamic period, i.e. during the 19th–20th centuries. The site appears to be a short-term campsite used by a small group of people, perhaps a family.

Tawi Beduwa Shwaiba

The site of Tawi Beduwa Shwaiba was first noted in 1992 and then an initial survey of the site was carried out in 1997 (see above for details). The joint Japanese–ADACH team re-surveyed Tawi Beduwa Shwaiba in early May 2008. Four different areas were identified within the site, designated as areas A, B, C and D.

Areas A and B were two scatters of ceramics and animal bones, about 25m apart, recorded towards the south-western part of the site (Fig. 17.8). Area A consisted of a surface of scatter of ceramics and dense fish bones, measuring 12m north–south by 9m east–west, spread across a flat area located besides a low sand dune about 7m high to the north.

Area B was a similar scatter of potsherds, mammalian and fish bones located about 50m to the north-east of area A (Fig. 17.9). The scatter was spread over an area measuring 20m north–south by 50m east–west, and was situated on the gently sloping side of the northern dune and its adjacent interdunal area. To the north of the scatter lay a high sand dune approximately 10–15m in height.

Area C was located about 200m east of area B beyond one large sand dune. This area was subsequently sub-divided into three sub-areas, designated as TBS-C1, TBS-C2 and TBS-C3. These three scatters of ceramics and fish bones were all located amongst an area of low sand dunes approximately 10–15m in height. Scatter TBS-C1 was located at the northern end of the cluster and measured about 9m north–south by 10m east–west. Scatter TBS-C2 was just 4m or so south of TBS-C1 and measured about 10m north–south by 12m east–west. Finally, scatter TBS-C3 was located at the southern end of the cluster, about 20m south-west of TBS-C2, and measured 9m north–south by 10m east–west.

Area D consisted of a rather sparse scatter of animal

Table 17.2 Fish anatomical elements represented at Muray and Tawi Beduwa Shwaiba

	Epinephelus sp.	*Lethrinus* sp.	*Acanthopagrus* sp.	*Argyrops spinifer*	*Rhabdosargus* sp.	*Thunnus* sp.	Indet	Total
MURAY SITE 1	*5*	*0*	*10*	*4*	*12*	*2*	*18*	*51*
Dentary	1		2	4	8		1	16
Maxilla			2		3			5
Premaxilla			6		1		4	11
Quadrate	1							1
Cleithrum	1							1
Spine frag							1	1
1st Vertebra							2	2
Abdominal vertebra	1						1	2
Caudal vertebra							2	2
Penultimate vertebra						1		1
Vertebra	1					1	6	8
Fragment							1	1
MURAY SITE 2	*1*	*0*	*15*	*3*	*3*	*0*	*11*	*33*
Neurocranium			1					1
Dentary				3	1		1	5
Maxilla					1			1
Premaxilla			6					6
Quadrate							1	1
Hyomandibular			1					1
Cleithrum	1		2					3
Opercula			3		1		2	6
Otolith			2					2
Spine frag							1	1
Scale							2	2
Abdominal vertebra							1	1
Caudal vertebra							1	1
Vertebra							1	1
Fragment							1	1
TAWI BEDUWA SHWAIBA	*52*	*1*	*8*	*2*	*5*	*1*	*19*	*88*
Vomer	2							2
Neurocranium	2							2
Articular	5							5
Dentary	8	1	1	2	2			14
Maxilla							1	1
Premaxilla	4		5		3			12
Quadrate	1		1					2
Opercular							1	1
Preopercular	1							1
Cleithrum	3							3
First vertebra			1				1	2
Abdominal vertebra	13						1	14
Caudal vertebra	13					1	4	18
Vertebra							2	2
Fragment							9	9

bones and sporadic ceramics, measuring 10m north–south by 30m east–west, located at the northern end of the site.

Overall, the analysis of the ceramic assemblage from Tawi Beduwa Shwaiba indicates that the site was occupied during the Late Islamic period, some time between the latter half of the 18th century and the mid-20th century. The site appears to be a large campsite of nomadic families and may have been used as an important way station on the transportation route between Abu Dhabi and Liwa.

Fish Bones

A total of 135 fish bones were examined from the three sites (Table 17.1). Fishes represented included some sort of shark, ray or skate, represented by indeterminate vertebra fragments, needlefish (Belonidae), grouper (Serranidae: *Epinephelus* sp.), the fish known locally in Arabic as '*hamoor*', emperors (Lethrinidae: *Lethrinus* sp.), known locally in Arabic as '*shaeri*', several types of seabream (Sparidae), including the genera *Acanthopagrus* and *Rhabdosargus*, as well as the king soldierbream (*Argyrops spinifer*), known locally in Arabic as '*kofar*'. These fish were in most cases represented by both cranial and vertebral elements (Table 17.2). This indicates that the majority of the fish were arriving to the site in a relatively complete whole state, inferring that the whole fish was salted and/or dried and brought to the site. Based on the relative size of diagnostic cranial fragments it was possible to reconstruct

Table 17.3 Size of the fishes at Muray and Tawi Beduwa Shwaiba, based on size reconstruction from diagnostic elements

SITE	*10–20 cm*	*20–30 cm*	*30–40 cm*	*40–50 cm*	*50–60 cm*	*60–70 cm*
MURAY 1						
Epinephelus sp.				1		2
Acanthopagrus sp.	2	8				
Argyrops spinifer		4				
Rhabdosargus sp.	5	7				
MURAY 2						
Epinephelus sp.						1
Acanthopagrus sp.		14	1			
Argyrops spinifer		3				
Rhabdosargus sp.		3				
TAWI BEDUWA SHWAIBA						
Epinephelus sp.			1	8	5	10
Lethrinus sp.			1			
Acanthopagrus sp.		7				
Argyrops spinifer		1	1			
Rhabdosargus sp.		5				

Table 17.4. Relative size of the fish vertebrae collected from Muray and Tawi Beduwa Shwaiba. Measurements taken in mm on the maximum diameter of the centrum of each vertebra

Size (mm)	*MURAY1*				*MURAY2*	*TAWI BEDUWA SHWAIBA*			
	Chondrichthyes	Serranidae	Sparidae	Scombridae	Sparidae	Chondrichthyes	Serranidae	Sparidae	Scombridae
5			4						
6	1		7					6	
7	1		3		2	1			
8	1				11		2		
9						1		2	
10								1	
11						1	6	1	
12		1					2		
13							5		
14		2					3		
15							5		
16				1			7		
17				1			5		
18							4		
19							6		
20							2		1
21							1		
22							1		

the size of these fish, based on relative comparisons with the primary author's personal osteological fishes reference collection of fishes collected in the region. The majority of these fish were medium in size, being from fishes 20–30cm in length (Table 17.3). There were some larger fish, in some cases up to around 70cm in size, but these were all examples of groupers ('*hamoor*'). Comparison of the relative size of fish vertebrae retrieved from all three sites showed a similar pattern of medium-sized fishes with the only larger fish represented being either groupers or tuna (Table 17.4). The site of Tawi Beduwa Shwaiba had a greater number of bones from larger groupers than at the two Muray sites. This difference may however be simply an effect of sample size.

Marine shells

The presence of moderate quantities of marine shells were also recorded, interspersed between the scatters of potsherds, mammal and fish bones at the three sites. These included cerithids, worm shells, cowries, murex, olive shells, Arcoidea shells, pearl oysters, rock oysters, cockles, Tellinoidea shells, as well as venus clams (Table 17.5). The range of habitats that these marine mollusca were originally collected from must have included both rocky coastal areas, as well as shallow sandy and muddy sandy areas situated along the coast. It seems likely that the majority of these marine mollusca would have been eaten by the inhabitants of these sites, although some of them may also have been a waste by-product of the manufacture of jewellery items, or may have been accidentally transported to the site, such as is perhaps the case for the small cerithid shells.

Table 17.5 Marine shells identified at Muray and Tawi Beduwa Shwaiba

Taxon	*Common name*	*Muray 1*	*Muray 2*	*Tawi Beduwa Shwaiba*	*Habitat*
Cerithidae					
Clypeomorus bifasciatus bifasciatus	Creeper/ceriths			1	Intertidal on & under rocks
Indeterminate		1		1	
Vermetidae					
Serpulorbis variabilis	Worm shell			1	On rocks & flat surfaces
Cypraeidae					
Cypraea sp.	Cowrie			3	Under rocks & coral
Muricidae					
Murex sp.	Murex			1	Offshore & beached in sand
Olividae					
Oliva bulbosa	Olive shell	2		4	Intertidal & offshore in sand
Ancilla farsiana	Olive shell	1		5	Intertidal & offshore in sand
Arcidae					
Anadara antiquata	Arcoidea	3			In muddy sand, intertidal & offshore
Anadara sp.	Arcoidea			2	
Pteriidae					
Pinctada margaritifera	Pearl oyster	2		2	Attached among rocks from lower shore & below
Pinctada radiata	Pearl oyster	2	1	2	Attached to rocks from lower shore & below
Pinctada sp.	Pearl oyster			2	
Ostreidae					
Saccostrea cucullata	Oyster	1			Covering rocks or in clumps in mangroves, upper middle shore
Cardiidae					
Acrosterigma lacunosa	Cockle	3		3	In sand, offshore
Semelidae					
Ervilia sp.	Tellinoidea	1			In sand, offshore
Veneridae					
Circenita callipyga	Venus clam	1		2	In sand and sandy mud, mud-tide levels
Total		17	1	29	

Mammal bones

The mammalian faunal remains are in general badly preserved. The loss of collagen has contributed in weakening the structure of the bones. Small bones have been preserved better than longbones. The only large mammal bones were found at Muray and these are two first phalanges of dromedary (*Camelus domedarius*) (Fig. 17.10). Caprini are the main taxonomic group represented in the assemblage (Table 17.6). Almost all the skeletal parts are present; the ones that are absent are due to taphonomic loss and/or recovery bias. The presence of the skull and mandibles indicates that animals were kept alive and that they were slaughtered at the site.

The distinction between sheep and goat is not easy on all bones. The criteria used here follow those defined by Boessneck 1969 and Clutton-Brock *et al.* 1990. Possible measurements were made on phalanges and on the more resistant parts of the body: Distal Tibia and Distal Humerus. The sheep and goat are generally young animals with reference to the fusion of the bones (Table 17.7).

The distinction of goat and sheep

Most of the bones could be allocated to goat according to the criteria of Boessneck (1969) and Clutton-Brock *et al.* (1990) (Figs 17.11–17.14). On the humerus, the distinction between sheep and goat was also verified with the measurements of the distal trochlea following Helmer & Rocheteau (1994); the indices below 60, obtained on the two measurable specimens are fully compatible with goat measurements and morphology. If most of the bones were allocated to goat some bones had an intermediary morphology. This was also the case in Tawi Beduwa Shwaiba of two unfused hemi-frontal bones of the skull with undeveloped horn cores (Fig. 17.15). Unimproved breeds of goat bear horn cores. The interfrontal fusion of

Table 17.6 Mammal bones represented at Muray and Tawi Beduwa Shwaiba

Site name	*Element*	*Camel*	*Large herbivore/ camel-horse*	*Capra*	*Ovis cf*	*Caprini*	*Unid*	*Total*
Muray								
	Horn core			2				2
	Skull					1		1
	Maxillary					2		2
	Vc atlas					1		1
	Vc axis					3		3
	Vcd					1		1
	Coxal					4		4
	Femora					1		1
	Tibia					2		2
	Talus			2				2
	Naviculo cuboide					1		1
	Metapodial					2		2
	Phalanx 1	2		2	1	2		7
	Phalanx 2				3	1		4
	Phalanx 3			3				3
	Unidentified						11	11
Total Muray		2		9	4	22	11	47
	Tawi Beduwa Shwaiba							
	Skull				2			2
	Mandible		1			2		3
	Rib					1		1
	Vc					2		2
	Vc atlas					1		1
	Vc6					1		1
	Vl					1		1
	Vt					2		2
	Humerus			2		2		4
	Radius					2		2
	Metacarpal					1		1
	Coxal					1		1
	Femor			1		1		2
	Tibia					2		2
	Talus				1			1
	Metapodial			1				1
	Phalanx 1			4				4
	Phalanx 3			1				1
	Total Tawi Beduwa Shwaiba	0	1	9	3	19	0	32
Total		2	1	18	7	41	11	79

the frontal bone is reported by Baron (1986) between 5 and 7 years. On the postcranial bones the information on age of death are scarce. Only one phalanx has an unfused proximal end indicating that it was killed between 7 and 10 months.

Discussion

The discovery of the Mari and Tawi Beduwa Shwaiba sites confirms one of the traditional routes followed by people and their camels between Abu Dhabi and Liwa. Many such sites are now known thanks to the earlier work carried out by the Abu Dhabi Islands Archaeological Survey (King and Hellyer 2003). The types of pottery and other possessions discovered at these desert campsites shows that some of the more affluent urban classes clearly moved between Abu Dhabi and Liwa.

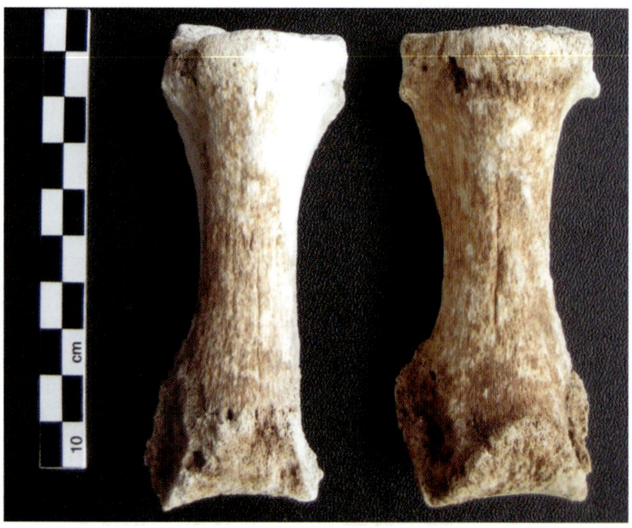

Fig. 17.10. Camel 1st phalanges from Muray site 2.

Fig. 17.11. Goat horncores from Muray site 1.

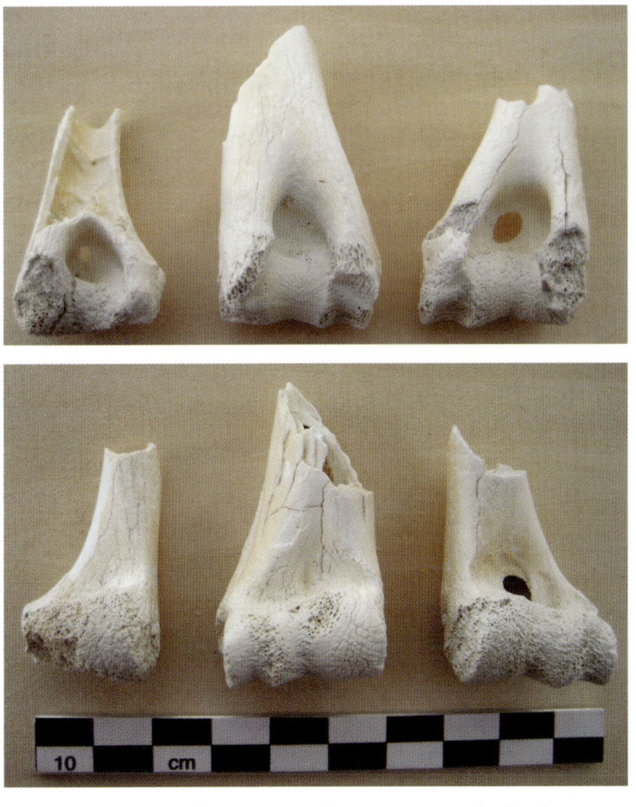

Fig. 17.12. Goat humeri: a. (posterior view); b. (anterior view) from Tawi Beduwa Shwaiba.

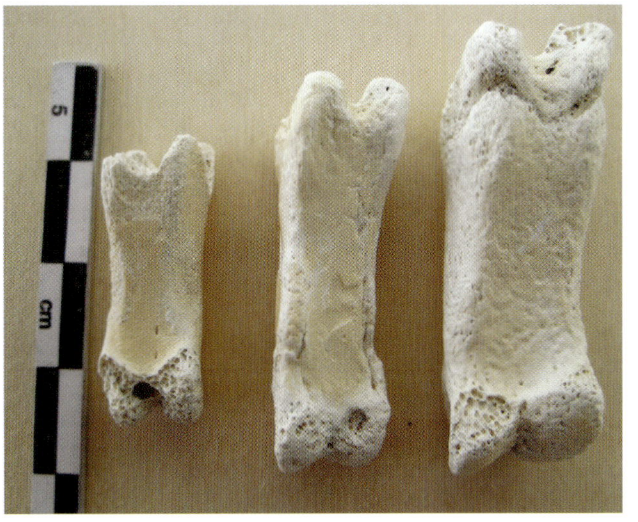

Fig. 17.13. Goat 1st phalanges from Tawi Beduwa Shwaiba.

Fig. 17.14. Goat 3rd phalanges from Tawi Beduwa Shwaiba.

Analysis of the pottery assemblages from all these sites confirms that the date when they were occupied was during the Late Islamic period, some time between the latter half of the 18th century until the middle–end of the 20th century. A similar range of types and shapes of ceramics were found at each site, including Iranian painted glazed ware, Chinese blue-and-white, European painted and blue-and-white wares, Omani brown glazed ware, as well as local earthenware vases and cooking pots.

Comparable pottery assemblages are known from the Bait Bin Ati al-Darmaki at Qattara oasis and the Bin Suroor House at Al Mutaredh oasis in Al Ain (Power & Al Kaabi in press). Similar ceramics are also found at a number of coastal sites such as Khor Kalba, Kalba, Fujairah, Khorfakkan, Madha, Dibba, Sharjah, Jumeirah, as well as the inland settlements at Masafi, Bithna, Awala and other locations (Sasaki & Sasaki 2012). Excavations at Jazirat al Hulaylah, Luluiyah and Julfar, however, have not recovered the same types of ceramics. This suggests that the period of occupation in this region of the Abu Dhabi desert was limited only to the Late Islamic period.

Why then are the early and middle Islamic occupation

Fig. 17.15. Sheep skull from Tawi Beduwa Shwaiba, site TBS-C2.

periods not represented in any of these desert sites? We know that numerous sites dating back to the Neolithic/Late Stone Age have been found in the deserts of Al Gharbia. Following the end of the so-called Holocene Climatic Optimum period, around 6000 years ago, there is a period of extreme aridity, marking what some authors have called the "Dark Millennium" (Uerpmann 2003) and this may be one of the reasons why we do not find any sites dating to the 4th millennium BC. It may also be a factor in explaining why do not find substantial archaeological material in the desert dating to the Bronze and Iron Ages. The few known sites in Al Gharbia are all located on the coast or on islands, where more favourable conditions may have existed for settlement and subsistence. If so, why then do we suddenly find evidence for settlement activities during the latter half of the 18th century? We know that this was a period which saw the emergence of the settlement of Abu Dhabi, the construction of Qasr Al Hosn and the Maqta watchtower, as well as the construction of a series of historic forts to safeguard water resources and date palm gardens in the oases of Liwa and Al Ain.

A number of questions, however, still remain. The sand dunes are highly mobile in this region of Abu Dhabi. Temporary camp sites utilised on a seasonal basis may have left only ephemeral traces in the archaeological record. The rapid movement of sands within this area can easily obscure or mask the traces of earlier activities at these locations. Sites such as Tawi Beduwa Shwaiba area A may soon disappear under the shifting sands of the Abu Dhabi desert. Similarly, new evidence of occupation areas may appear as the interdunal areas change in size and form.

What do these scatters of finds deep in the desert actually represent? Were there any traces of structures associated with these pottery and animal bone scatters? Are these find scatters associated with the location of former wells? Do the larger examples of these scatters (like perhaps in the case of Tawi Beduwa Shwaiba) remain *in situ* for longer periods of time than some of the more ephemeral find scatters, since they are located along favoured travelling routes? Did the tribe in this area come to the Tawi Beduwa Shwaiba camp site looking for pasture and water for their camels during the winter? Why were ceramics and fish bones found in such profusion at these sites?

Analysis of the fish bones discovered at these sites provides evidence of the types of dried fish that were consumed by the nomadic people, as well as which resources were imported to the oases in the interior. The fish remains recovered included bones from shark/ray/skate, needlefish, grouper, emperor, several types of seabream and tuna.

Marine resources exploited and transported to the interior was not only restricted to fish; marine shellfish were also recovered at these sites and would have been consumed. This demonstrates the general importance of marine resources to the nomadic people and the relations between coastal communities and those living within the desert interior. Perhaps those living on some of the major transportation/migration routes between the coast and Liwa, or who had a taste of urban life on the coast in the newly emerging settlement of Abu Dhabi, were more likely to eat fish than some of their other nomadic counterparts who spent less time in the coast and more time wandering within the desert interior.

Salted and dried fish are a valuable resource, which can be kept for many months, even within the desert. It is known that people who lived in Al Ain oasis visited the coast selling charcoal, returning with salted and dried fish (Mohammed Al Neyadi, pers. comm.). Coastal inhabitants also penetrated the desert with camels to sell dried fish. These accounts from the pre-oil era give us an idea why there are dense scatters of fish bones at some of these desert sites.

Various domestic activities clearly took place at these desert sites. Other finds recovered included stone pestles and mortars, used for milling flour and/or grinding other products, whetstones for sharpening iron knives, as well as flint fragments, used as strike-a-lights. Small chips of calcite, haematite and limestone were also noted. Small finds recovered from the survey included glass bottles and glass fragments, glass beads, as well as a few metal fragments, including a finger ring base and several coins.

Mammalian bones recovered from these sites demonstrate which animals accompanied the people as they made their seasonal migration between the coast and the interior. Mammal bones recovered from the sites included camels, sheep, goats as well as desert hare. The domestic animals would have accompanied the people as they moved between the coast and the inland oasis, whilst the latter was probably hunted locally in the vicinity of the site. Camels would have been also used to help transport water at wells and oases.

Further research

Tawi Beduwa Shwaiba was the largest amongst the three sites examined and appears to be located on one of the regular transportation/migration routes used by migrating tribal groups between the coast and the desert interior, heading to and from the Liwa oasis. Muray was a more ephemeral nomadic desert camp site, which was perhaps less frequently utilised, judging from the relative volume of finds recovered from the sites.

Further research is clearly required to investigate further examples of such sites. It would be profitable to map the locations of known historic wells to then target new areas for future archaeological surveys in their surrounding dunes. As the sands of Al Gharbia shift year by year who knows what new secrets may be revealed in the future.

In conclusion, it is hoped that this work demonstrates the value of undertaking archaeological investigations of historical sites from the pre-oil era. The analysis of both ceramic assemblages and environmental finds can provide new insights into both the life and daily diet of pre-oil era populations in Abu Dhabi emirate.

Acknowledgements

Thanks go to Marjan Mashkour for assistance with the analysis of the mammal bones. This work would not have been possible without the logistic support of the Historic Environment Department, as well as the help of local Abu Dhabi people who assisted with the original identification and documentation of these sites, as well as the difficulty of navigating through the sands to access these difficult locations.

References

ADIAS. 1997. *Abu Dhabi Islands Archaeological Survey Project – Occasional Newsletter, 15th September 1997*. Online source: http://www.adias-uae.com/newsletter/newsletter-sept-97.pdf

Barone, R. 1986. *Anatomie Comparée des Mammifères domestiques*. Vigot Frères, Paris.

Beech, M. J. 2004. *In the Land of the Ichthyophagi: Modelling fish exploitation in the Arabian Gulf and Gulf of Oman from the 5th millennium BC to the Late Islamic period.* Abu Dhabi Islands Archaeological Survey Monograph 1/British Archaeological Report S1217, Oxford.

Boessneck, J. 1969. Osteological differences between Sheep (*Ovis aries*) and goat (*Capra hircus)*, in D. R. Brothwell & E. S. Higgs (eds), *Science in Archaeology*, 331–358, 378. Thames & Hudson, London.

Clutton-Brock, J., Dennis-Bryan, P. L. Armitage & Jewell, P. A. 1990. Osteology of the Soay sheep. *Bulletin of the British Museum (Natural History)* 56(1), 1–56.

Codrai, R. 1990. *The Seven Shaikhdoms. Life in the Trucial States before the Federation of the United Arab Emirates*. Stacey International, London.

Cordes, R. & Scholz, F. 1980. *Bedouins, wealth and change : a study of rural development in the United Arab Emirates and the Sultanate of Oman*. United Nations University, Tokyo. URL: http://archive.unu.edu/unupress/unupbooks/80143e/80143E00.htm

Driesch, A. von den. 1976. *A Guide to the Measurement of Animal Bones from Archaeological Sites* (11th edn). Peabody Museum Bulletin 1, Harvard University, Cambridge MA.

Al-Fahim, M. 1998. *From Rags to Riches: A Story of Abu Dhabi*. Centre for Arab Studies & I. B. Taurus & Co. Ltd, London (new edition).

Heard-Bey, F. 1996. *From Trucial States to United Arab Emirates*. Longman, London (2nd enlarged edition).

Hellyer, P. 1998. *A Preliminary Note on an Archaeological Site at Tawi Beduwa Shwaiba, Emirate of Abu Dhabi, UAE. A Report to H.E. Sheikh Mohammed Bin Butti Al Hamed, Ruler's Representative in the Western Region – June 1998*. Unpublished report: Abu Dhabi Islands Archaeological Survey Project.

Helmer, D. and M. Rocheteau. 1994. *Atlas du squelette appendiculaire des principaux genres holocènes de petits ruminants du nord de la méditérranée et du Proche-Orient (Capra, Ovis, Rupicapre, Capreolus, Gazella)*. Fiches d'ostéologie animale pour l'Archéologie. APDCA, Juan les Pins.

Iddison, P. 1997. *Arabian Traveller's Observations on Bedouin Food. Paper Presented at the Oxford Symposium on Food and Cookery: Food and Travel*. Privately published booklet – Philip Iddison, Al Ain, UAE (2nd edition). Also available online: http://enhg.4t.com/iddison/oxsymp96.htm.

King, G. R. D. & Hellyer, P. 2003. Islamic archaeology in the deep sands of Abu Dhabi Emirate, U.A.E., in D. T. Potts, H. Naboodah and P. Hellyer (eds), *Archaeology of the UAE: Proceedings of the First International Conference on the Archaeology of the UAE,* 264–276. Trident Press, London.

Power, T. & Al Kaabi O. In press. A preliminary chronology for the al-'Ain (Buraimi) oases based on the quantified study of ceramics from the Bayt Bin 'Ati al-Darmaki, Qattara Oasis. *Archaeology in the United Arab Emirates* 6.

Sasaki, H. & Sasaki, T. 2008. *Nomadic Occupations in the Desert, Abu Dhabi emirate in the United Arab Emirates*. Unpublished preliminary report submitted to Abu Dhabi Authority for Culture and Heritage, 8 May, 2008.

Sasaki, H. & Sasaki, T. 2012. Trade ceramics from East Asia to the Arabian peninsula, in D. T. Potts and P. Hellyer (eds.), *Fifty Years of Emirates Archaeology – Proceedings of the Second International Conference on the Archaeology of the United Arab Emirates,* 223–233. Ministry of Culture, Youth and Community Development & Motivate Publishing, Abu Dhabi, Dubai and London.

Thesiger, W. 1959 (2007 new edithion). *Arabian Sands*. Penguin Classics, Harmondsworth.

Uerpmann, M. 2003. The Dark Millennium: remarks on the Final Stone Age in the Emirates and Oman, in D. Potts, H. Al Naboodah and P. Hellyer (eds), *Archaeology of the United Arab Emirates – Proceedings of the First International Conference on the Archaeology of the U.A.E.*, 73–81. Trident Press, London.

18. Nomads, horses and mobility: an assessment of geographic origins of Iron Age horses found at Tsengel Khairkhan and Baga Turgen Gol (Mongolian Altai) based on oxygen isotope compositions of tooth enamel

R. Bendrey, S. Lepetz, A. Zazzo, M. Balasse, T. Turbat, P.-H. Giscard, D. Vella, G. I. Zaitseva, K. V. Chugunov, J. Ughetto, K. Debue and J.-D. Vigne

The ephemeral nature of pastoral nomadism leaves few traces in the archaeological record with which to study the functioning of these communities. Although the Iron Age peoples of the Altai region did not build permanent settlements, they did construct kurgans (funerary barrows), which on occasion include depositions of horses. As horse teeth grow, enamel records isotope compositions related to water and food consumed during the period of its mineralization as a continuous record covering several years. This paper discusses results of a project to explore the movements and social connections of these nomadic groups through isotopic analyses of tooth enamel from horse burials excavated from first millennium BC funerary barrows. It employs the phenomenon that the oxygen isotope composition ($\delta^{18}O$) of meteoric precipitation varies geographically, and presents tooth enamel $\delta^{18}O$ values for Iron Age horses excavated from two sites in the Mongolian Altai and an assessment of differentiating between 'local' and 'non-local' values for the sites as a first step in building a framework within which to analyse the past geographical movement of the horses.

Keywords Pastoralism; Mobility; Central Asia; Iron Age; Mongolia; Horse; Oxygen isotope ratios; Tooth enamel

Introduction

Migration is frequently cited as a key feature of the later prehistoric societies of the Eurasian steppe, with two forms of practice being invoked: the regular movement of people with their herds between different environments at specific times of the year (e.g. Khazanov 1984; Koryakova & Hanks 2006), and the colonising migration of groups of people from one region to permanently settle in another (e.g. Anthony 2007; Koryakova 2002, 276; Scott *et al.* 2004). Although the development of mobile forms of pastoralism had a long trajectory in this region (Allard & Erdenebaatar 2005; Anthony 2007; Legrand, 2006), it is at the start of the 1st millennium BC, with increased evidence for mobility in the steppe, that it is often agreed that fully-fledged pastoral nomadism emerged (Koryakova & Hanks 2006; Renfrew 2002). The archaeological record also witnesses the concurrent spread of some common cultural elements throughout the steppe (e.g. Bashilov & Yablonsky 1995; Christian 1998, 124; van Geel *et al.* 2004). Within such a context, of increased mobility and of extensive long-distance interactions, the transport capabilities of the horse emerge as a vital factor. The great social importance of horses during the Iron Age is clearly visible in the rich archaeological finds of this period, in

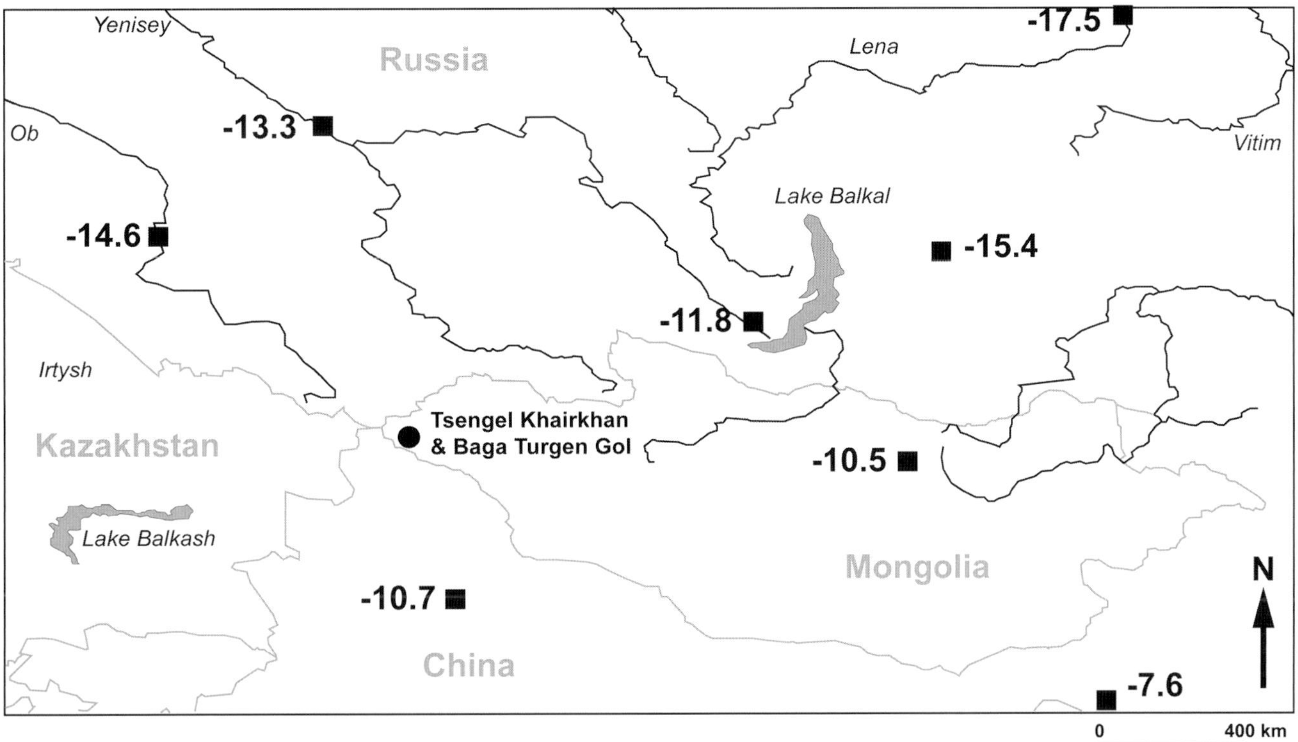

Fig. 18.1. Map showing location map of Tsengel Khairkhan and Baga Turgen Gol and modern precipitation $\delta^{18}O$ values (‰): long-term weighted means from the Global Network of Isotopes in Precipitation (GNIP) database are plotted according to the positions of the GNIP stations (black squares) (IAEA/WMO, 2006).

particular the multiple horse burials at high status barrows in the region (Alexeev *et al.* 2001; Čugunov *et al.* 2003; Francfort *et al.* 2000; Rudenko 1970).

The available means of transport would have been a key factor in the dynamics of any colonising migration (Anthony 2007 111), and horses would have formed an essential component of nomadic lifestyles as movement to seasonal pastures would have been important for the success of subsistence strategies (Koryakova & Hanks 2006; Shnirelman *et al.* 1996; Vainhstein 1980). In contemporary and historic pastoral movement systems, distances and patterns of seasonal movements vary considerably across Central Asia, ranging from a few kilometres to over 1000 (Ferret 2009; Sneath 1999; Shnirelman *et al.* 1996). The latter distances are known from (meridional) migrations of Kazakhs in the 19th century AD across the steppe (Shnirelman *et al.* 1996). Vertical migrations of groups exploiting uplands and mountains tend to follow shorter routes. During the 1930s, for example, nomadic groups of the Sayan-Altai uplands in Tuva followed average annual migrations of 25–30km a year for a poor *aal*, and 50–60km for a rich *aal* (Vainshtein, 1980, 83–103) [an *aal* is a unit of a few households that migrated together, equivalent to the Mongol *ail* and the Kazakh *aul* (Humphrey 1980, 13)]. Some modern mobile pastoralists in the Mongolian Altai are known to cover annual migrations in excess of 200km (Foggin *et al.* 1997).

In addition to migrations, there are other possible reasons for journeys of humans with their horses, over shorter or longer distances, for example in acts of trade, hunting, conflict, and communication. There is evidence for long-distance movements of commodities in the later prehistoric steppe (e.g. Bokovenko 2004; Koryakova 2002), and horses could have been traded between communities. The movement of horses could also have occurred via gift-giving, both to the living and the dead. Ethnographic information on recent practices in Central Asia indicates that gifts of livestock to newly married couples are an important part of the marriage rite (Vainshtein 1980, 101; Shnirelman *et al.* 1996, 147–148). Horses may also have been given as sacrifices, to be buried with the deceased. Iconographic analysis of the tack worn by thirteen horses buried at the rich Iron Age kurgan excavated at Berel, in eastern Kazakhstan, indicates that they may have come from diverse and distant sources, perhaps as gifts from allied tribes (Francfort *et al.* 2000; Keyser-Tracqui *et al.* 2005). Various forms of social conflict could also have resulted in the acquisition and movement of horses. Shnirelman *et al.* (1996, 148–149), for example, describe historic practices of horse theft amongst Kazakh nomadic peoples. Warfare is often seen as an important element of Iron Age life on the steppe, with the appearance of fully-fledged pastoral nomadism at the start of the 1st millennium BC being linked with a militaristic component in the society of the Iron Age steppe (mounted warriors) (Anthony 2007 322; Koryakova & Hanks 2006; Hanks 2002).

Despite the considerable research which has focused on the Iron Age societies and economies of Central

Fig. 18.2. Human and horse skeleton in tomb 12 at Baga Turgen Gol, Mongolia.

Asia, identifying the mobility, seasonal migrations, and interactions of these groups has proven a considerable challenge. Problems in studying these cultures reside, in particular, in the mobile and ephemeral nature of nomadic occupation (Cribb 1991). The approach taken in this paper is based on measurements of isotope ratios in the tooth enamel of Iron Age horses. Mammalian tooth enamel preserves a record of dietary and environmental changes for the part of the animal's life during which the tooth was growing. Stable and radiogenic isotopes in the enamel can vary according to the locations where the animal lived while the tooth was growing, and can thus be used to study past movements (e.g. Balasse *et al.* 2002; Bendrey *et al.* 2009; Evans *et al.* 2006). This paper employs the phenomenon that the oxygen isotope composition ($\delta^{18}O$) of meteoric precipitation varies geographically (e.g. Bowen *et al.* 2005), and presents tooth enamel $\delta^{18}O$ values for Iron Age horses excavated from two sites in the Mongolian Altai (Fig. 18.1) and an assessment of differentiating between 'local' and 'non-local' values for the sites as a first step in building a framework within which to analyse the past geographical movement of the horses.

Table 18.1. Horse skeletons excavated from Baga Turgen Gol and Tsengel Khairkhan: location, age at death, and tooth/column sampled. Ages attributed on the basis of incisor wear following Cornevin and Lesbre (1894). All horses are male

Site	*Location*	*Horse (code)*	*Age (years)*	*Tooth / column sampled*
Baga Turgen Gol	tomb 3	B3	+ 20	M_3/anterior
	tomb 8	B8	±4	M_3/central
	tomb 10	B10	+ 20	M_3/anterior
	tomb 11	B11A	±18	M_3/anterior
	tomb 11	B11B	±17	M_3/anterior
	tomb 12	B12	+ 20	M_3/central
	tomb 13	B13	±1	not sampled (too young)
Tsengel Khairkhan	tomb 1	T1	±5	M_3/anterior
	tomb 2	T2	±19–20	M_3/central
	tomb 3	T3A	2–4	M_3/anterior
	tomb 3	T3B	±10	M_3/anterior

Methods, materials and background

Study sites

This paper considers $\delta^{18}O$ values of tooth enamel from ten Iron Age horses from Mongolia. Samples were collected from lower third molar (M_3) tooth enamel of horses excavated from burial mounds at Tsengel Khairkhan and Baga Turgen Gol, two neighbouring sites that lie some 40km apart in the Altai Mountains, Mongolia (Figs. 18.1 and 18.2). The two sites are located in Bayan-Ölgiy aimag (province) in the west of the country at an elevation of a little over 2000m and date to the 'Pazyryk culture' (the 5th–3rd centuries BC) (Jordana *et al.* 2009). Tsengel Khairkhan produced four horse skeletons, all of which have been sampled. Baga Turgen Gol produced seven horses. Six of the seven horses excavated have been sampled; one horse, B13, was only ~1 year old at death, and consequently the M_3 had not yet begun to form (Table 18.1). Three of the horses – B8, T1 and T3A – died before the M_3 had fully formed [M_3 tooth enamel mineralization lasts from 21±3 months to 55±3 months in horses (Hoppe *et al.* 2004b)]. Further, from the time at which a tooth comes into occlusal contact it is progressively worn down (from the occlusal surface) due to contact with the opposite teeth, food, and other extraneous matter in the mouth (Levine, 1982; Kilic *et al.*, 1997), and consequently the teeth of the older animals have significantly shorter crowns present for analysis than the younger horses (Table 18.2).

Oxygen isotope compositions of horse tooth enamel

The oxygen isotope composition of meteoric precipitation ($\delta^{18}O_{ppt}$) varies widely and systematically across the globe, and can be used for determining the origin and movement of animals (Bowen *et al.* 2005; Bowen & Wilkinson 2002). This variation is controlled by Rayleigh distillation of atmospheric vapour, driven primarily by air-mass temperature which is largely dependent on latitudinal and altitudinal variations (Bowen & Wilkinson 2002). Mapping of modern oxygen isotope content of precipitation by the Global Network of Isotopes in Precipitation (GNIP) database confirms the general latitudinal gradient for the oxygen isotope composition of meteoric precipitation across the region (Fig. 18.1). The positions of the stations collecting data on the isotope content of precipitation do not allow a detailed consideration of the modern variability in precipitation $\delta^{18}O$ values at Tsengel Khairkhan and Baga Turgen Gol. However, the long-term weighted means of the GNIP stations plotted in Figure 18.1 indicate the magnitude of the variation in mean values across the region (IAEA/WMO, 2006).

As ingested local surface waters constitute the primary source of oxygen in the body water of a mammal, and tooth enamel bioapatite is formed in isotopic equilibrium with body water, then $\delta^{18}O$ values of tooth enamel may be used to infer $\delta^{18}O$ values of local waters (Fricke *et al.* 1998; Longinelli 1984; Luz *et al.* 1984). Horses are obligate drinkers, and their $\delta^{18}O$ values are considered to be more closely linked to drinking water values than in the case of other animals that obtain water mainly from plant food (which is enriched in $\delta^{18}O$ relative to precipitation due to evapotranspiration) (Hoppe *et al.* 2004a; 2005; Levin *et al.* 2006). Furthermore, tooth enamel bioapatite has been shown to be resistant to diagenetic changes (alterations due to post-depositional interactions between the tooth and the burial environment) (Lee-Thorp & Van der Merwe 1991; Wang & Cerling 1994; Zazzo *et al.* 2004). In a recent study, Bell *et al.* (2006) confirmed that latitudinal variation in precipitation $\delta^{18}O$ values are recorded in horse tooth enamel.

Mongolia has a continental climate of extreme cold and heat, and absolute temperatures can vary between –50°C in winter to 40°C in the summer (Pyankov *et al.* 2000). Modern records indicate seasonal temperature variation for Ulan Bator, Mongolia, of between around –30°C to 20°C, and that precipitation $\delta^{18}O$ values vary between around –30 to –4‰ (IAEA/WMO, 2006). Precipitation $\delta^{18}O$ values are positively correlated with temperature, with higher values occurring in warmer seasons and lower values in colder seasons (Fricke *et al.* 1998), and we would thus expect to see pronounced seasonal oscillations in the $\delta^{18}O$ values of sequential microsamples taken along a tooth of a Mongolian horse. As stated above, the process of mineralisation of enamel as the tooth grows preserves a record of $\delta^{18}O$ values of body water for this period of the animal's life; however, the delayed process of enamel mineralisation means that $\delta^{18}O$ values from enamel microsamples return a time averaged signal, which has the effect of attenuating the input signal in its amplitude (Balasse 2002; Hoppe *et al.* 2004b; Passey & Cerling 2002; Zazzo *et al.* 2005). The amplitude of seasonal variation in $\delta^{18}O$ values in tooth enamel will therefore be less than that seen in meteoric precipitation.

In addition, precipitation and surface waters become increasingly depleted in $\delta^{18}O$ with increasing altitude. In a compilation of studies from throughout the world, Poage and Chamberlain (2001) identify a linear relationship between elevation and the isotopic composition of precipitation. They identify an isotopic lapse rate of ~2.8‰/1000m applicable for most regions of the world (except the Himalayas and at elevations of over 5000m). In studies of the 'altitude effect' on the isotopic composition of tropical rains in Cameroon and Bolivia, Gonfiantini *et al.* (2001) identified isotopic lapse rates of 1.4‰ to 2.6‰/1000m altitude; while Rietti-Shati *et al.* (2000) identified no significant 'altitude effect' on Mt Kenya between 4200m and 5200m. Tsengel Khairkhan and Baga Turgen Gol are located at a little over 2000m in the Altai mountains, which reach a maximum of ~4000m, although not all of this range would have been exploited, for example modern mobile pastoralists in Hovd sum (district), Uvs aimag (province) follow seasonal routes moving between altitudes of ~1000m and ~2400m (Sneath 1999, 236–239). Thus temperature-related seasonal variation in precipitation $\delta^{18}O$ values (with a range of >20‰, as discussed above) is expected to have a greater influence on recorded values in the horse tooth enamel than altitudinal movements that could have been made.

Table 18.2. Summary statistics of stable oxygen isotope measurements: minimum (min), maximum (max), and amplitude of variation (range). Two average values are given: mean-1 is the arithmetic mean; mean-2 has been calculated by first separately calculating the means of peak values (interpreted as summers) and trough values (interpreted as winters) in the isotopic data series from each tooth, then the average was taken of these two values (see text)

Horse (code)	*No. samples*	*Distance sampled (mm)*	$\delta^{18}O$ (‰)				
			min	*max*	*range*	*mean-1*	*mean-2*
B3	4	8.2	−13.2	−10.0	3.2	−11.7	−11.6
B8	8	26.4	−12.6	−10.7	1.9	−11.6	−11.7
B10	8	23.7	−12.5	−10.7	1.8	−11.5	−11.4
B11A	8	23.8	−12.0	−9.5	2.5	−11.1	−11.0
B11B	14	44.7	−14.6	−10.9	3.7	−12.8	−12.6
B12	5	14.0	−11.2	−9.3	1.9	−10.4	–
T1	15	46.6	−13.4	−10.2	3.1	−11.7	−12.0
T2	6	14.2	−15.5	−11.6	3.8	−13.5	−13.0
T3A	13	35.2	−12.4	−9.7	2.7	−11.1	−11.0
T3B	14	44.1	−14.4	−10.7	3.7	−11.9	−12.1

Comparative data and caveats

To provide an estimate of geographical variation in horse tooth enamel $\delta^{18}O$ values, a small quantity of data from other sites in the region are used to contextualise the Mongolian results. These include modern and archaeological horse M_3 tooth enamel $\delta^{18}O$ values: a modern domestic horse specimen collected in 2008 from near Tsengel Khairkhan, Mongolia; a Przewalski's horse mandible collected in 1896 from Lake Ulungar, Xinjiang State, northwest China; and 13 horses buried at Arzhan 2, Tuva Republic, dating to the end of the 7th century BC (Čugunov *et al.* 2003; Zaitseva *et al.* 2007). Lake Ulungar is situated at an elevation of ~500m, and Arzhan 2 is at a little over 800m.

There are several caveats that must be borne in mind with the dataset. Mean $\delta^{18}O$ values are employed in the paper to provide an indication of the average environment in which the horse lived during the period of tooth formation, with which to compare to an assessment of 'local' range for the sites, from which animals of 'non-local' origin may be distinguished. The number of sites and samples analysed, and the absence of previously published comparative data for the region, limits our ability to identify geographical movements from these results. There is also uncertainty over the locations at which the horses had lived: we know where they were excavated from, but do not have details of where they had lived. Further, inter-annual temperature variations could create differences in $\delta^{18}O$ values between teeth and inter-individual variation between horses due to differences in ingested water from plants and individual metabolism could affect the variability of the $\delta^{18}O$ values (Hoppe *et al.* 2004a).

Analytical methods

Dental cementum was removed on one of the columns of each M_3 using a tungsten drill bit in order to access the enamel. A series of enamel samples was then taken along each crown, in a sequence perpendicular to the growth axis of the column of the tooth, using a diamond-encrusted drill bit. Each sample consisted of a horizontal band (~1mm wide and 3–5mm long) through the depth of the enamel.

Samples were taken from the buccal side of the M_3 anterior column, except in cases where this column was damaged in which case they were taken from the central column. Positions of the samples on the crown were measured from the most apical part of the enamel on the specimens, at the junction between the enamel and the root. For teeth in which the root was absent – those from the younger horses – the sample positions were measured from the first drilled sample adjacent to the maturing enamel.

After a protocol defined in Balasse *et al.* (2002), the pre-treatment of the sampled enamel powder included: treatment with sodium hypochlorite (NaOCl, 2–3%, 0.1ml/mg) for 24 hours to remove organic matter; then with 0.1 M acetic acid for 4 hours (0.1ml/mg) to remove exogenous carbonate.

Bioapatite samples weighing around 560–640µg were reacted with 100% phosphoric acid at 70°C in individual vessels in an automated cryogenic distillation system (Kiel IV device) interfaced with a Delta V Advantage isotope ratio mass spectrometer. Over the period of analysis of the bioapatite samples from Tsengel Khairkhan and Baga Turgen Gol, the analytical precision for $\delta^{18}O$ estimated from 65 analyses of the laboratory internal carbonate standard (Marbre LM, normalised on the international standard NBS19) is 0.06‰ (1σ). All $\delta^{18}O$ values are expressed versus VPDB.

Results and discussion

Oxygen isotope compositions

In total, 95 serial enamel samples were obtained from the teeth of the ten horses excavated from Baga Turgen Gol and Tsengel Khairkhan. The $\delta^{18}O$ values measured in the horse tooth enamel vary from −15.5‰ to −9.3‰ (Table 18.2). The amplitude of variation within a single tooth is

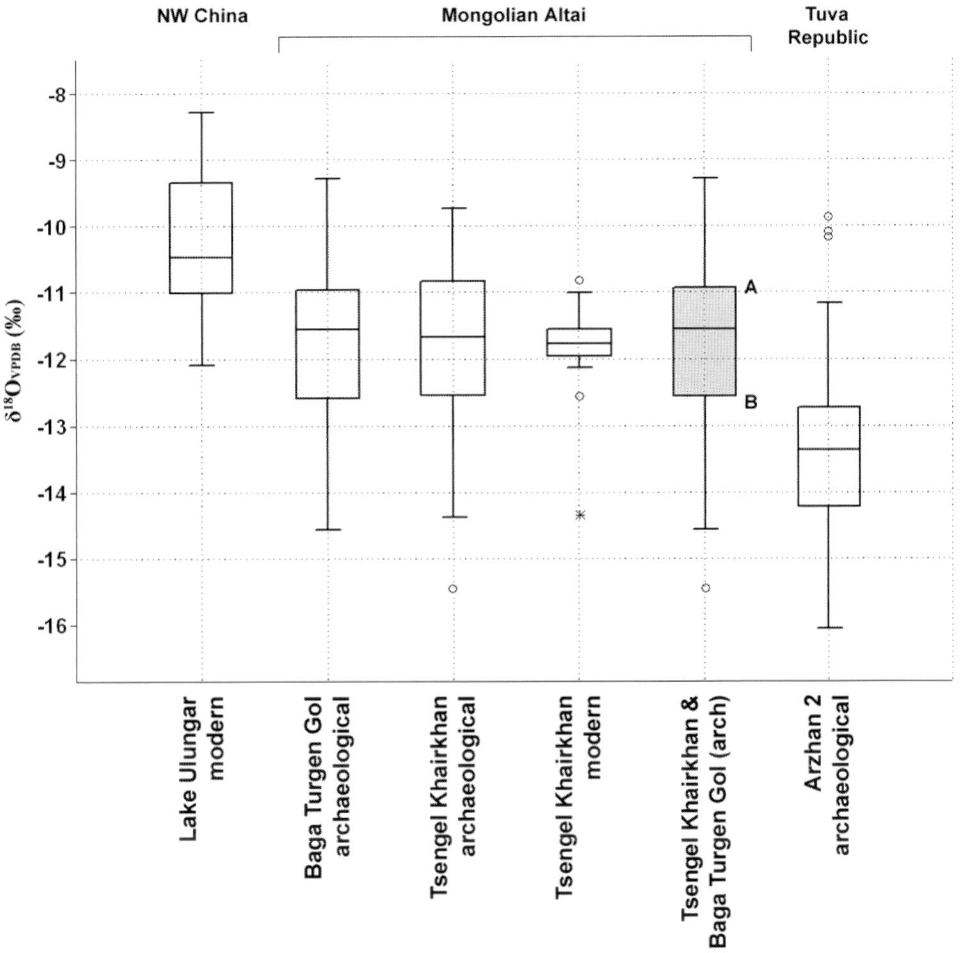

Fig.18.3. Box plots of $\delta^{18}O_{VPDB}$ values of horse tooth enamel from certain sites in Central Asia: data sets described in Figure 18.5, and site locations shown in Figure 18.1. The box plots divide a distribution according to the inter-quartile range, with the box containing 50% of the values. Possible outliers are marked by circles, probable outliers by asterisks. 'A' and 'B' mark the upper and lower limits of the central 50% of the distribution of values from the Mongolian archaeological samples, which includes both Baga Turgen Gol and Tsengel Khairkhan plotted together (greyed box).

between 1.8‰ and 3.8‰. The amplitude of variation for precipitation $\delta^{18}O$ values quoted above is 26‰ (for modern Ulan Bator), and thus the enamel values can be seen to be significantly attenuated relative to likely precipitation values (see discussion above and Balasse 2002; Hoppe *et al.* 2004b; Passey & Cerling 2002; Zazzo *et al.* 2005).

Geographical variation in oxygen isotope compositions of horse tooth enamel

To provide an assessment of geographical variation in oxygen isotope composition of horse tooth enamel we present a comparison of the range and distribution of the $\delta^{18}O$ values from the Mongolian archaeological horse tooth enamel against some other Central Asian samples (described above) from modern Lake Ulungar (China), modern Tsengel Khairkhan (Mongolia) and archaeological Arzhan 2 (Tuva) (Fig. 18.3 and Table 18.3). The locations of these sites are shown in Figure 18.4. Figure 18.3 presents the data in the form of box plots: this divides the distribution of values into the interquartile range and identifies possible outliers (circles) and probable outliers (asterisks). Relating the $\delta^{18}O$ values to the relative geographical locations of the sites (Figs. 15.3 & 18.4) confirms that the $\delta^{18}O$ values recorded in Central Asian horse tooth enamel of the region vary according to a large-scale latitudinal gradient (Bell *et al.* 2006; Bowen *et al.* 2005), as seen in the modern precipitation values (Fig. 18.4).

Table 18.3. Samples analysed to provide data used in Table 18.2

Location	*Site*	*No. horses*	*No. samples*
NW China modern	Lake Ulungar	1	12
Mongolian archaeological	Baga Turgen Gol and Tsengel Khairkhan	10	95
Mongolian modern	Tsengel Khairkhan	1	13
Tuva archaeological	Arzhan 2	13	128

Modern patterns in precipitation values may differ

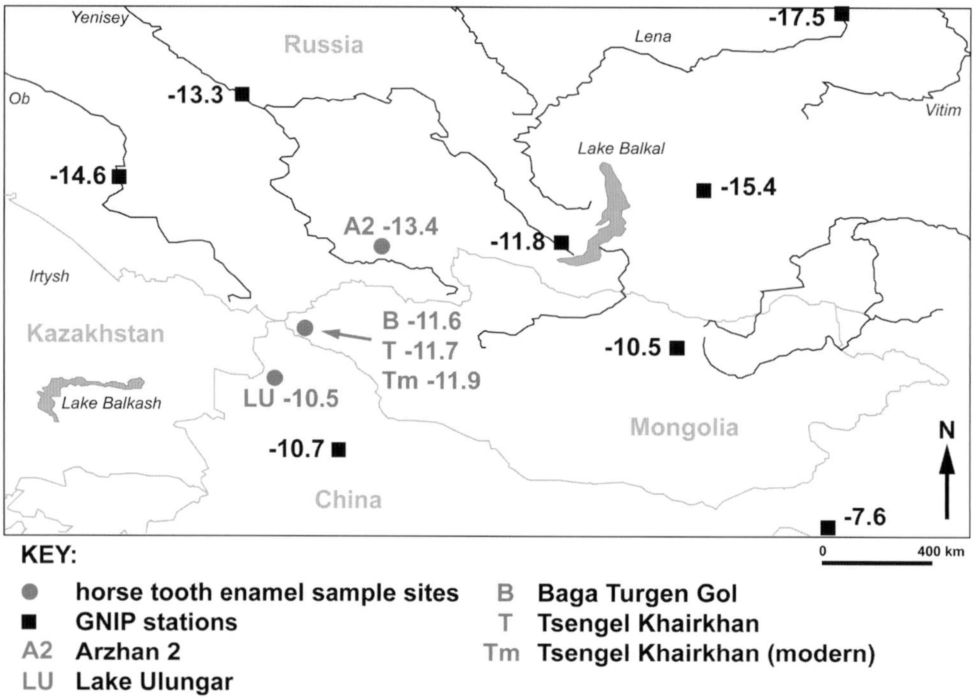

Fig. 18.4. Map showing median of the horse tooth enamel values (‰) from each of the sample sites plotted for comparison with the modern precipitation $\delta^{18}O$ values (‰) (Fig. 18.1): long-term weighted means from the Global Network of Isotopes in Precipitation (GNIP) database are plotted according to the position of the GNIP stations (IAEA/WMO, 2006).

from ancient ones and thus cannot themselves be used to define variation across the study area for periods in the past. However, the consistency between the regional geographical variation of the tooth enamel values and modern precipitation in terms of both direction and magnitude of variation supports the premise that the $\delta^{18}O$ values of horse tooth enamel can be used to identify long-distance movements (*sensu* Bell *et al.* 2006).

Are the Tsengel Khairkhan and Baga Turgen Gol 'local' or 'non-local' horses?

In order to identify long-distance movements of horses, and consider the question of whether the Tsengel Khairkhan and Baga Turgen Gol horses are 'local' or 'non-local', we shall: first, calculate a mean $\delta^{18}O$ value for each tooth, in the assumption that this reflects the averaged intake of oxygen during the period of mineralisation sampled [see Hoppe and Koch (2006) and Evans *et al.* (2006)]; second, estimate a 'local' range for Tsengel Khairkhan and Baga Turgen Gol, to which we can compare the individual horse mean values.

MEAN $\delta^{18}O$ VALUES FOR EACH TOOTH

In taking the mean $\delta^{18}O$ value from each tooth we are combining the values recorded from throughout the period of enamel mineralisation to produce a measure of the averaged intake of oxygen, and so are not considering the movements between the different locations during any seasonal nomadic migration. The length of each tooth sampled varies due to the ages of the horses (Tables 18.1 & 18.2) and represents different quantities of time, and we cannot thus use the arithmetic means (Table 18.2, mean-1) of each tooth dataset to calculate the averaged intake of oxygen. Rather, the average value for each tooth has been calculated in the following way. First, the means of peak values (interpreted as summers) and trough values (interpreted as winters) were calculated separately; then an overall tooth mean was calculated from the means of the peaks and the troughs (see Table 18.2, mean-2).

The adjusted means (mean-2) of the Mongolian horses are plotted in Figure 18.5. Horses B8, T1 and T3A were slaughtered during the period of formation of the M_3 (i.e. before the root had developed), and the values recorded in these teeth represent those of the regions where the horses lived in the months prior to their deaths. The consistency of these mean values and their agreement with the modern horse data from Tsengel Khairkhan (Figs 18.4 & 18.5) support the interpretation that these three horses were living in the region of the excavated tombs at the time of their deaths, and that these horses were thus 'local'.

IDENTIFYING A 'LOCAL' RANGE OF $\delta^{18}O$ VALUES?

The calculated mean values have a range of ~3‰ (Fig. 18.5). Horse B12 only produced a short data sequence from the M_3 consisting of a single 'trough' (interpreted as a winter signal). As it does not show a full seasonal cycle, no mean could be calculated, although its position is

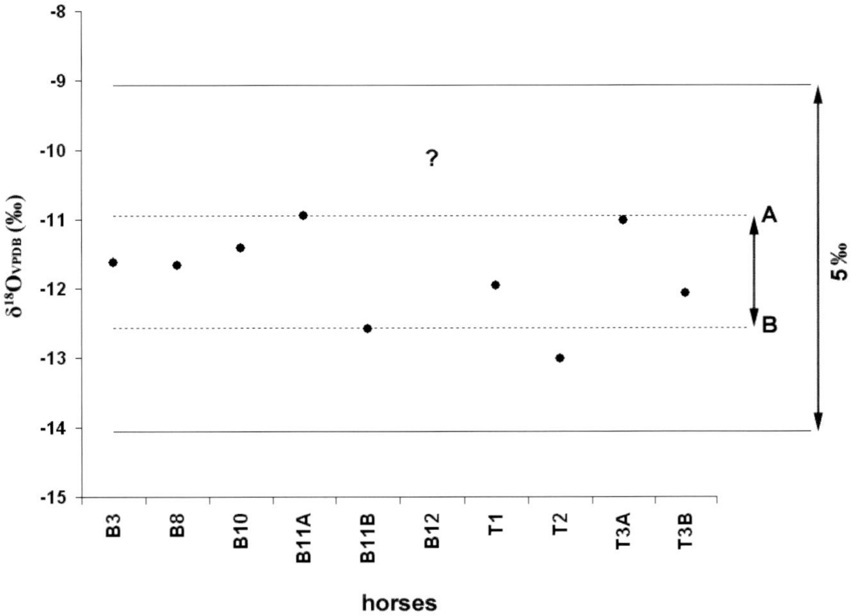

Fig. 18.5. Mean $\delta^{18}O_{VPDB}$ value for each horse M_3 from Baga Turgen Gol and Tsengel Khairkhan (mean-2 from Table 18.2). Dotted-lines 'A' and 'B' mark the limits of the central 50% of the distribution of values from the Mongolian archaeological samples (from Fig. 18.3). Solid lines mark a range of 5‰ centred on the median of the Mongolian archaeological values.

estimated in Figure 18.5. Exclusion of the data from horse B12 would give a range of 2‰ for the other nine horses.

The range of mean $\delta^{18}O$ values for the Mongolian horses is relatively small when compared to other published modern equid isotopic values. Hoppe *et al.* (2004a) demonstrated that in desert horses $\delta^{18}O$ values can vary by >6.5‰, although the extreme conditions of such arid environments may produce large fractionations among environmental sources and among different individuals. Hoppe *et al.* (2005) calculate a range of 5‰ for $\delta^{18}O$ values for horses in a single feral population in a temperate climate from the barrier island of Shackleford Banks, which lies off the coast of North Carolina (USA). Such variation is confirmed by Wang *et al.* (2008), who found a maximum range of 5.2‰ for horses living at high altitudes on the Tibetan plateau (at Gyirong, 4100m above sea level).

To distinguish between 'local' and 'non-local' horses it would be necessary to set limits for what is considered a 'local' range. If we assume a 5‰ range, as that which would be expected to be seen in a single population living in one area [following Hoppe *et al.* (2005) and Wang *et al.*. (2008)], then it would follow that only mean values outside this range could be suggested as being 'non-local'. A 5‰ range is plotted in Figure 18.5 for the Mongolian horses, centred on the median value (–11.56‰) from the Mongolian archaeological values (Fig. 18.3). Above, we argue that the data from horses B8, T1 and T3A are those that can most clearly be interpreted as 'local' from the two sites: the arithmetic mean for the data from these three horses is –11.55‰, which supports the selection of the overall median value as the centre of the local range. Using the 5‰ range to define the distribution of expected $\delta^{18}O$ mean values in single populations, all ten horses would be interpreted as being 'local' (Fig. 18.5). Indeed, this range also covers the great majority of the enamel $\delta^{18}O$ values recovered from both Arzhan 2 and Lake Ulungar (Fig. 18.3), and would identify all as 'local'. This would thus not allow differentiation of origins of horses from sites across this geographic distance (in this case ~700 km from Lake Ulungar to Arzhan 2); although it would be possible to identify horses originating from outside this area.

The narrower range of the Mongolian horse $\delta^{18}O$ mean values than those animals in the studies of Hoppe *et al.* (2005) and Wang *et al.* (2008), however, raises the question whether the modern horses from these other climatic regions provide appropriate comparative data for the horses and the environment of the Mongolian Altai. For example, as stated by Hoppe *et al.* (2005) different water sources and feeding environments could contribute to the variability of the $\delta^{18}O$ values of the Shackleford Banks horses, as might sampling error if some of the bulk samples represent less than a year of growth. Wang *et al.* (2008) do not publish details of the life-histories of their horses: they have sample collection locations, and it would seem to be (at least) possible that animals may have moved within the region. In contrast, Shackleford Banks is an island of $12km^2$, and the geographic origins of the horses in the study by Hoppe *et al.* (2005) can be known with certainty.

Without further modern horse tooth enamel data from the region it will not be possible to definitively propose a more appropriate 'local' range for Tsengel Khairkhan and Baga Turgen Gol. However, consideration of the archaeological data suggests that: 1) the 'local' ranges of means, for a population living in a single area, may be less than 5‰ [the Arzhan 2 data presents a similar range and distribution to that seen in the Mongolian data (Fig. 18.3)]; 2) it should be possible to distinguish between horses whose geographic origins are separated by several hundred kilometres – the Mongolian and Tuvan (Arzhan 2) sites are situated ~500km apart and it can be seen that the central 50% of the distributions of the archaeological tooth enamel $\delta^{18}O$ values from the two areas have a distinct range (Fig. 18.3). Assuming that precipitation $\delta^{18}O$ values change gradually and continuously over the region, it is expected that it would be increasingly difficult to identify movements as the distances between sites is reduced, i.e. the shorter the distance between two sites, the more the ranges of 'local' values will over lap. Tsengel Khairkhan and Baga Turgen Gol, for example, are ~40 km apart, and the ranges of $\delta^{18}O$ values are almost identical.

If we compare the upper and lower limits of the central 50% of the Mongolian archaeological data to the individual horse means, it can be seen that all but two (horses T2 and B12) plot within these limits (Fig. 18.5). Relatively short isotopic sequences were derived from these two animals – they have short crown heights due to their advanced ages at death (Table 18.2) – which along with the speculative nature of the limits set means that these two horses cannot be identified as 'non-local' ('local' limits may in fact lie somewhere between the two ranges illustrated in Fig. 18.5). However, it can be concluded that all the Mongolian horses, with the possible exception of B12 and T2, exhibit results consistent with having lived in the locality of Tsengel Khairkhan and Baga Turgen Gol during the period of M_3 formation.

Conclusions

This paper represents a first step in building a framework within which to consider the movements of horses in the study area. It considers mean oxygen isotope compositions of tooth enamel from horses buried as part of the funerary rites for Iron Age people in the Mongolian Altai, within the context of a number of modern and archaeological horse tooth enamel $\delta^{18}O$ values and modern oxygen isotope composition of meteoric precipitation for the region.

The horse tooth enamel values from the sites studied vary with latitude, and agree in terms of the direction and magnitude of changes with the pattern of mean $\delta^{18}O$ values of modern precipitation. The range of mean $\delta^{18}O$ values from the Tsengel Khairkhan and Baga Turgen Gol horses is relatively small compared to the ranges (~5‰) of isotopic values demonstrated for modern horse populations from an island site in the USA (Hoppe *et al.* 2004a; 2005) and at high altitudes on the Tibetan Plateau (Wang *et al.* 2008). It is suggested that these horses are not appropriate comparative populations for horses in the environment of the Mongolian Altai, that mean $\delta^{18}O$ values from single Mongolian populations will have smaller ranges than this, and that with the collection of a greater quantity of appropriate modern control data from across the region it will be possible to more accurately define 'local' limits for particular sites. We conclude, on the basis of the evidence to hand, that all the Mongolian horses, with the possible exception of two (B12 and T2), lived in the locality of Tsengel Khairkhan and Baga Turgen Gol during the period of (sampled) M_3 enamel formation.

Acknowledgements

Thanks are due to Jacques Cuisin (Muséum national d'Histoire naturelle, Paris) for permission to sample the Lake Ulungar Przewalski's horse mandible, Delphine Frémondeau (MNHN) for useful discussions, Rob Batchelor (University of Reading, UK) for help with figures, and Jane Evans (NIGL, Nottingham, UK) for critical comments on an earlier draft that greatly helped to improve this paper. We would also like to thank Marjan Mashkour and Mark Beech for their organization of the 9th ASWA conference. This research was funded by the CNRS. The isotope analyses were carried out in the Service de Spectrométrie de Masse Isotopique of the Muséum national d'Histoire naturelle, Paris (SSMIM). The SSMIM is financially supported by the Région Ile de France.

References

Alexeev, A. I., Barkova, L. L. & Galanina, L. K. 2001, *Nomades des steppes: Les Scythes, VIIe–IIIe siecle av. J.–C.* Éditions Autrement, Paris.

Allard, F. & Erdenebaatar, D. 2005, 'Khirigsuurs, ritual and mobility in the Bronze Age of Mongolia'. *Antiquity* 79, 547–563.

Anthony, D. W. 2007, *The Horse, the Wheel and Language.* Princeton University Press, Princeton & Oxford.

Balasse, M. 2002, 'Reconstructing dietary and environmental history from enamel isotopic analysis: time resolution of intra–tooth sequential sampling'. *International Journal of Osteoarchaeology* 12, 155–165.

Balasse, M., Ambrose, S. H., Smith, A. B. & Price, T. D. 2002. 'The seasonal mobility model for prehistoric herders in the south–western Cape of South Africa assessed by isotopic analysis of sheep tooth enamel'. *Journal of Archaeological Science* 29, 917–932.

Bashilov, V. A. & Yablonsky, L. T. 1995, 'Some current problems concerning the history of Early Iron Age Eurasian Steppe nomadic societies', in J. Davis-Kimball, E. M. Murphy, L. Koryakova & L. T. Yablonsky (eds), *Kurgans, Ritual Sites, and Settlements: Eurasian Bronze and Iron Age*, 9–12. British Archaeological Report S890, Archaeopress, Oxford.

Bell, L. S., Lee-Thorp, J. A. & Dobney, K. 2006, 'Mapping human movement using stable oxygen isotopic ratio mass spectrometry: potential application to forensic science demonstrated by a modern horse–human study'. *Canadian Society of Forensic Science Journal* 39, 47–54.

Bendrey, R., Hayes, T. E. & Palmer, M. R. 2009, 'Patterns of Iron Age horse supply: an analysis of strontium isotope ratios in teeth'. *Archaeometry* 51, 140–150.

Bokovenko, N. 2004, 'Migrations of early nomads of the Eurasian steppe in a context of climatic changes', in E. M. Scott, A.

Y. Alekseev & G. Zaitseva (eds), *Impact of the Environment on Human Migration in Eurasia*, 21–33. Kluwer, Dordrecht.

Bowen, G. J, Wassenaar, L. I. & Hobson, K. A. 2005, 'Global application of stable hydrogen and oxygen isotopes to wildlife forensics'. *Oecologia* 143, 337–348.

Bowen, G. J. & Wilkinson, B. 2002, 'Spatial distribution of $\delta^{18}O$ in meteoric precipitation'. *Geology* 30, 315–318.

Christian, D. 1998, *A History of Russia, Central Asia and Mongolia, Volume 1, Inner Eurasia from Prehistory to the Mongol Empire.* Blackwell, Oxford.

Cornevin, C. & Lesbre, X. 1894, *Traité de l'age des animaux domestiques: d'après les dents et les productions épidermiques*. Librairie J.-B. Baillière et Fils, Paris.

Cribb, R. 1991, *Nomads in Archaeology.* Cambridge University Press, Cambridge.

Čugunov, K. V., Parzinger, H. & Nagler, A. 2003, 'Der skythische Fürstengrabhügel von Aržan 2 in Tuva: Vorbericht der russisch-deutschen Ausgrabungen 2000–2002'. *Eurasia Antiqua* 9, 113–162.

Evans, J. A., Chenery, C. A. & Fitzpatrick, A. P. 2006, 'Bronze Age childhood migration of individuals near Stonehenge revealed by strontium and oxygen isotope tooth enamel analysis'. *Archaeometry* 48, 309–321.

Ferret, C. 2009, *Une civilisation du cheval: Les usagers de l'équidé, de la steppe à la Taïga.* Belin, Paris.

Foggin, P., Farkas, O., Shiirev-Adiya, S. & Chinbat, B. 1997, 'Health status and risk factors of seminomadic pastoralists in Mongolia: A geographical approach'. *Social Science and Medicine* 44, 1623–1647.

Francfort, H.-P., Ligabue, G. & Samashev, Z. 2000, 'La fouille d'un kourgane scythe gelé du IVe siècle av. notre ère à Berel dans l'Altaï (Kazakhstan). *Académie des Inscriptions et Belles-Lettres*, 775–806.

Fricke, H. C., Clyde, W. C. & O'Neil, J. R. 1998, 'Intra–tooth variations in $\delta^{18}O(PO_4)$ of mammalian tooth enamel as a record of seasonal variations in continental climate variables'. *Geochimica et Cosmochimica Acta* 62, 1839–1850.

Geel, B. van, Bokovenko, N. A., Burova, N. D., Chugunov, K. V., Dergachev, V. A., Dirksen, V. G., Kulkova, M., Nagler, A., Parzinger, H., Plicht, J. van der, Vasiliev, S. S. & Zaitseva, G. I. 2004, 'Climate change and the expansion of the Scythian culture after 850 BC: a hypothesis'. *Journal of Archaeological Science* 31, 1735–1742.

Hanks, B. 2002, 'The Eurasian Steppe 'nomadic world' of the first millennium BC: Inherent problems within the study of Iron Age nomadic groups', in K. Boyle, C. Renfrew, & M. Levine, *Ancient Interactions: east and west in Eurasia*, 183–197. McDonald Institute for Archaeological Research, Cambridge.

Hoppe, K. A. & Koch, P. L. 2006, 'Reconstructing the migration patterns of late Pleistocene mammals from northern Florida, USA'. *Quaternary Research* 68, 347–352.

Hoppe, K. A., Stushka, S. & Amundson, R. 2005, 'The implications for paleodietary and paleoclimatic reconstructions of intrapopulation variability in the oxygen and carbon isotopes of teeth from modern feral horses'. *Quaternary Research* 64, 138–146.

Hoppe, K. A., Amundson, R., Vavra, M., McClaran, M. P. & Anderson, D. L. 2004a, 'Isotope analyses of equid teeth from modern North American feral horses: implications for paleoenvironmental reconstructions'. *Palaeogeography, Palaeoclimatology, Palaeoecology* 203, 299–311.

Hoppe, K. A., Stover, S. M., Pascoe, J. R. & Amundson, R. 2004b, 'Tooth enamel biomineralization in extant horses: implications for isotopic microsampling'. *Palaeogeography, Palaeoclimatology, Palaeoecology* 206, 355–365.

Humphrey, C. 1980, 'Introduction', in S. I. Vainshtein, *Nomads of South Siberia: the pastoral economies of Tuva*, 1–36. Cambridge University Press, Cambridge.

IAEA/WMO, 2006. *Global Network of Isotopes in Precipitation. The GNIP Database*. Accessible at: http://isohis.iaea.org

Jordana, X., Galtés, I., Turbat, T., Batsukh, D., García, C., Isidro, A., Giscard, P.-H. & Malgosa, A. 2009, 'The warriors of the steppes: osteological evidence of warfare and violence from Pazyryk tumuli in the Mongolian Altai'. *Journal of Archaeological Science* 36, 1319–1327.

Kilic, S., Dixon, P. M. & Kempson S. A. 1997, 'A light microscopic and ultrastructural examination of calcified dental tissues of horses: 1. The occlusal surface and enamel thickness'. *Equine Veterinary Journal* 29, 190–197.

Legrand, S. 2006, 'The emergence of the Karasuk culture'. *Antiquity* 80, 843–879.

Levine, M. A. 1982, 'The use of crown height measurements and eruption-wear sequences to age horse teeth', in B. Wilson, C. Grigson & S. Payne, (eds), *Ageing and Sexing Animal Bones from Archaeological Sites*, 223–250. British Archaeological Report 109. British Archaeological Reports, Oxford.

Keyser-Tracqui, C., Blandin-Frappin, P., Francfort, H.-P., Ricaut, F.-X., Lepetz, S., Crubezy, E., Samashev, Z. & Ludes, B. 2005, 'Mitochondrial DNA analysis of horses recovered from a frozen tomb (Berel site, Kazakhstan, 3rd century BC)'. *Animal Genetics* 36, 203–209.

Khazanov, A. M. 1984, *Nomads and the Outside World.* Cambridge University Press, Cambridge.

Koryakova, L. 2002, 'Between Steppe and forest: Iron Age societies of the Urals', in K. Boyle, C. Renfrew & M. Levine (eds), *Ancient Interactions: east and west in Eurasia*, 265–292. McDonald Institute for Archaeological Research, Cambridge.

Koryakova, L. 1995, 'Some notes about the material culture of Eurasian nomads', in J. Davis-Kimball, E. M. Murphy, L. Koryakova & L. T. Yablonsky (eds), *Kurgans, Ritual Sites, and Settlements: Eurasian Bronze and Iron Age*, 13–18. Britiah Archaeological Report S890, Archaeopress, Oxford.

Koryakova, L. & Hanks, B. 2006, 'Horse husbandry among the Early Iron Age Trans-Ural Societies', in S. L. Olsen, S. Grant, A. M. Choyke, & L. Bartosiewicz (eds), *Horses and Humans: the evolution of human–equine relationships*, 275–287. British Archaeological Report S1560, Archaeopress, Oxford.

Lee-Thorp, J. A. & Merwe, N. J. van der 1991, 'Aspects of the chemistry of modern and fossil biological apatites'. *Journal of Archaeological Science* 18, 343–354.

Levin, N. E., Cerling, T. E., Passey, B. H., Harris, J. M. & Ehleringer J. R. 2006, 'A stable isotope aridity index for terrestrial environments'. *Proceedings of the National Academy of Sciences* 103, 11201–11205.

Longinelli, A. 1984, 'Oxygen isotopes in mammal bone phosphate: A new tool for paleohydrological and paleoclimatological research?'. *Geochimica et Cosmochimica Acta* 48, 385–390.

Luz, B., Kolodny, Y. & Horowitz, M. 1984, 'Fractionation of oxygen isotopes between mammalian bone–phosphate and environmental drinking water'. *Geochimica et Cosmochimica Acta* 48, 1689–1693.

Passey, B. H. & Cerling, T. E. 2002, 'Tooth enamel mineralization in ungulates: implications for recovering a primary isotopic time–series'. *Geochimica et Cosmochimica Acta* 66, 3225–3234.

Pyankov, V. I., Gunin, P. D., Tsoog, S. & Black, C. C. 2000, 'C_4 plants in the vegetation of Mongolia: their natural occurrence and geographical distribution in relation to climate'. *Oecologia* 123, 15–31.

Renfrew, C. 2002, 'Pastoralism and Interaction: some introductory questions', in K. Boyle, C. Renfrew, & M. Levine (eds), *Ancient Interactions: east and west in Eurasia*, 1–10. McDonald Institute for Archaeological Research, Cambridge.

Rudenko, S. I. 1970, *The Frozen Tombs of Siberia.* Dent, London.

Scott, E. M., Alekseev, A. Y. & Zaitseva, G. (eds) 2004, *Impact of the Environment on Human Migration in Eurasia.* Kluwer, Dordrecht.

Shnirelman, V. A., Olsen, S. L. & Rice, P. 1996, 'Hooves across the steppe: the Kazak life–style', in S. L. Olsen (ed.), *Horses Through Time*, 129–152. Roberts Rinehart Publishers, Boulder, CO.

Sneath, D. 1999, 'Spatial mobility and Inner Asian pastoralism', in C. Humphrey & D. Sneath (eds), *The End of Nomadism?: society, state, and the environment in Inner Asia*, 218–277. Duke University Press, Cambridge.

Vainshtein, S. I. 1980. *Nomads of South Siberia: the pastoral economies of Tuva.* Cambridge University Press, Cambridge.

Wang, Y. & Cerling, T. E. 1994, 'A model of fossil tooth and bone diagenesis: Implications for paleodiet reconstruction from stable isotopes'. *Palaeogeography, Palaeoclimatology, Palaeoecology* **107**, 281–289.

Zaitseva, G. I., Chugunov, K. V., Alekseev, A. Y., Dergachev, V. A., Vasiliev, S. S., Sementsov, A. A., Cook, G., Scott, E. M., Plicht, J. van der, Parzinger, H., Nagler, A., Jungner, H., Sonninen, E. & Bourova, N. D. 2007, 'Chronology of key barrows belonging to different stages of the Scythian period in Tuva (Arzhan-1 and Arzhan-2 Barrows)'. *Radiocarbon* 49, 645–659.

Zazzo, A., Balasse, M. & Patterson, W. P. 2005, 'High–resolution $\delta^{13}C$ intra-tooth profiles in bovine enamel: implications for mineralization pattern and isotopic attenuation'. *Geochimica et Cosmochimica Acta* 69, 3631–3642.

Zazzo, A., Lécuyer, C. & Mariotti, A. 2004, 'Experimentally–controlled carbon and oxygen isotope exchange between bioapatites and water under inorganic and microbially mediated conditions'. *Geochimica et Cosmochimica Acta* 68, 1–12.

19. Zooarchaeological evidence for pastoralism in the Early Transcaucasian Culture

Jennifer J. Piro and Pam J. Crabtree

The papers presents the Early Transcaucasian Culture complex (ETC) subsistence economy during the 4th millennium BC based on the zooarchaeological evidence from four sites: Sos Höyük in northeastern Anatolia, Mokhra Blur and Gegharot in Armenia, and Godin Tepe in west-central Iran. ETC sites appear to be influenced not simply by ecological conditions, but by a range of social, economic, political, and cultural factors with varying degrees of archaeological visibility. This study indicates the emphasis on highly mobile forms of pastoralism both as a defining feature and as a mechanism for the creation and spread of the Early Transcaucasian Culture must be reconsidered.

Keywords Early Transcaucasian Culture complex, 4th millennium, pastoralism, Anatolia, Armenia, Iran

Introduction

The Early Transcaucasian Culture complex (ETC) emerged during the mid-4th millennium BC and later spread throughout the southern Caucasus and its adjacent regions. While the architecture and material culture from ETC sites have been well documented, little is known about the pastoral activities of these communities, particularly in terms of herding goals and strategies. Many researchers have suggested that mobile forms of pastoralism likely characterised the ETC subsistence economy, but the nature and extent of this mobility is much debated, ranging from long-distance seasonal transhumance (Kushnareva 1997, 194), to increasing mobility over time (Kushnareva 1997, 208, 230), to alternating forms of nomadisation and permanent settlement (Cribb 1991, 221), to the varying role of nomadic elements in the development of ETC communities (Sagona 1993, 453–454). However, the notion that mobile pastoralism figured prominently in the ETC is problematic for two reasons. First, we know little about ETC pastoral practices from their primary source, animal bone remains, since detailed faunal reports are largely unavailable or unpublished (Kohl 2007, 95; Sagona 1993, 453). Second, our current understanding of ETC pastoral activities is largely inferred from the portability of material culture and the impermanent or composite nature of dwelling forms, which resemble those used by contemporary nomadic groups (Burney & Lang 1971, 57; Cribb 1991, 221; Kiguradze & Sagona 2003; Sagona 1993; Whallon & Kantman 1969, 103). The rapid spread of the ETC horizon during the 3rd millennium BC is assumed to be linked, in part, to the seasonal movement of ETC populations in search of pasturage for their flocks (Kohl 1992, 122; Rothman 2003, 109).

Recent studies of excavated faunal remains provide new insight into the role of mobile pastoralism in the Early Transcaucasian Culture (cf. Bartosiewicz 1998; Bökönyi 1983; Crabtree 2011; Howell-Meurs 2001a; 2001b; Monahan 2005; Piro 2008; 2009). Although the existing dataset is still quite small, we can begin to test the premise that the economic base of ETC communities rested primarily on mobile pastoralism, and furthermore, that such mobility helped stimulate the migrations of ETC populations into neighboring regions. This paper presents zooarchaeological evidence from four ETC sites (Figs. 19.1 & 19.2): Sos Höyük in northeastern Anatolia, Mokhra Blur and Gegharot in Armenia, and Godin Tepe in west-central Iran.[1]

Fig. 19.1. Map of Sos Höyük, Mokhra Blur, and Gegharot in the region (created by Chandra Jayasuriya)

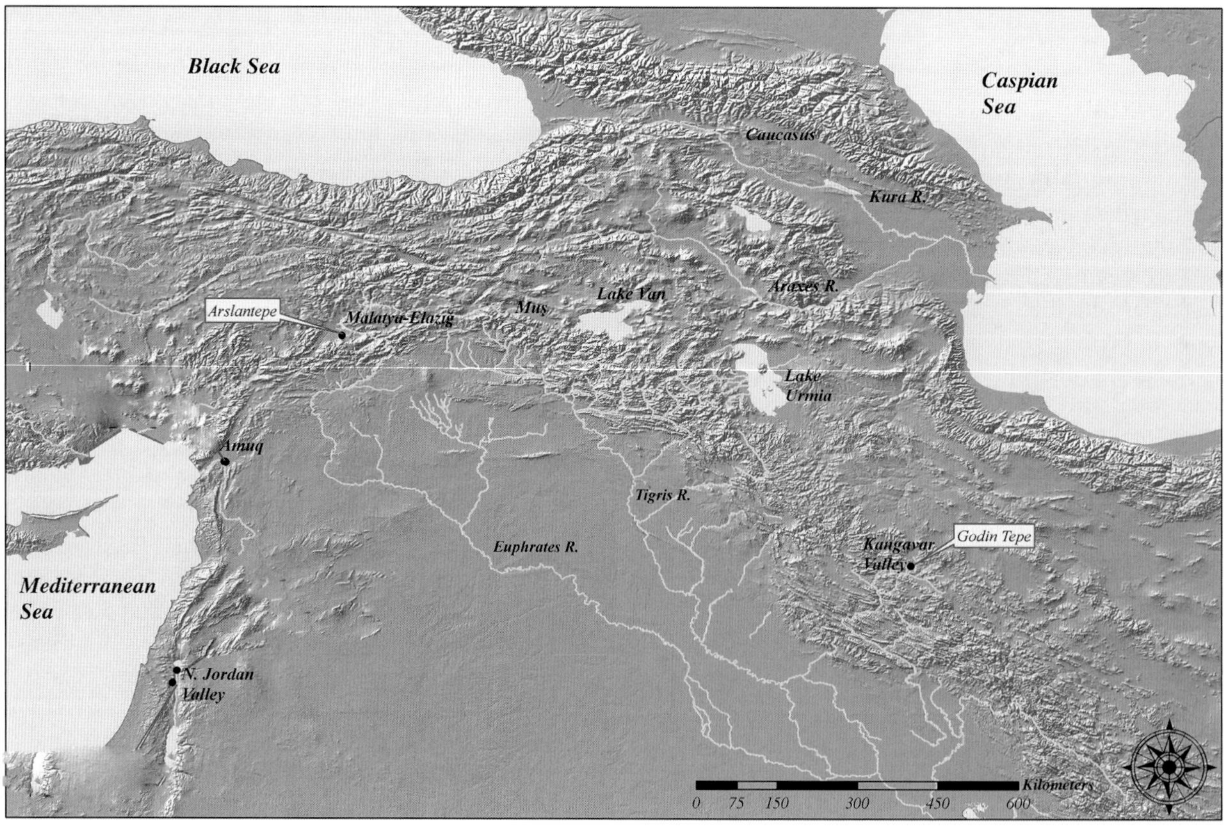

Fig. 19.2. Map of Godin Tepe in the region (adapted from a map created by Stephen Batiuk)

Sos Höyük faunal assemblage

The archaeological site of Sos Höyük is located in the modern village of Yiğittaşı, approximately 24km east of the city of Erzurum in the highlands of northeastern Turkey (1800m a.s.l.). Both the Late Chalcolithic and Early Bronze Age periods at the site yielded architectural forms and artefacts that resemble the ETC horizon style: round and rectangular houses constructed from wattle and daub or mud brick; figured hearth supports; black and red burnished wares and other related suites of ceramics; tools manufactured from bone, obsidian, and arsenical bronze; and horned animal figurines (Sagona *et al.* 1995; 1996; 1997; Sagona & Erkmen 1998; Sagona & Sagona 2000). At present, Sos is atypical of ETC sites due to its Middle Bronze Age settlement phase – rarely found elsewhere in Transcaucasia – and the continued use of the ETC ceramic suite, along with new stylistic and technical innovations, throughout this time period.

To investigate the nature of pastoral activities at Sos, approximately 16,000 faunal specimens were analysed from archaeological contexts containing ETC materials.[2] These contexts were chronologically divided into Late Chalcolithic (3500/3300–3000 cal BC), Early Bronze I (3000–2800 cal BC), Early Bronze II–III (2800–2200 cal BC), and Middle Bronze Age (2000–1500 cal BC) – almost 2000 years of the occupational sequence at the site (Sagona 2000). The vast majority of the faunal assemblage in each phase at Sos consists of domesticated species: caprines are the most abundant followed by cattle, while pigs, horses, and dogs are very rarely represented (Fig. 19.3 and Table 19.1). Wild animals constitute less than 8% of the sample in each period. Worth noting is the richness in species diversity found in the later Early Bronze Age, which yielded red deer, a variety of birds and small mammals, and the wild versions of cattle, sheep, goats, and pigs.[3]

Figure 19.4 presents the species ratios of the main domesticates from the Late Chalcolithic through the Middle Bronze Age at Sos. Overall, the relative abundance of cattle, caprines, and pigs is largely comparable between the Late Chalcolithic and Early Bronze I phases. During the later Early Bronze Age, however, the proportion of cattle rises, while that of caprines declines by approximately 13%. This reverses in the Middle Bronze Age as the number of cattle decreases and caprines once again dominate the assemblage. These fluctuations in the ratio of cattle to caprines over time are to some extent related to biases inherent to the NISP method of quantification. However, the analysis of specimen fragmentation rates in the EBII–III sample indicates that the sizeable proportion of cattle is not due to high levels of fragmentation (Piro 2009, 176–177, 184–185). Thus, while sheep and goats are the predominant species in each phase, cattle appear to have a particularly strong secondary role in terms of relative abundance during the later Early Bronze Age.

Ovis/capra: mortality data

Caprine herding goals and strategies are the focus of

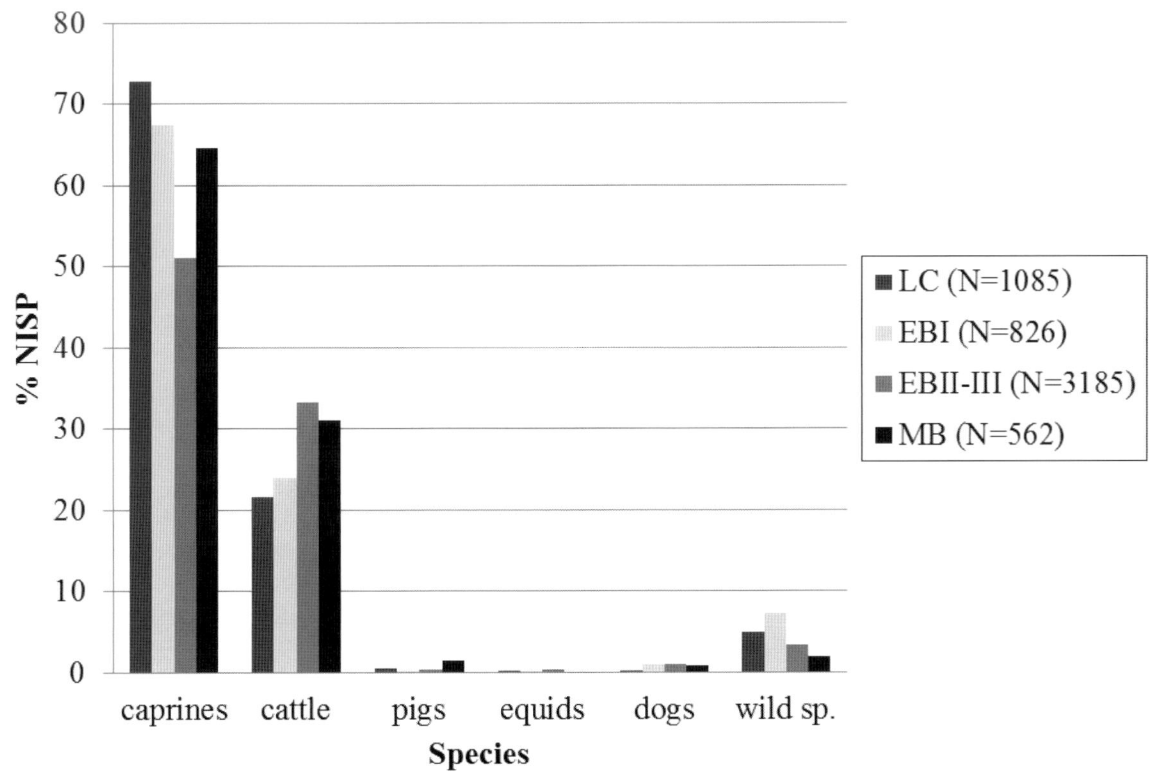

Fig. 19.3. Species representation at Sos Höyük. EBII–III phase includes data analysed by Howell-Meurs (2001b, 117)

Table 19.1.Number of Identified Specimens (NISP) assigned to genus from Sos

SPECIES	*LC*	*%*	*EBI*	*%*	*EBII-III*	*%*	*MB*	*%*
Domesticates								
Bos taurus (cow)	234	21.6	199	24.1	181	25.6	175	31.1
Ovis/Capra (sheep/goat)	789	72.7	560	67.8	476	67.2	363	64.6
including:								
Ovis aries (sheep)	211		150		124		109	
Capra hircus (goat)	54		36		53		20	
Sus scrofa dom. (pig)	5	0.5	2	0.2	2	0.3	8	1.4
Equus sp. (horse)	1	0.1	2	0.2	1	0.1	0	0.0
Canis familiaris (dog)	3	0.3	8	1.0	10	1.4	5	0.9
Subtotal	*1032*	*95.1*	*771*	*93.3*	*670*	*94.6*	*551*	*98.0*
Wild Species								
Capreolus capreolus (roe deer)	0	0.0	0	0.0	1	0.1	0	0.0
Cervus elaphus (red deer)	1	0.1	13	1.6	11	1.6	3	0.5
Castor fiber (beaver)	0	0.0	0	0.0	2	0.3	0	0.0
Lepus europaeus (hare)	3	0.3	6	0.7	4	0.6	0	0.0
Meles meles (Eurasian badger)	3	0.3	0	0.0	0	0.0	3	0.5
Mustela nivalis (common weasel)	1	0.1	2	0.2	2	0.3	0	0.0
Ursus Arctos (brown bear)	0	0.0	0	0.0	1	0.1	0	0.0
Mauremys caspica cas. (Caspian turtle)	0	0.0	0	0.0	1	0.1	0	0.0
Aves								
Alectoris chukar (chukar partridge)	0	0.0	2	0.2	0	0.0	0	0.0
Anas platyrhynchos (mallard)	0	0.0	0	0.0	0	0.0	1	0.2
Aquila chrysaetos (golden eagle)	0	0.0	0	0.0	1	0.1	0	0.0
Athene noctua (little owl)	0	0.0	0	0.0	1	0.1	0	0.0
Columba livia (rock pigeon)	0	0.0	0	0.0	1	0.1	0	0.0
Coturnix coturnix (common quail)	1	0.1	0	0.0	0	0.0	0	0.0
Otis tarda (great bustard)	1	0.1	1	0.1	0	0.0	0	0.0
Passer domesticus (house sparrow)	0	0.0	2	0.2	0	0.0	0	0.0
Perdix perdix (grey partridge)	1	0.1	0	0.0	0	0.0	0	0.0
Phasianus colchicus (common pheasant)	0	0.0	0	0.0	0	0.0	1	0.2
Podiceps cristatus (great crested grebe)	1	0.1	0	0.0	0	0.0	0	0.0
Other								
Unidentified fish	1	0.1	0	0.0	0	0.0	0	0.0
Unidentified amphibian	0	0.0	1	0.1	0	0.0	0	0.0
Gastropod (land snail)	1	0.1	0	0.0	1	0.1	0	0.0
Unionidae (bivalves).	39	3.6	28	3.4	12	1.7	3	0.5
Subtotal	*53*	*4.9*	*55*	*6.7*	*38*	*5.4*	*11*	*2.0*
TOTAL IDENTIFIED	1085	100.0	826	100.0	708	100.0	562	100.0
Intrusive								
Nannospalax nehringi (mountain mole rat)	2		3		1		0	
Mesocircetus brandti (Turkish hamster)	4		2		0		0	
Apodemus sylvaticus (wood mouse)	9		4		0		0	
Crocidura sp. (shrew)	1		0		0		0	
Microtus sp. (vole)	1		1		0		0	
Subtotal	*17*		*10*		*1*		*0*	
TOTAL	1102		836		2		0	

this section due to the substantial quantity of sheep and goat remains in the assemblage. Mortality profiles were constructed from tooth eruption and wear using mandibles, the lower 4th deciduous premolar, and the lower 3rd molar. Due to the limited dental data available for this study, a combined caprine sample was analysed from each phase (Fig. 18.5).[4] Findings indicate that caprine survivorship patterns are highly comparable throughout the Early Transcaucasian period at the site (Kolmogorov-Smirnov two-sample test, $p>0.05$). In each phase, a substantial proportion of animals were slaughtered between 1 and 3 years of age – a period that includes adolescence and early adulthood. This mortality pattern suggests that animals, perhaps subadult males, were killed for meat before reaching full maturity. A much smaller percentage of the herd, probably females, survived well into old age, provided that they continued to breed and provide milk. In order to infer production goals and management strategies, the caprine tooth eruption and wear data were then compared to Payne's (1973) and Redding's (1981; 1984) idealised kill-off patterns. The resulting survivorship pattern shows a close fit with Redding's (1981) herd security model and Payne's (1973) meat model in each phase (Kolmogorov-Smirnov two-sample test, $p>0.05$) (Fig. 19.5). Furthermore,

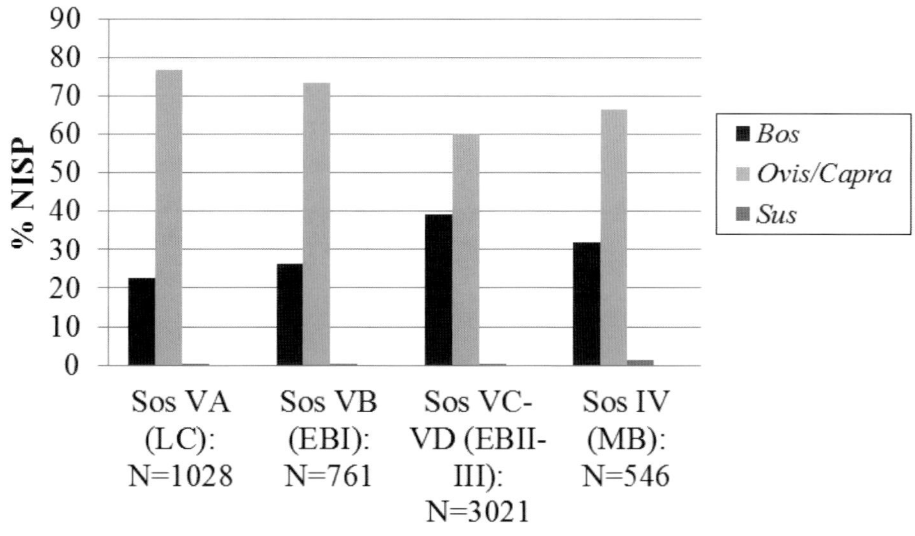

Fig. 19.4. Relative abundance of the main domesticates from Sos Höyük (% NISP). EBII–III phase includes data analysed by Howell-Meurs (2001b, 117)

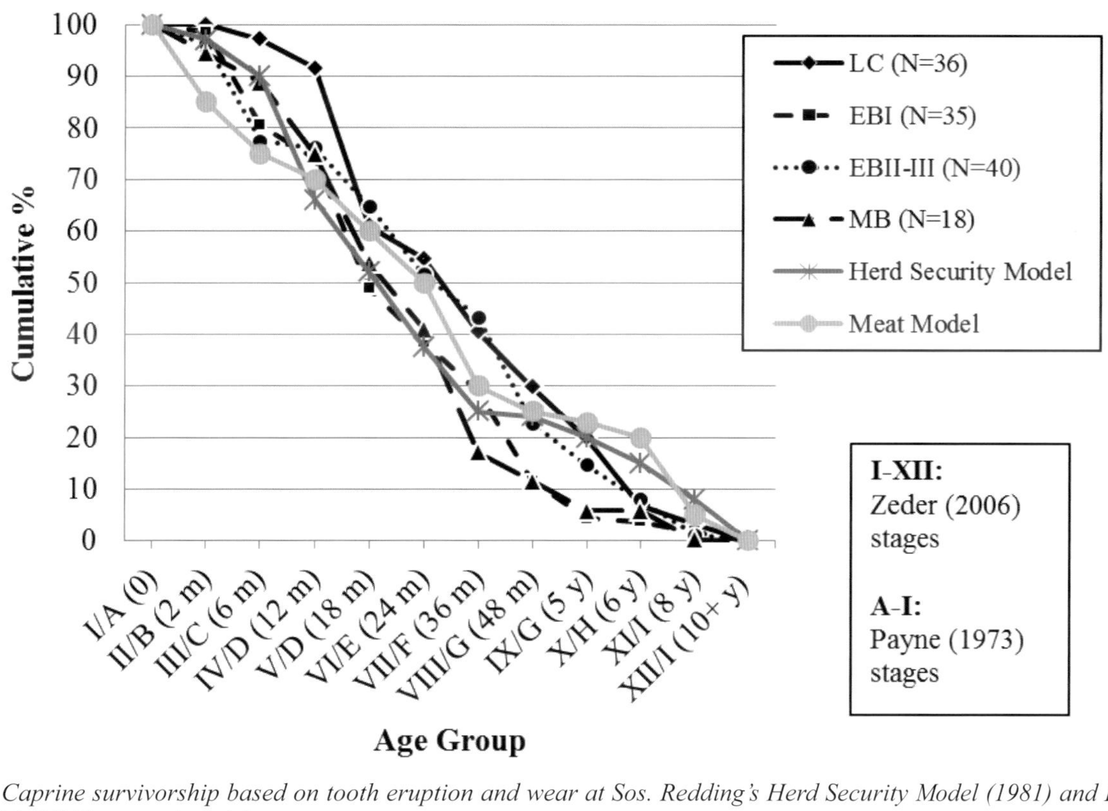

Fig. 19.5. Caprine survivorship based on tooth eruption and wear at Sos. Redding's Herd Security Model (1981) and Payne's Meat Model (1973) are included for comparison

the survivorship curves point to a stability and resilience in herd management practices throughout the Early Transcaucasian period at the site.

Pastoral production and seasonal mobility at Sos

Strategies centred on risk avoidance likely served as the guiding principle for subsistence practices from the Late Chalcolithic to the Middle Bronze phase at the site. First, the caprine mortality data indicate that production

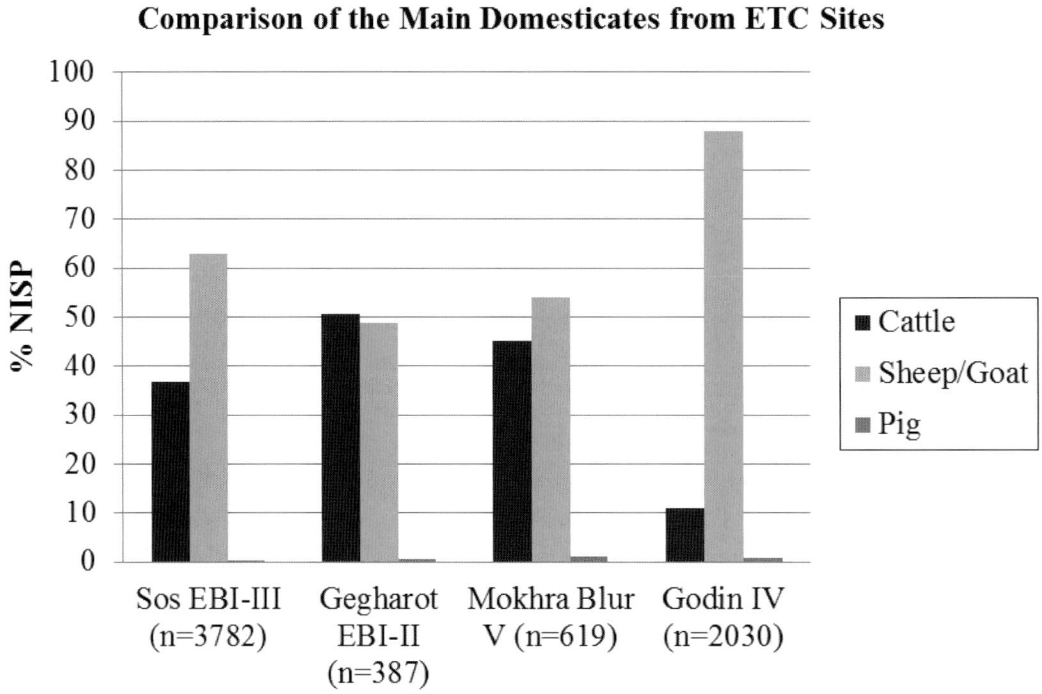

Fig. 19.6. Percentages of the main domesticates from ETC sites (NISP). EBII–III phase includes data analysed by Howell-Meurs (2001b, 117); Gegharot data from Monahan (2005, 386); Mokhra Blur data from Piro (2009, 279–282); Godin data from Crabtree (in press)

strategies focused on risk reduction, herd security, and the exploitation of a variety of products to meet subsistence needs. Second, the presence of wild taxa in all four phases of the assemblage, even at low levels, may reflect a strategy of resource diversification intended to minimise subsistence risk. While domestic livestock provided the bulk of consumable products in the form of meat and milk, an extensive variety of wild species (particularly the wild cattle, caprines, and pigs noted by Howell-Meurs (2001b, 117)) could be exploited as additional meat sources, when necessary. Third, the herding of domestic cattle, caprines, and pigs may indicate strategies aimed at the diversification of livestock production by exploiting a wide range of environmental niches in the site's vicinity.[5] A broad-based approach to husbandry practices, together with small-scale hunting, trapping, and agricultural activities, would help minimise subsistence risk and uncertainty by increasing predictability and stability in the subsistence economy (Halstead & O'Shea 1989, 3–4; Stein 1989).[6]

In terms of seasonal exploitation of livestock, analysis of tooth eruption data and foetal/neonatal remains suggests that caprines were present at the site throughout the year. The eruption time of the 1st and 2nd molars – at approximately 5–6 months and 9–12 months, respectively – may help estimate the season of death. In the Sos assemblage, the presence of specimens in the younger age groups, particularly the 6 month and 12 month categories, raises the question as to whether this pattern represents year-round mortality or specific seasonal culling. The birthing season of domestic caprines from Sos was likely more comparable to that of wild sheep (*Ovis orientalis*) and wild goats (*Capra aegagrus*), which rut in October/November and give birth in April/May (MacDonald & Barrett 1993, 219), rather than the extended birthing season associated with the intensive management of sheep and goats by contemporary pastoralists in the Middle East (Redding 1981, 86; Epstein & Herz 1964, 240; Rottensten & Ampy 1971, 371). If lambing occurs in the spring, and mortality occurs among juveniles at 5–6 and 9–12 months of age, this implies the presence of animals in the autumn, late winter, and early spring at the site.[7] In addition, the recovery of foetal and neonatal bones shows that pregnant ewes and very young lambs were present at the site during the spring and summer. Together, these data indicate that caprines were exploited at the site throughout the year and, therefore, refute previous interpretations of Sos as a summer encampment associated with transhumant or nomadic forms of pastoralism (Sagona & Sagona 2000, 65). Furthermore, if Sos was occupied only in the summer months, this would be reflected by gaps or peaks in the mortality data. The fact that all age categories are fairly well-represented in the assemblage argues against seasonal herding at the site.

Regional comparison of the main domesticates

Bone remains of the main domesticates from Early Bronze Sos were compared to those from Mokhra Blur and Gegharot, two Early Bronze Age settlements in northwestern

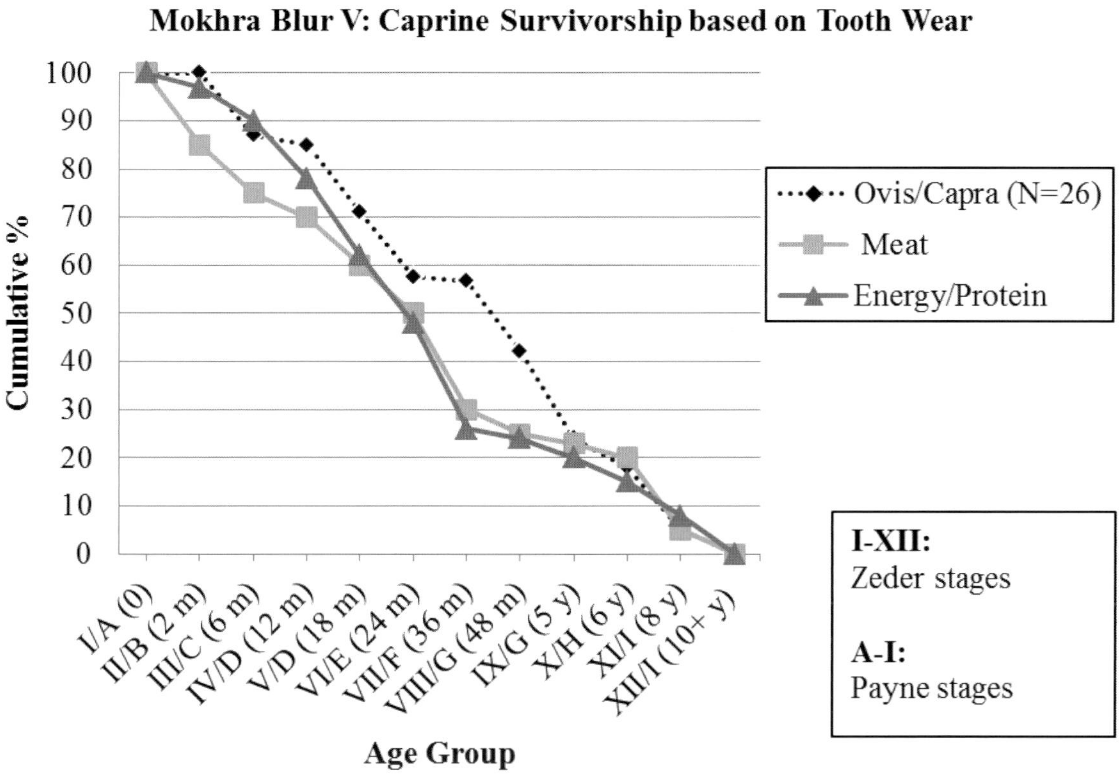

Fig. 19.7. Mokhra Blur V: caprine survivorship based on tooth eruption and wear. Redding's (1981; 1984) energy/protein model and Payne's (1973) meat model are included for comparison

Table 19.2. Number of Identified Specimens (genus) from Mokhra Blur V

Identification	*Common Name*	*No.*	*Percentage*
Bos	Cattle	279	42.0
Ovis/Capra	Sheep/goat	334	50.2
Sus	Pig	6	0.9
Cervus elaphus	Red deer	16	2.4
Ovis orientalis	Wild sheep	19	2.9
Canis familiaris	Dog	11	1.6
Total		665	100.0

Armenia, in order to provide insight into the nature of ETC herding economies on a broader regional scale (Fig. 19.1). These sites are located in very different ecological zones: Mokhra Blur lies in the Ararat Valley, while Gegharot is situated on an upland plain, about 2100m a.s.l.

Despite the small size of the Gegharot faunal assemblage, the high proportion of cattle suggests that this site was a sedentary village during the early 3rd millennium BC. The assemblage is dominated by roughly equal proportions of cattle (49% NISP, genus) and caprines (47.3%), followed by red deer (2.5%) and very rarely, pigs, equids, gazelle, and weasel (Monahan 2005, 386, table 15.4). Sheep greatly outnumber goats in the assemblage, but this should be taken with caution due to the high fragmentation of bones and the fact that goats were represented by a single specimen. Caprine mortality profiles based on tooth wear are not yet available from Gegharot due to the small sample size.

Similar to Sos, the zooarchaeological evidence from Gegharot questions the commonly held view, supported primarily by ethnographic accounts, that ETC sites in high elevations were linked with patterns of transhumance in the region (cf. Kushnareva 1997, 192). The high proportions of cattle and caprines at both sites, particularly at Gegharot, do not coincide with specialised forms of pastoralism that focus on the production of one species and that may involve the seasonal movement of herds (Fig. 19.6). Moreover, the long duration of Early Bronze Age occupation at Gegharot further challenges the notion of an overall migration of people from the lowlands and valleys to the foothills and mountains during the 3rd millennium BC (Smith 2005, 259; cf. Kushnareva 1997, 188).

The analysis of faunal remains from Mokhra Blur was limited to level V – dated approximately to the first half of the 3rd millennium BC. Interestingly, the relative proportion of species is broadly comparable to that from Gegharot (Fig. 19.6): cattle and caprines dominate the assemblage in roughly equal percentages, pigs are very rare, and red deer are present in small amounts (Table 19.2). Low percentages of wild sheep and dogs complete the assemblage. In terms of productive goals, the pattern of sheep and goat herding at Mokhra Blur resembles both Redding's (1981; 1984) energy/protein maximization model and Payne's (1973) meat model (Fig. 19.7).

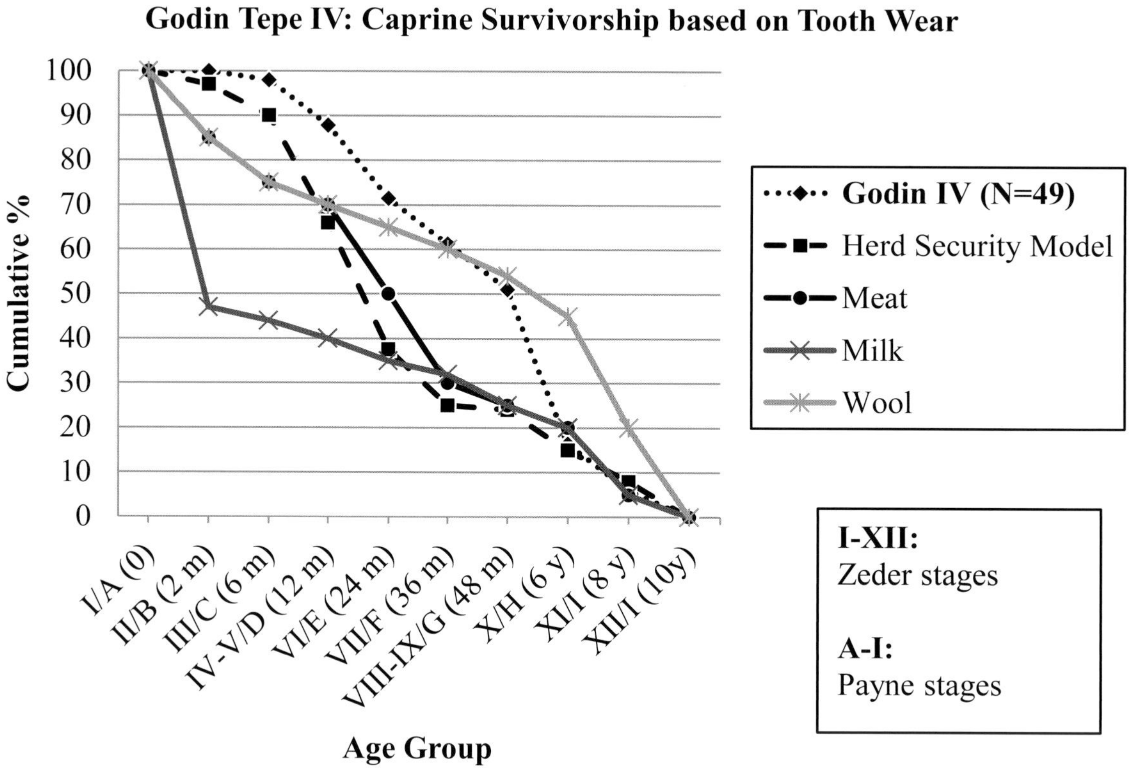

Fig. 19.8. Godin Tepe IV: caprine survivorship based on tooth eruption and wear. Redding's (1981; 1984) herd security model and Payne's (1973) meat, milk, and wool model are included for comparison

However, this interpretation is tentative due to the limited tooth wear data available for analysis.

Similar to Sos, the pattern of animal use from Mokhra Blur points to resource diversification and conservative production strategies: 1) the broad-based exploitation of domestic livestock, instead of specialised herding; and 2) caprine meat production supplemented by the small-scale hunting of wild caprines and red deer. Moreover, the presence of all age groups at the site indicates that caprines were raised and consumed locally. Overall, the faunal data are consistent with the archaeological evidence (see Areshian & Kafadaryan 1975; Areshian 2005), which together portray Mokhra Blur V as an extremely stable, sedentary agro-pastoral village in the Early Bronze Age.

While the zooarchaeological evidence revealed Sos Höyük, Mokhra Blur, and Gegharot as stable village communities, in which the ETC horizon persisted for over a millennium, animal bone remains from the site of Godin Tepe, located in the Kangavar Valley in west-central Iran, may offer insight into the movement of ETC populations into neighboring regions (Fig. 19.2). The faunal remains from Godin IV (ca. 3000–2400 BC) reveal striking differences from those in the ETC levels at Sos, Gegharot, and Mokhra Blur (Fig. 19.6). In terms of species composition, caprines represent over 80% of the identified remains, followed by cattle at a distant second at just over 10% (Table 19.3) (Crabtree 2011). Caprine mortality data from Godin IV indicate an emphasis on secondary product exploitation (i.e. wool, hair, and possibly dairy products) due to the sizeable proportion of older animals in the assemblage (Fig. 19.8). Nevertheless, the kill-off pattern does not correspond closely to Payne's meat, milk, and wool models for specialised production, nor does it resemble Redding's herd security model (Kolmogorov-Smirnov two-sample test, $p<0.01$). The low representation of young animals, particularly below 6 months of age, likely affected the shape of the survivorship curve and may suggest that caprines were raised outside the site's vicinity.

Table 19.3. Number of Identified Specimens (genus) from Godin IV

Identification	*Common Name*	*No.*	*Percentage*
Bos	Cattle	225	10.6
Ovis/Capra	Sheep/goat	1786	83.8
incl. *Ovis*	Sheep	333	
Capra	Goat	134	
Sus	Pig	19	0.9
Equus sp.	Horse	76	3.6
Canis familiaris	Dog	4	0.2
Cervus elaphus	Red deer	7	0.3
Capreolus capreolus	Roe deer	9	0.4
Lepus capensis	Brown hare	4	0.2
Meles meles	Badger	1	–
Total		2131	100.0

In terms of cattle management, the paucity of mortality

data limits the interpretation of production goals (Crabtree 2011). Available dental and fusion data indicate that most of the cattle (69%) survived beyond 3 years of age, with no evidence for juvenile mortality. This culling pattern, with its focus on adolescents and older adults, points to the importance of cattle secondary products, such as dairying and traction. Pathological conditions (i.e. lipping and exostosis) on cattle phalanges and the occurrence of plow marks provide further evidence of traction activities.

The zooarchaeological evidence from Godin provides insight into the degree of mobility and the type of pastoralism practiced at the site. It appears that the inhabitants of Godin IV raised caprines for meat and secondary products, particularly wool, within a broad subsistence base that included agriculture and hunting. In this respect, Godin IV may be considered a sedentary village, with a number of the inhabitants living at the site year-round. Clear evidence of pathologies associated with penning indicates that caprines were kept at the site (Crabtree 2011). Yet, the overwhelming proportion of caprines in the Godin IV sample, along with the absence of the younger age categories in the mortality profiles, may indicate that some animals were kept away from the settlement (Crabtree 2011). It is possible that some members of the site may have moved the caprine herds on a seasonal basis, perhaps by practicing vertical transhumance – a form of pastoralism still visible today in the Zagros region. Alternatively, local pastoral groups in the area may have provisioned the settlement. Although secondary products were important, the low sheep to goat ratio (2.5:1) and the caprine mortality pattern argue against highly specialized forms of pastoral production.

In terms of ETC migrations, the zooarchaeological data from Godin appear to corroborate Mitchell Rothman's model describing the gradual infiltration of new populations from Transcaucasia into west-central Iran (Batiuk & Rothman 2007; Rothman 2003; 2005). According to this model, the increase in the number of cave sites, open air sites, and isolated cemeteries in highland areas during the Late Chalcolithic – and the occurrence of small amounts of ETC ceramics at some of these sites – suggests the arrival of ETC pastoral nomads. Following this period, many new sites with exclusively ETC ceramics were established on the valley bottom, while a few abandoned Late Chalcolithic sites, such as Godin, were reoccupied. Thus, Godin IV would represent the reoccupation of the site by incoming ETC settlers. On the whole, faunal evidence demonstrating sedentary or transhumant agro-pastoralism at Godin is consistent with this view. However, it is also possible that pastoralists, who were part of the first wave of ETC migration, or even local Late Chalcolithic pastoralist groups who may have existed on the margins of the site, may have interacted with the Godin IV settlement (Young 2004, 257). More analysis of faunal assemblages from postulated ETC migration sites is necessary to continue testing this model and to clarify the role of pastoral movements in the spread of the Early Transcaucasian Culture.

Conclusion

The comparison of faunal data from sites with ETC occupation phases indicates that herding strategies may not coincide with expectations based on settlement type and location. The nature and extent of pastoral activities at ETC sites appears to be influenced not simply by ecological conditions, but by a range of social, economic, political, and cultural factors with varying degrees of archaeological visibility. Thus, it is challenging to define what constitutes a 'typical' ETC pastoral economy, especially given the general paucity of carefully analysed faunal data from ETC sites. What is clear, however, is that the emphasis on highly mobile forms of pastoralism both as a defining feature and as a mechanism for the creation and spread of the Early Transcaucasian Culture must be reconsidered. As larger faunal samples from additional ETC sites become available, we can examine this heterogeneity in herding economies more closely and assess whether they are indeed part of a broader trend of agro-pastoralism across the region or whether mobile forms of pastoralism were the dominant strategy at certain sites. Such analyses are the necessary foundation for understanding the processes behind the spread of the ETC phenomenon across the Near East.

Acknowledgements

Analysis of the fauna from Sos Höyük was funded by grants from the Sigma Xi Scientific Research Society and the Center for Ancient Studies at New York University. Analysis of the fauna from Mokhra Blur V was funded by a Fulbright IIE grant to Armenia (2004–5). Additional thanks go to Dr Tony Sagona (co-director of the Sos excavations) for the opportunity to analyze the animal bone remains from Sos; Dr Hakob Simonyan, (Center for Armenian Culture and Historical Heritage, Armenia) for permission to study the faunal remains from Mokhra Blur; Dr Nina Manaseryan (Institute of Zoology, National Academy of Sciences, Armenia) for assistance and guidance with the animal bones from Mokhra Blur; and Dr Mitchell Rothman, who arranged for the Godin fauna to be analysed at the NYU Faunal Laboratory. Final responsibility for content rests with the authors.

Notes

1 The zooarchaeological evidence from Sos and Mokhra Blur (Level V) was analysed by Piro (2009) (see also Howell-Meurs 2001a; 2001b). Faunal remains from Gegharot were studied by Monahan (2005), while those from Godin were examined by Crabtree 2011.

2 Previous analysis of faunal remains conducted by Howell-Meurs (2001a; 2001b) from the later Early Bronze Age at Sos is included in this study.

3 See Howell-Meurs (2001b, 117) for additional identified species from the EBII–III phase.

4 See Piro (2009, 345–360) for tables and figures detailing caprine mandibular data by phase and Howell-Meurs (2001b,

171) for additional caprine dental data from the EBII–III phase.

5 See Longford *et al.* (2009) and Newton's section in Sagona & Sagona (2004, 109–110) for in-depth discussions of the ancient landscape at Sos.

6 Several varieties of domesticated wheat and barley – both hulled and naked types - were found in the Late Chalcolithic and Middle Bronze levels at Sos (Longford *et al.* 2009, table 2).

7 This pattern was also documented by Howell-Meurs (2001b, 99) in her analysis of caprine tooth eruption from the EBII–III phase at the site.

References

Areshian, G. 2005, 'Early Bronze Age settlements in the Ararat Plain and its vicinity'. *Archäologische Mitteilungen aus Iran und Turan*, 37, 71–88.

Areshian, G. & Kafadaryan, K. K. 1975, 'Rozhdeniye mounumentalnoy arkhitektury na territorii Armenii: 1-ya polovina IIIt tys. do n.e. (The rise of monumental architecture within the territory of Armenia: the first half of the 3rd millennium B.C.)'. *Pamyatniki Kultury: Novye Otkrytiya 1974*, 397–403.

Bartosiewicz, L. 1998, 'Interim report on the Bronze Age animal bones from Arslantepe (Malatya, Anatolia)', in H. Buitenhuis, L. Bartosiewicz, & A. M. Choyke (eds), *Archaeozoology of the Near East III,* 221–232. ARC Publicaties, Groningen.

Batiuk, S. & Rothman, M. S. 2007, 'Early Transcaucasian Cultures and their neighbors' *Expedition* 49(1), 7–17.

Bökönyi, S. 1983, 'Late Chalcolithic and Early Bronze I animal remains from Arslantepe (Malatya), Turkey: a preliminary report', in M. Frangipane & A. Palmieri, 'Perspectives on protourbanization in Eastern Anatolia: Arslantepe (Malatya). An interim report on 1975–1983 campaigns', *Origini XII*, 581–598.

Burney, C. A. & Lang D. M. 1971, *The Peoples of the Hills: ancient Ararat and Caucasus*, Weidenfeld & Nicolson, London.

Crabtree, P. J. 2011, 'The animal remains from Godin Period IV', in H. Gopnik & M. S. Rothman (eds), *On the High Road: the history of Godin Tepe, Iran*, 178. Mazda, New York.

Cribb, R. 1991, *Nomads in Archaeology*. Cambridge University Press, New York.

Epstein, H. & Herz, A. 1964, 'Fertility and birth weights of goats in a subtropical environment'. *Journal of Agricultural Science* 62, 237–244.

Halstead, P. & O'Shea, J. 1989, 'Introduction: cultural responses to risk and uncertainty', in P. Halstead & J. O'Shea (eds), *Bad Year Economics: cultural responses to risk and uncertainty*, 1–7. Cambridge University Press, New York.

Howell-Meurs, S. 2001a, 'Archaeozoological evidence for pastoral systems and herd mobility: the remains from Sos Höyük and Büyüktepe Höyük'. *International Journal of Osteoarchaeology* 11, 321–328.

Howell-Meurs, S. 2001b, *Early Bronze and Iron Age Animal Exploitation in Northeastern Anatolia: the faunal remains from Sos Höyük and Büyüktepe Höyük.* British Archaeological Report S945, Archaeopress, Oxford.

Kiguradze, T. B & Sagona, A. G. 2003, 'On the origins of the Kura–Araxes cultural complex', in A. T. Smith & K. S. Rubinson (eds), *Archaeology of the Borderlands: investigations in Caucasia and beyond*, 38–94. Cotsen Institute of Archaeology, University of California, Los Angeles.

Kohl, P. L. 1992, 'The Transcaucasian 'periphery' in the Bronze Age: a preliminary formulation', in E. M. Schortman & P. A. Urban (eds), *Resources, Power, and Interregional Interaction*, 117–137. Plenum Press, New York.

Kohl, P. L. 2007, *The Making of Bronze Age Eurasia*. Cambridge University Press, New York.

Kushnareva, K. K. 1997 , *The Southern Caucasus in Prehistory: stages of cultural and socioeconomic development from the eighth to the second millennium B.C.*. University Museum, University of Pennsylvania, Philadelphia.

Longford, C., Drinnen, A. & Sagona, A. G. 2009, 'Archaeobotany of Sos Höyük, northeast Turkey', in A. Fairburn, S. O'Connor & B. Marwick (eds), *New Directions in Archaeological Science*, 121–136. Australian National University E Press, Canberra.

MacDonald, D. & Barrett, P. 1993, *Mammals of Britain and Europe*. Harper Collins, London.

Monahan, B. H. 2007, 'Nomadism in the Early Bronze Age southern Transcaucasia: the faunal perspective', in L. M. Popova, C. Hartley & A. T. Smith (eds), *Social Orders and Social Landscapes. Second University of Chicago Conference on Eurasian Archaeology*, 379–392. Cambridge Scholars Publishing, Chicago.

Payne, S. 1973, 'Kill-off patterns in sheep and goats: the mandibles from Asvan Kale'. *Anatolian Studies* 23, 281–303.

Piro, J. J. 2008, 'Pastoral economies in Early Transcaucasian communities from the mid-IVth to IIIrd millennium B.C.', in E. Vila, L. Gourichon, A. M. Choyke & H. Buitenhuis (eds), *Archaeozoology of the Near East VIII: proceedings of the eighth international symposium on the archaeozoology of southwestern Asia and adjacent areas, 2006*, 451–463. CNRS, Lyon.

Piro, J. J. 2009, 'Pastoralism in the Early Transcaucasian Culture: the faunal remains from Sos Höyük', Ph.D. dissertation, New York University, New York.

Redding, R. 1981, 'Decision making in subsistence herding of sheep and goats in the Middle East', Ph.D. dissertation, University of Michigan, Ann Arbor.

Redding, R. 1984, 'Theoretical determinants of a herder's decisions: modeling variation in the sheep/goat ratio, in J. Clutton–Brock & C. Grigson (eds), *Animals and Archaeology: early herders and their flocks* 3, 223–241. British Archaeological Report S202, British Archaeological Reports, Oxford.

Rothman, M. S. 2003, 'Ripples in the stream: Transcaucasia–Anatolian interaction in the Murat/Euphrates Basin at the beginning of the third millennium B.C.', in A. T. Smith & K. S. Rubinson (eds), *Archaeology of the Borderlands: investigations in Caucasia and beyond*, 95–110. Cotsen Institute of Archaeology, University of California, Los Angeles.

Rothman, M. S. 2005, 'Transcaucasians: settlement, migration, and trade in the Kura Araks periods'. *Archäologische Mitteilungen aus Iran und Turan* 37, 1–10.

Rottensten, K. & Ampy, F. 1971, 'Studies on Awassi sheep in Lebanon, I. Production traits of a flock'. *Journal of Agricultural Science* 77, 371–373.

Sagona, A. G. 1993, 'Settlement and society in late prehistoric Trans–Caucasus', in M. Frangipane (ed.), *Between the Rivers and Over the Mountains*, 453–474. La Sapienza, Rome.

Sagona, A. G. 2000, 'Sos Höyük and the Erzurum region in late prehistory: a provisional chronology for northeast Anatolia', in C. Marro & H. Hauptmann (eds), *Chronologies des pays du Caucase et de l'Euphrate aux IVe–IIIe millenaires*, 329–373. De Boccard, Paris.

Sagona, A. G. & Erkmen, M. 1998, 'Excavations at Sos Höyük, 1997: fourth preliminary report'. *Anatolica* 24, 31–64.

Sagona, A. G., Erkmen, M., Sagona, C., McNiven, I. & Howells, S. 1997, 'Excavations at Sos Höyük, 1996: third preliminary report'. *Anatolica* 23, 181–226.

Sagona, A.G., Erkmen, M., Sagona, C., Thomas, I. & Howells, S. 1996, 'Excavations at Sos Höyük, 1995: second preliminary report'. *Anatolian Studies* 46, 27–52.

Sagona, A. G. & Sagona, C. 2000, 'Excavations at Sos Höyük 1998–2000: fifth preliminary report'. *Ancient Near Eastern Studies* 37, 56–127.

Sagona A. G. & Sagona, C. 2004, *Archaeology at the North-east Anatolian Frontier, I: an historical geography and a field survey of the Bayburt province*. Peeters, Louvain.

Sagona, A. G., Sagona, C. & Ozkorucuklu, H. 1995, 'Excavations at Sos Höyük: first preliminary report 1994'. *Anatolian Studies* 45, 193–218.

Smith, A. T. 2005, 'Prometheus unbound: southern Caucasia in prehistory'. *Journal of World Prehistory* 19, 229–279.

Stein, G. 1989, 'Strategies of risk reduction in herding and hunting systems of Neolithic southeast Anatolia', in P. Crabtree, D. Campana & K. Ryan (eds), *Early Animal Domestication and its Cultural Context*, 87–97. MASCA, University of Pennsylvania, PA.

Whallon, R. & Kantman, S. 1969, 'Early Bronze Age development in the Keban Reservoir, east central Turkey'. *Current Anthropology* 10, 128–183.

Young, T. C. 2004, 'The Kangavar Survey – Periods VI to IV', in A. G. Sagona (ed.), *A View from the Highlands: archaeological studies in honor of Charles Burney*, 645–660. Peeters, Louvain.

Zeder, M. A. 2006, 'Reconciling rates of long bone fusion and tooth eruption and wear in sheep (*Ovis*) and goat (*Capra*)', in D. Ruscillo (ed.), *Recent Advances in Ageing and Sexing Animal Bones*, 87–118. Oxbow, Oxford.

Part 5

Exploitation of animals in the Arabian Peninsula

20. New data on domestic and wild camels (*Camelus dromedarius* and *Camelus* sp.) in Sabaean and Minaean Yemen

F. G. Fedele

Recently studied faunas from controlled, dated Sabaean archaeological contexts provide information on the presence of both domestic and contemporary wild camel in the 9th–7th century BC time range in Yemen. Although still very limited this evidence may be of interest. The wild or domestic status was determined on the basis of both osteometrical data and archaeological context. Domestic Camelus dromedarius *was occasionally butchered for household meat consumption at Yalā (Ad-Durayb), southwest of Marib, in two subsequent occupations that can be dated within the 8th–7th century BC, if not slightly earlier (A. de Maigret's excavations). At Barāqish, in the Wadi al-Jawf region of northeastern Yemen, deep stratigraphic testing outside the walls in 2005–2006 has revealed a long sequence of occupations spanning the whole 1st millennium BC (F. G. Fedele's excavations). The domestic camel is frequent throughout the series, increasing in numbers during the post-Sabaean phases. Its earliest presence can again be dated to the 8th century BC, in association with probable evidence of early trading activities. A unique find of this period, an isolated humerus deriving from a caravan 'campsite' context on the outskirts of the Sabaean town, suggests the existence of wild* Camelus *populations in the area.*

Keywords Domestic dromedary, wild dromedary, Sabaean, Yemen, Yalā

Introduction and background

In a wide-ranging reappraisal of *Camelus* domestication in the Arabian Peninsula, and discussion of camel domestication in general, Uerpmann and Uerpmann (2002) have argued for the relatively late appearance of the domesticated dromedary in southeast Arabia. On present zooarchaeological evidence they suggest that all the remains of domestic camels date from the Iron Age II onwards, and particularly from after 900 BC. This is at least a century later than the assumed earliest appearance of domestic camels in the Levant, or arguably in the general area of the northern Arabian fringe. In these latter regions, reliable historical sources in Assyrian records (e.g. Liverani 1992; Na'aman 2007) point to the presence of domestic dromedaries in the 9th century BC, or somewhat earlier by implication, while osteological evidence from the southern Levant (e.g. Horwitz & Rosen 2005; Wapnish 1984) documents a pronounced increase in their utilisation from about 800 BC. The sociocultural context appears to be provided by a combination of rapidly evolving caravan trade and a new political assertiveness by 'aggressive' pastoral nomads of the Arabian area. Plausible evidence for a much earlier domestication in the 13th–12th centuries BC will be discussed below.

The late appearance of osteologically domestic camels in the Gulf, as particularly recorded at the large stratified settlement of Tell Abraq and at Muweilah in the U.A.E. (Magee 2007; Potts 1991; 1993) (Fig. 20.1), can be taken as evidence against an independent domestication of the dromedary in southeast Arabia. This very much reopens a debate, as noted by Magee (2007, 94), the details of which are beyond the limits of this paper. In addition, this might lead to the rather surprising conclusion that southern Arabia should be excluded from any role in camel

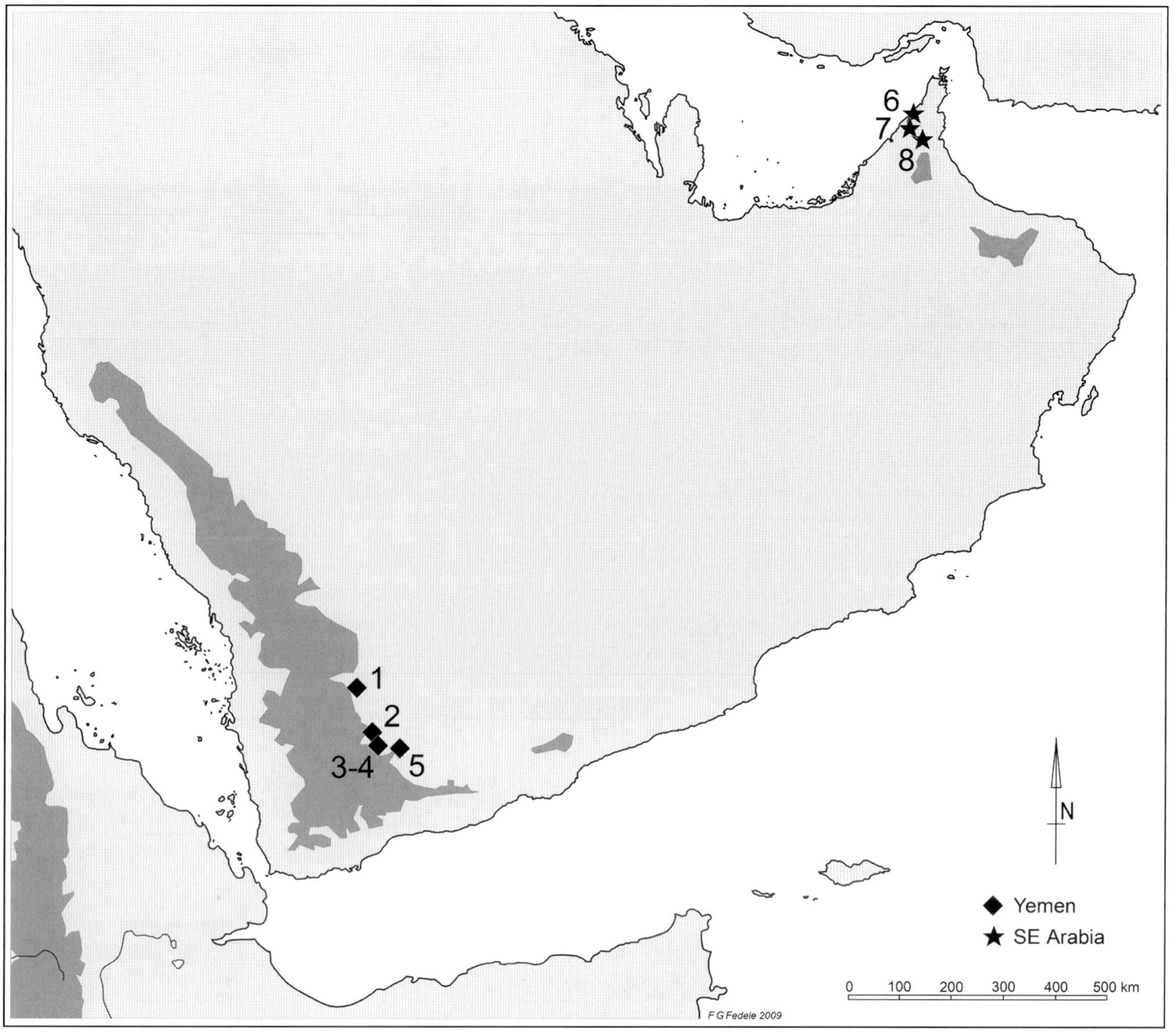

Fig. 20.1. Locations of the sites mentioned in the text (Yemen and Southeast Arabia): 1. Barāqish, 2. Yalā, 3. Hajar ar-Rayhānī, 4. Hajar at-Tamra, 5. Hajar Ibn Humayd, 6. Tell Abraq, 7. Muweilah, 8. Mleiha.

domestication. But in fact most of the Peninsula has yet to be investigated. Uerpmann and Uerpmann (2002, 258) are careful to conclude that 'clear biological evidence of when and where the Arabian camel was first domesticated is still missing', and go on to suggest that 'future research should concentrate on northern Arabia, although southern Arabia is not ruled out as a potential area of origin'. To the extent that direct bone evidence is considered of primary importance to recognise domestication, such statements represent the most that can be said at the moment concerning the Arabian camel.

As a subject of broad archaeological and anthropological interest, the origins and early developments of domesticated camels in the Old World have generated a considerable amount of literature and debate. The discussion has often suffered from ineffective dialogue between textual and osteological analysts, and even from misunderstandings on the part of some culture historians of the fundamental differences between the two species concerned, the Bactrian camel in cold-winter central Asia and the dromedary in hot, dry Arabia (Payne & Wilson 1999). In principle, ecologically distinct species would call for distinct domestication histories. This is true even if they are considered to be subspecies (Köhler 1981; Wapnish 1983, 104 n. 6; Wilson 1997). Equally underestimated by less zoologically-focused scholars has been the fact that until very recently the *wild* dromedary was unknown, thus negating the possibility of documenting a direct relationship between a wild ancestor and the domestic descendant in Arabia or elsewhere. The distinction of *Camelus dromedarius* and *Camelus* sp. in the title of this paper reflects this problem (cf Uerpmann & Uerpmann 2008, 102–103).

In the field of osteology, quantitative studies of body size and proportions conducted from a zooarchaeological

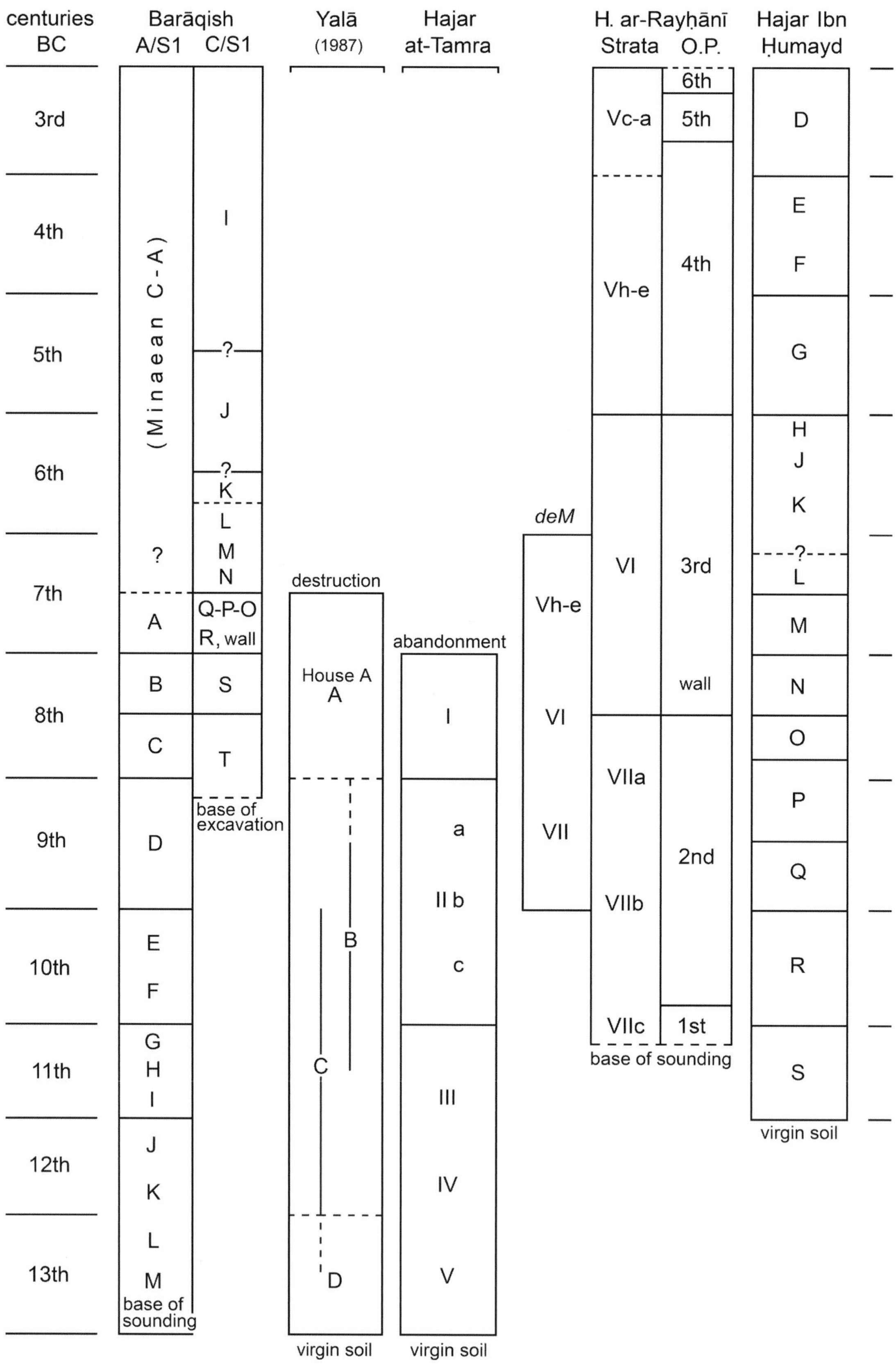

Fig. 20.2. Comparative stratigraphy and radiocarbon chronology of the five Yemeni Iron Age sites discussed in this paper. See text for relevant details. Abbreviations in the Hajar ar-Rayḥānī column: O.P. = Occupational phases, deM = A. de Maigret's (2010) correlation.

perspective only began a decade ago (Uerpmann 1999; Uerpmann & Uerpmann 2002), in spite of size indexes having been applied to species other than camels for a considerable time. Renewed criticisms of the value of size index scaling techniques to detect early animal domestication (e.g. Lesur 2007, 56; Vigne, Helmer & Peters 2005, 6–7, with references) do not detract from the fact that such morphometric statistics can greatly help in comparing size when dealing with small ungulate assemblages. One such device, the logarithmic size index (LSI), will be consistently employed below. LSIs provide an objective method for scaling measurements of different skeletal elements in order to make them comparable (see Uerpmann & Uerpmann 2002, 252). Recent examples, notably Beech *et al.* (2009) working on *Camelus* sp. remains in Abu Dhabi, suggest that the LSI continues to have good comparative value.

The still fragmentary knowledge of early dromedary control in the Arabian Peninsula confers potential interest on new archaeofaunal evidence from Yemen. This paper reports information on the early presence of both the domestic and wild camel in southwest Arabia in association with the Sabaean culture of the 9th–7th centuries BC, if not slightly earlier. These data primarily derive from the walled cities of Yalā and Barāqish, both excavated by the Italian Archaeological Mission (Fig. 20.1). Faunal collections from both sites have recently been studied, and work at Barāqish is expected to continue. Although small, the assemblages come from stratified, controlled archaeological contexts precisely defined by diagnostic pottery and a preliminary radiocarbon chronology. At Barāqish the *Camelus* record expands during the subsequent Minaean period until the end of the city, round the beginning of the current era. On morphological grounds (Köhler-Rollefson 1989; Steiger 1990; Wapnish 1984), all camel remains discussed in this paper come from the dromedary or its wild form. Their cultural status – whether domestic or wild – was determined on the basis of osteometric and LSI values independent of archaeological context, however important this latter for socio-cultural interpretation.

The study of Yalā and Barāqish provides a basis for appraising additional material from Yemen. In particular, an attempt is made to reassess the camel samples that were obtained from three other settlements of the 'early' South Arabian culture, or the Sayhad Iron Age (as called by Edens & Wilkinson 1998, 95). Listed according to their date of excavation, these samples derive from Hajar Ibn Humayd, Hajar at-Tamra and Hajar ar-Rayhānī (Fig. 20.1). The sites were investigated by the American Foundation for the Study of Man (henceforth AFSM) with a 30-year interval between the sounding at Hajar Ibn Humayd in 1950–1 and the subsequent two (Glanzman 1994, 8–27). Due to their geographic and environmental proximity the AFSM sites will be considered in detail after examination of Yalā. The camel remains from the northern, desert-edge city of Barāqish will be presented in a further section.

Fundamental of course to the historical evaluation of the bone assemblages are the stratigraphy and associated dating of the archaeological sequences from which they derive. This point only needs to be emphasised because both the pottery sequences and their intersite correlation, on one side, and their radiometric chronology on the other, still represent work in progress as far as the Sayhad Iron Age is concerned (e.g. de Maigret 2003; 2010; Fedele 2010; Glanzman 1994). For the purposes of the present paper the chronology of the relevant sites has been compiled in Figure 20.2. The figure provides an interim correlation-chronological chart based on comparative stratigraphy and calibrated radiocarbon determinations, and must be understood as a critical synthesis of partially controversial information. Several underlying assumptions will be discussed on a site by site basis beginning with Yalā.

Yalā (or Hajar ad-Durayb)

The ruins locally known as Hajar ad-Durayb near the present-day village of Yalā, or Yalā' (Tiede & Overstreet 1988, xxxviii), were reported by Alessandro de Maigret in 1985. A 2.3 ha walled settlement measuring 230 × 170m across, the city was identified epigraphically with ancient Hafarī (cf Robin & Brunner 1997). A topographic survey of the Yalā area led to recognise that the city was only part of a larger and remarkable archaeological complex of the early Sabaean period. This included a dam, an agricultural estate composed of farms, and a rocky gorge where engraved inscriptions recorded the performance of ceremonial ibex hunts (de Maigret 1988; 2002, 173–186, 273–285, with references). The locality lies in Wādī Yalā, 30 km southwest of Mārib, at an altitude of 1280m above sea level, on the eastern border of the upland region of Khawlān at-Tiyāl.

Because of local difficulties, an excavation at Yalā only became possible in 1987. A well-preserved, two-storey house in the upper part of the tell was excavated in its entirety (House A; de Maigret & Robin 1989). This building had been destroyed by fire and its occupancy represents Stratum A of the city's sequence as currently known. Its destruction broadly corresponds with the end of Yalā. Testing beneath the floor revealed earlier layers and structures which were grouped as Strata B, C and D, with D probably resting on virgin deposits (Fig. 20.3). Animal remains were present in all layers from A to C, totalling about 3000 specimens (Fedele 2009). The occupation sediments within the house were particularly rich, and included the stratigraphically sealed ground floor and a collapsed upper floor. A spatial-functional study of the house has jointly been made from the pottery assemblage (de Maigret 2003; Loreto 2009) and the faunal collection (Fedele 2009), on the premise that they plausibly represent the domestic refuse from a single household during a defined period of time.

The presence of domesticated camels in the fauna, whose significance appeared to be crucially contingent on their temporal placement, has prompted a critical reappraisal of

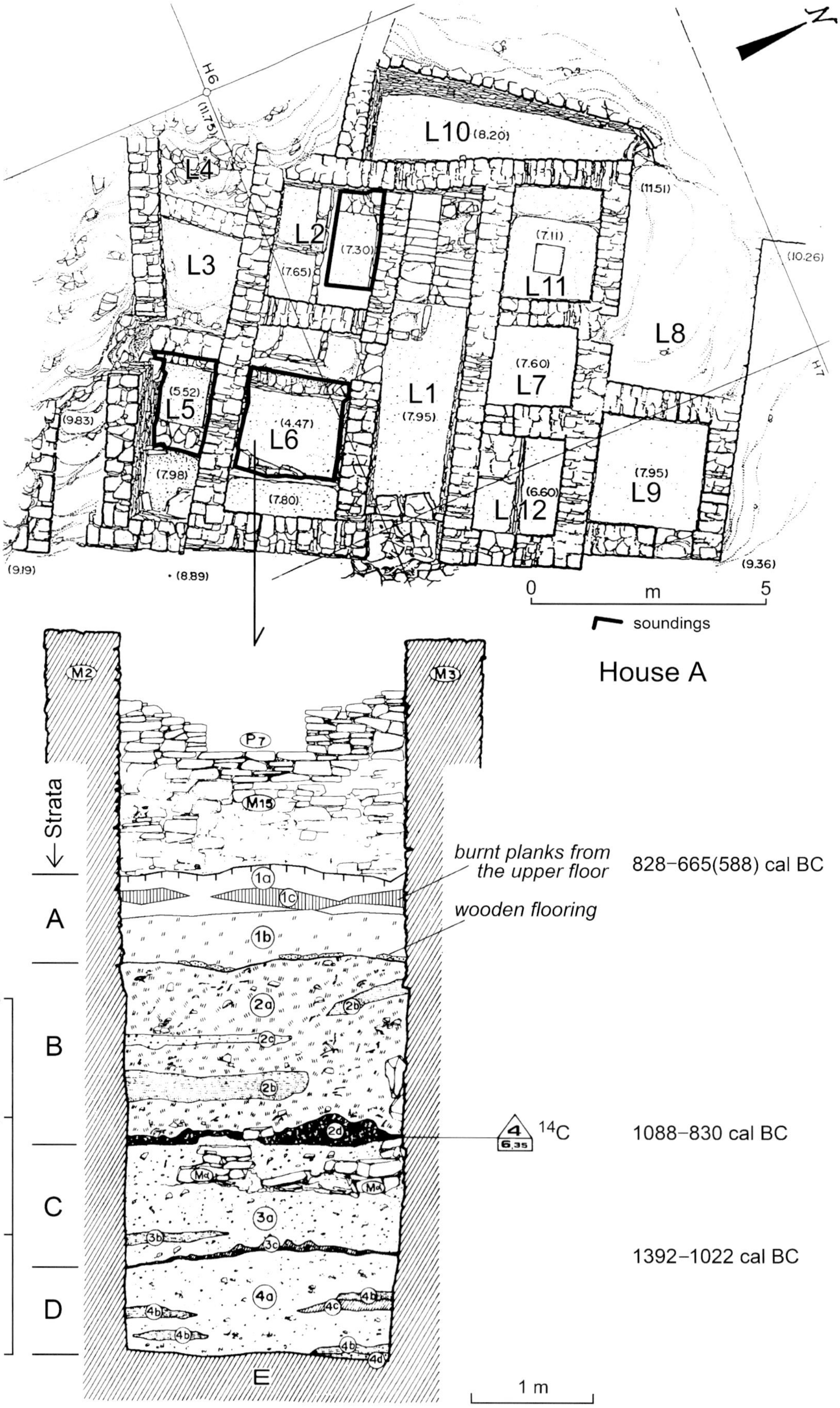

Fig. 20.3. Yalā, House A: a complete stratigraphic profile obtained from room L6, with plan above. Adapted from de Maigret (2002, pl. 6) (plan) and de Maigret & Robin (1989, fig. 3) (profile), with added information, after Fedele (2009, fig. 2).

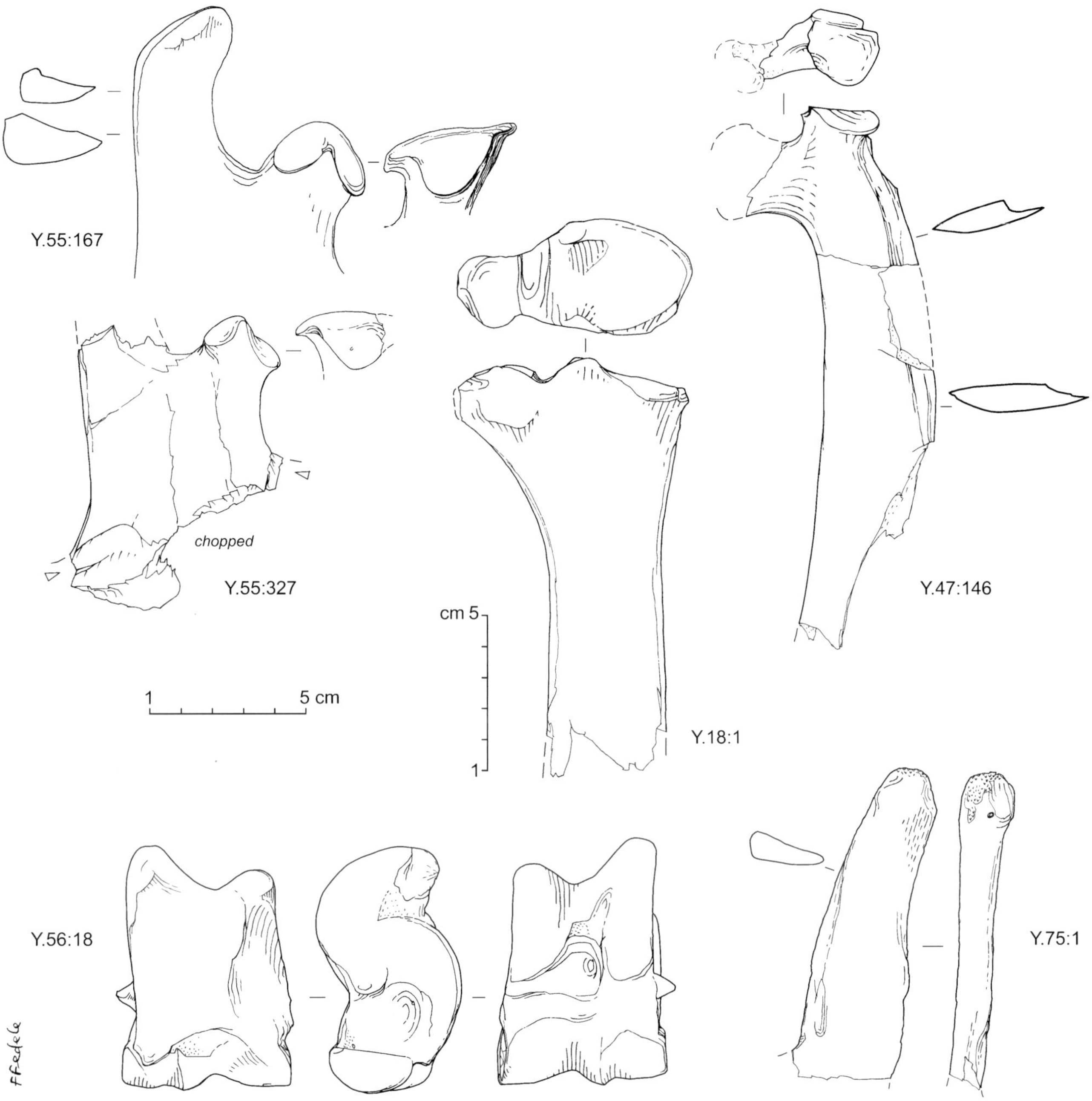

Fig. 20.4. Yalā. Camelus dromedarius, a selection of bone finds.

Yalā's stratigraphy and chronology (de Maigret & Robin 1989). Recalibration of the five radiocarbon measurements (Edens & Wilkinson 1998, 94; Fedele 2009, 138, fig. 3, and unpublished documents; Manzo 2005), pottery parallels (de Maigret 2010) and epigraphic evidence suggesting a continuance of Yalā into the reign of *mukarrib* Karib'īl Watār I, ca. 700–680 BC (e.g. Robin 2002), support the following dates for the sequence (see Fig. 20.2):

- End of the house and city ca. 650 BC;
- Stratum A ca. (850)800–650 BC (occupation of the house ≤ two centuries);
- Stratum B ca. 1050–(850)800 BC;
- Stratum C ca. (1300?)1200–900 BC.

Sixteen *Camelus* bone finds were identified from the house floor and the underlying upper part of Stratum B, level 2a; there are in addition four less certain fragments (Fig. 20.4, Table 20.2). An astragalus, Y.56:18, represents *C. dromedarius* on the basis of its distinctive morphology and proportions (Steiger 1990, figs 50–53; Uerpmann 1999, figs 6–7; Wapnish 1984, figs 5–6). On visual impression it seems likely that all the identifiable bones are from dromedaries. Only four bones from the house provide usable measurements, of which three allow logarithmic size indices to be calculated (Tables 20.1 & 20.2) and checked against Arabian biometric records (Fig. 20.5). The comparisons show that the dromedary bones from

Table 20.1. List of abbreviations for non-standard bone measurements (not defined in von den Driesch 1976) used in Tables 20.2, 20.4 and 20.6. BLF according to Uerpmann (1999, fig. 6, 'BAQ')

Skeletal element	*Abbreviation*	*Definition*
Mandible	Wram	minimum width of mandibular vertical ramus (at constriction)
Mandible	Bcon	mediolateral breadth of articular condyle
Mandible	$H_{P/M}$	height of mandibular body at the P/M contact, lingual side (cf no. 15b von den Driesch for *Bos*)
Mandibular cheekteeth	La, Lo	alveolar, occlusal length
Mandibular cheekteeth	Ba, Bo	alveolar, occlusal breadth
Humerus, distal	HT	maximum height of trochlea, medial side (Payne)
Humerus, distal	HTC	height of trochlea at main constriction (Davis)
Humerus, distal	Dd	greatest depth of the distal end, by analogy to Dd of tibia
Astragalus	BLF	breadth of the lateral articular surface of distal trochlea
Rib	GLV	greatest length of the dorsal (vertebral) end, head to tubercle

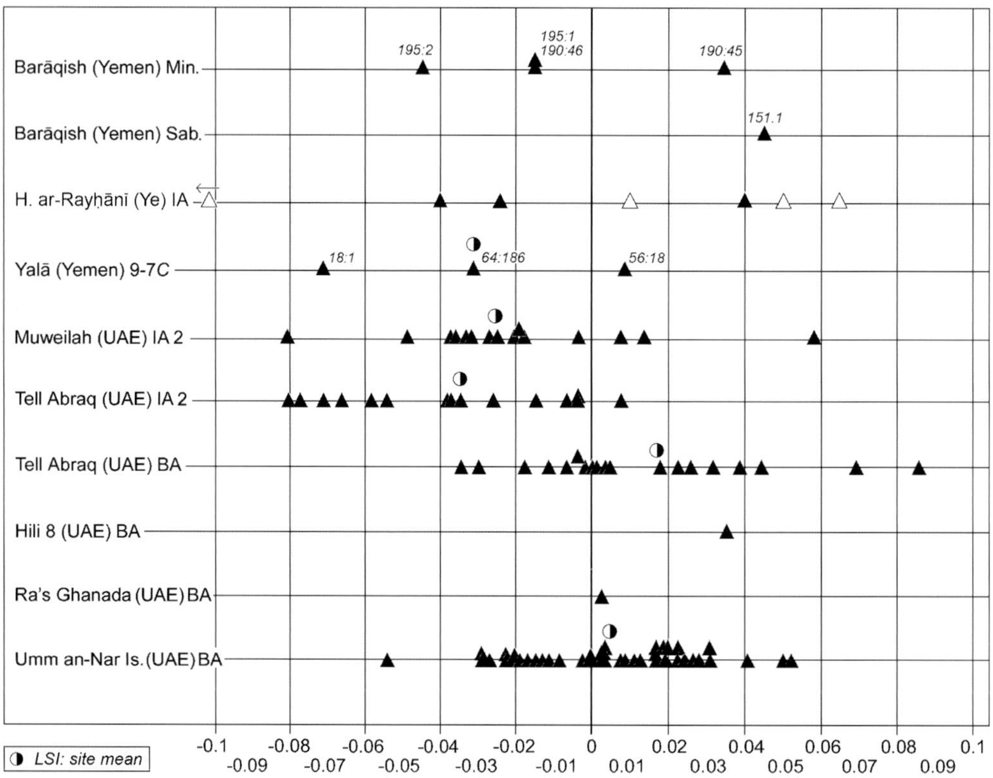

Fig. 20.5. Camelus: distribution of logarithmic size indices (LSI) of bone finds from Yalā, Barāqish, and Hajar ar-Rayhānī compared to background information for Southeast Arabia (after Uerpmann & Uerpmann 2002, as redrawn in Beech et al. 2009, fig. 9). The white triangles indicate approximate values.

Yalā have a relatively small size and clearly fall within the domestic range (radius = –0.072; astragalus = 0.008; first anterior phalanx = –0.032). Furthermore, taken together, they provide an osteometric mean of –0.032 (or –0.012 if the slightly immature radius is excluded): a mean closely similar to the values for the Iron Age II 'reference' sites of Tell Abraq and Muweilah in southeast Arabia (Uerpmann & Uerpmann 2002). The LSI total ranges are also similar.

The earlier fragments from upper Stratum B, level 2a, include two mandibular rami from which non-standard measurements can be taken (Fig. 20.4; Tables 20.1 & 20.2). Such measurements have little comparative value, but in conjunction with bone morphology and appearance they suggest that we are dealing with domestic animals well within the variation observed later in the house. A domestic status is equally suggested by context. This is of particular interest in the light of Yalā's chronology. The upper levels of Stratum B show masonry features of an earlier phase of the house and can probably be dated before 800 BC. These bones thus establish the presence of dromedaries in

Table 20.2. Yalā, House A and earlier layers: list of identified Camelus bone finds. Measurements in mm, with estimates in parentheses, as defined in von den Driesch (1976) and in Table 20.1 (accessory measurements). LSI is the averaged logarithmic size index

Yalā: *Camelus dromedarius*

			House A						
Find no.	*Locus*	*Skeletal element*			*Measurements*				*LSI*
			GL	*SD*	*CD*	*Bp*	*BFp*	*Dp*	
Y.18:1	GF:*L2*	Radius$_d$ px (J)	–	(37)	–	72.6	66.5	(37)	–0.072
			GLl	*GLm*	*Dl*	*BLF*	*Bd*	*GB*	
Y.56:18	GF: *L10*	Astragalus$_s$ A*	77.2	69.1	45.7	20	52.2	54.9	+0.008
			GL	*GLpe*	*SD*	*Bp*	*Dp*		
Y.64:186	GF: *L11*	Phalanx 1_3 an, px half A	–	–	(20)	36.1	–		–0.032
			GLV						
Y.47:146	GF: *L10*	Mid-thoracic rib fr A	>55						
Non-measurable finds									
Y.18:2	GF: *L2*	Long-bone diaphysis fr A							
Y.49-V:1	GF: *L6* ck	Thoracic vertebra J		whole, shattered					
Y.75:1	GF: *L12*	Thoracic vertebra spin.		103mm long					
Y.75:33	GF: *L12*	Thoracic vertebra spin. fr A							
Y.35:100	UF *(L6)*	Thoracic vertebra spin. A		155mm long					
Y.75:40	GF: *L12*	Rib median fr A							
Y.75:39	GF: *L12*	Pubis$_s$ acetabular fr J		from ashes, just fused					
Y.56:19	GF: *L10*	Astragalus px fr A							
? Camelus									
Y.67:144	GF: *L9*	Flat-bone fr A							
			Below house: Stratum B, level 2a						
			Wram	*Bcon*					
Y.55:167	*(L6)*	Mandible$_s$ ramus fr A	62	39.6		chopped			
Y.55:327	*(L6)*	Mandible$_s$ ramus fr A	55.2	–		chopped			
Y.55:168	*(L6)*	Mandible$_s$ fr + rM A		charred, gray					
Y.28:12	*(L5)*	Radius$_d$ sliver A		chopped					
? Camelus									
Y.31:1	*(L5)*	? Mandible fr A		weathered					
Y.55:370	*(L6)*	Epiphyseal fr J		camel size					
Y.55	*(L6)*	Heavy & dense frs of long-bone diaphyses and flat bones A		4–5 frs are probably camel					

* In addition to the formal measurements here reported, an estimate of the medial depth Dm ('impossible to measure precisely'; von den Driesch 1976, 89) gave a value of 36.5 mm.

Locus: GF = ground floor; UF = collapsed upper floor; ck = cooking structure (oven?).
Osteology codes in alphabetic order: A = full adult; an = anterior; $_d/_s$ = right/left; fr/frs = fragment/s; J = juvenile; px = proximal; rM = root of molar tooth; spin. = spinous process; $_3$ = third or axial toe (phalanges)

a domestic context at Yalā during the second half of the 9th century BC, or at a cautious minimum date of about 800 BC. Nondiagnostic 'large-size' fragments from lower levels of thick Stratum B (Fedele 2009, fig. 6) may in part be derived from camels. A clay figurine of a camel is among the surface finds of the 1985 survey at Yalā (de Maigret 1988, fig. 27).

The dromedary appears at Yalā in a faunal context characterised by a limited range of species, with overwhelming domestic stock throughout the sequence (97.1% of identified specimens). Overall, sheep and goats are by far the most common livestock (82%), followed by taurine cattle (11.5%) and in much smaller numbers by the dromedary (1.6%) and the occasional donkey (1%). The dog and a few wild game animals – gazelle, ibex, hare – are documented in trace amounts. To some extent this faunal composition reflects food consumption. No doubt, domestic dromedaries were occasionally butchered for their meat, not unlike the standard meat animals. This is indicated by chopping traces on most of their bones (very obvious in

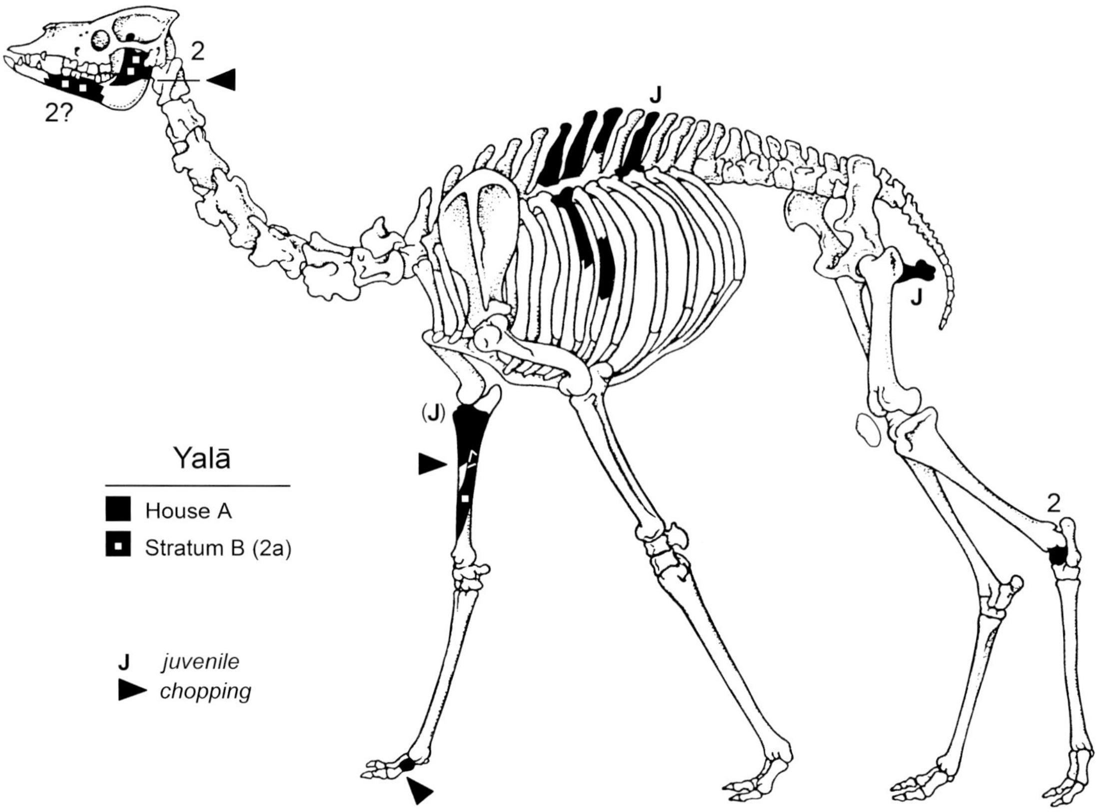

Fig. 20.6. Yalā. Camelus dromedarius, representation of skeletal parts. Camel skeleton redrawn from Kingdon (1988).

eight cases) as well as modifications from burning (one bone from the house and two from Stratum B). A limited food bias may perhaps be perceived in the representation of skeletal parts (Fig. 20.6), where vertebrae and ribs – almost certainly under-diagnosed – tend to outnumber the other bones. Also indicated in the figure with the letter 'J' are the elements from immature animals, three bones from the house and a '*?Camelus*' fragment from Stratum B. If dromedaries were mostly butchered as full adults, this might imply that they were primarily used for milk, fibre and transport – as working and producing animals. The variation in the adult mandibles from Stratum B would suggest that both sexes are represented. Obviously, biological statistics have limited value on such a small sample, and clues to the herd composition cannot reliably be derived.

The above data cast some light on two complementary, poorly known facets of an Early Sabaean settlement: its animal economy and the use of animals in household consumption. At Yalā the faunal record is probably too biased in a functional sense – meat processing for domestic use – to fully portray the significance of the dromedary at the regional level. However, the results from Yalā do not seem to be only local (Fedele 1992, 74–75; 2009; see also Edens & Wilkinson 1998, 99–100). The faunal picture closely resembles that obtained from the Wādī al-Jūbah lowlands to the southeast, see next section, while indirect evidence for the role of camels, suggested by the settlement organisation, is available from the city of Mārib (Japp 2008). The message is clear: between about 850 and 650 BC dromedaries had become common animals in the domestic stock of this part of Yemen, even at a distance from the famed Frankincense Road. Yalā is interesting in this context precisely because it lies away from the routes employed by long-distance caravan trade. By the late 9th century BC the dromedary, although on occasion appreciated at the table, had in fact acquired the complete multiplicity of roles that was to be its lasting distinction (e.g. Gauthier-Pilters & Dagg 1981; Köhler-Rollefson 1991; 1996; Wilson 1984).

Wādī al-Jūbah and Wādī Bayhān sites

a. Hajar ar-Rayhānī (HR3) and Hajar at-Tamra (HT12)

Hajar ar-Rayhānī, ancient Marda', is the largest Iron Age settlement in Wādī al-Jūbah, 40km south of Mārib (Fig. 20.1). A relatively low tell, it is located on a branch of the main trade route that linked the ancient Sabaean capital with Wādī Bayhān, hence with the major site of Hajar Ibn Humayd which will be considered below. Hajar at-Tamra (this spelling after Robin & Brunner 1997) is a smaller mound located 7km south-southeast of ar-Rayhānī. Both sites were tested by the renewed AFSM mission, at-Tamra in 1983 (Blakely 1985) and ar-Rayhānī in 1984 and 1987 (see Glanzman & Ghaleb 1987 and Glanzman

Table 20.3. Stratigraphic distribution of camel finds at Hajar ar-Rayhānī (after Hesse 1996a, table 1; 1996b, table 8.05) and Hajar at-Tamra (Toplyn 1985, corrected). NISP = number of specimens identified to species or genus, mammals only, per layer

Hajar ar-Rayhānī	*Camel bones*	*Total NISP*
mixed Va–IIa	**16**	760
Stratum II	–	101
Total III	**11**	724
Stratum IIIa	2	333
Stratum IIIb	5	221
Stratum IIIc	4	170
Total IV	**7**	354
Stratum IVa	–	91
Stratum IVb	4	135
Stratum IVc	3	128
Total V	**30**	2097
Stratum Va	12	376
Stratum Vc	13	708
Stratum Vd	–	82
Stratum Ve	1	198
Stratum Vf	–	66
Stratum Vg	1	88
Stratum Vh	3	492
Stratum Vi	–	87
Total VI	**1**	115
Stratum VIa	–	2
Stratum VIb	1	73
Stratum VIc	–	40
Total VII	**3**	758
Stratum VIIa	1	115
Stratum VIIb	–	75
Stratum VIIc	2	568
Total	**68**	4909

Hajar at-Tamra	*Camel bones*	*Total NISP*
Surface	–	9
Stratum I	**6**	91
Total II	**6**	168
Stratum II indet.	2	30
Stratum IIa	–	88
Stratum IIb	3	21
Stratum IIc	1	29
Stratum III	*	35
Stratum IV	*	57
Stratum V	–	84
Pre–Stratum V	–	1
Total	**12**	445

* Hajar at-Tamra, Strata III–IV: trace amounts as 'large mammal' fragments?

1994, respectively). A detailed stratigraphic sounding was conducted at this latter site, and virgin soil was reached at a depth of approximately 9m, reportedly corresponding to the mid- or late 2nd millennium BC and to some earlier 'pre-site' deposits (Glanzman 1994; Blakely 1996). The 1987 campaign is *de facto* unpublished, however, apart from Glanzman's (1994, 51–97) important updating on stratigraphic ordering and chronology, complemented by some subsequent refinements (Glanzman 1987b; 2004a; 2005). Nothing is known of the faunal finds that were obtained.

Working on Hajar ar-Rayhānī's stratigraphy is complicated by the fact that the two successive reports use different phasing schemes, one based on 'strata' (Glanzman & Ghaleb 1987) and the second on 'occupational phases' (Glanzman 1994). The stratigraphy presented in Figure 20.2 takes into account both reports, because, as explained below, 'strata' are still relevant for the needs of archaeofaunal analysis, although occupational phases are meant to supersede the previous scheme. Chronometric control is provided by ten radiocarbon determinations for Hajar ar-Rayhānī (Glanzman 1987a), some of them inconsistent because of their large standard deviations, and five for Hajar at-Tamra (Blakely 1985). A comprehensive catalogue with extended discussion is to be found in Glanzman (1994, 71–74, appx 2, and tabs 2.1–2.5); the datings for both sites are listed in a handy format in Edens and Wilkinson (1998, tab. viii and 94–95), who provide a useful commentary. However, intersite correlations based on pottery parallels raise questions. In a recent reappraisal de Maigret (2010) would equate Hajar ar-Rayhānī's Strata VII–Ve with the 9th–7th centuries BC, a much shorter interval than Glanzman's. This will affect the chronology and duration of the 3rd and 4th Occupational Phases, or Strata VI and lower V. In Figure 20.2, a compromise has been attempted between the demands of stratigraphic inference, pottery correlation, and radiometric dating, but inconsistencies remain and cannot presently be resolved. Furthermore, de Maigret (2010) would put the end of Hajar at-Tamra's sequence at 700 BC, explicitly admitting the considerable change in interpretation inherent in such a revision. Any temporal evaluation of the faunal finds will depend on such divergences.

The 1984 faunal collection from Hajar ar-Rayhānī was studied in detail by Brian Hesse (1996a; 1996b; these papers are to some extent complementary). It totals 25,540 bones of which 19.2% were identified to species or genus. Overall, a fundamental reliance on sheep and goats is apparent (97% of the identifiable pieces), with cattle and camel each about 1–2% and gazelle and uncertain *Equus* (donkey?) in trace amounts. A much smaller sample from Hajar at-Tamra shows the same (see below). The stratigraphic distribution of camel finds is summarised for both sites in Table 20.3. It needs be noted that ar-Rayhānī's

Table 20.4. Hajar ar-Rayhānī: summary measurements in mm, after Hesse (1996a, table 4, with corrections from his original manuscript dated May 1991; 1996b, table 8.04). Inferred anatomical specifications in square brackets, non-standard measurements as defined in Table 20.1. Added references are to Steiger (1990)

Hajar ar-Rayhānī: *Camelus dromedarius* and *Camelus* sp.

Skeletal elements						*LSI and size evaluation*
			Single bones (n = 1)			
	DC					
Femur	48.8					–0.040
	GLl	*GLm*	*Dl*	*Dm*	*Bd*	
Astragalus	70.7	64.2	40.5	40.5	48.0	–0.024
	GB					
Os carpi intermedium	25.0					Within Steiger's (GBCi) 24.8±1.8
Os tarsi centrale	34.0					Very small
Os tarsale 2+3	39.3					–
'Fused os tarsale'	69.8					–
	Bp					
Phalanx 1 [an]	44.5					+0.040
			*Grouped bones (n = 2)**			
	GB					
Os carpi ulnare	48.0 ± 4.2					Range exceeds Steiger's (GTCu) 47.8±2.9
	GB	*GL*				
Patella	43.2 ± 2.5	91.8 ± 2.0				LSI from the means (43.2, 91.8) = +0.030; LSIs for individual bones (inferred size): +0.01 and +0.05, approximately
	GL	*Bp*	*Dp*	*SD*		
Phalanx 2 [an/po?]**	60.9	26.2 ± 8.6	28.3	34.9		LSI from Dp [anterior] = +0.037
						LSI from SD [anterior] = +0.174
						LSI from Bp [large, anterior] ~ –0.02
						LSI from Bp [small, posterior] ~ –0.2+
						LSI from GL [posterior] = –0.006

* Reported as mean value ± standard deviation.
** Limb position and individual measurements of each phalanx unspecified. The abnormally high standard deviation for Bp (if correct) points to extreme size difference between the bones. The reported measurements are compatible with the presence of a large anterior phalanx (averaged LSI ~ +0.064) and a small posterior phalanx (averaged LSI ~ -0.11).

fauna was published in terms of strata, not occupational phases. Metrical data are only available for ar-Rayhānī and comprise a mix of single-specimen measurements and group 'bulk' values (Table 20.4).

Within these limits, a biometric evaluation of Hajar ar-Rayhānī's measurements is presented in Table 20.4. Only three bones can be assigned a precise LSI, but some further bones allow approximate size estimates. An interesting result, alongside the expected presence of domestic stock, is the appearance of animals that patently exceed the size range of the domestic dromedary (Fig. 20.5). A first anterior phalanx is probably from a wild animal (Bp = 44.5; LSI = 0.040), but a patella (LSI ~0.05) and especially a second phalanx (LSI ~0.064) might in theory indicate camel hybrids (see Köhler-Rollefson 1989 and Uerpmann 1999 for comparative data, and discussion under 'Barāqish' below). Unfortunately, a historical assessment is impossible because the stratigraphic provenance of the specimens is unpublished. Based on the urban context and a single astragalus with '*dromedarius*' morphology all camel material was attributed domestic status (Hesse 1996b, 272). Some foetal bones were present (B. Hesse pers. comm., October 1989).

Never a major faunal component (site frequency 1.39%), the dromedary increases from trace amounts (0.4%–0.9%) in Strata VII, VI and lower V, equivalent to the 2nd, 3rd and 4th Occupational Phases (9th–7th centuries BC, de Maigret's correlation), to 1.5%–2.3% from upper Stratum V (Vc) onwards, peaking at 2.3% in level Vc. The relative abundance of camel bones was calculated as a percentage of the pooled cattle-camel subsample, on the assumption that the two species do not co-vary; in another calculation the relative frequency was used (Hesse 1996b, 275). These statistics, and a consideration of the unidentified large-mammal fraction, lead Hesse to suggest that there was a shift in camel frequencies towards the end of Stratum V and a threefold increase at the transition to Stratum IV (see also Edens & Wilkinson 1998, 106). Hesse is tempted to link this shift in relative representation to the advent of extensive overland transport based on camels. Hajar ar-Rayhānī was supposedly 'tied into the caravan network leading from Timna [Tamna'] toward Marib and points north' (Hesse 1996b, 263; cf Toplyn 1988, 102–103, for field data describing a relevant upland pass, with further comments in Glanzman 1994, 31–33).

A sudden increase from trace amounts to significant frequency appears at Hajar at-Tamra (Toplyn 1985; see also Fedele 2009, 147–149) (Table 20.3). The published data suggest that camel might be represented among the large-mammal fragments of Strata IV–III, but its presence is only demonstrated from the base of Stratum II onwards, with a strong representation (3.6% of identified bones in Stratum II, 6.6% in Stratum I). The site's total shows that camel (n=12) comes second in frequency after the domestic caprines (n=163) and before cattle (n=9); however, as mentioned, the sample is small. Accepting de Maigret's correlation, the dromedary makes a signal appearance at Hajar at-Tamra as early as the 10th century BC (Fig. 20.2), a date which would be of remarkable interest if the domestic status of these few bones could be established.

Before conclusions are reached, see below, one also needs to determine how representative are the collections from Hajar ar-Rayhānī and Hajar at-Tamra. Concerning at-Tamra no information on recovery procedures was given; at ar-Rayhānī, on the other hand, the problem of 'collection intensity and bias' was carefully addressed (Hesse 1996b, 267–272). The overall sampling bias was examined in terms of size bias, taphonomic pressure, and differential bone destruction. Hesse concluded with confidence that 'the larger identifiable bones of small mammals and all the identifiable bones of medium-sized mammals and larger' were collected (Hesse 1996b, 269). However, some caution must be used in interpreting the relative frequency of species of significantly different size, and concurrently, 'the temporal integrity of bone deposits has been significantly distorted' due to the melting of mud-bricks containing old bone (Hesse 1996b, 272).

b. Hajar Ibn Humayd

This mound located in Wādī Bayhān, known in the literature as 'Hajar bin Humeid' (HBH), was the object of a pioneering excavation directed by William Foxwell Albright in 1950–1, as part of Wendell Phillips' AFSM expedition to southern Arabia. Unfortunately, disregarding more objective stratigraphic methods already practised elsewhere, HBH was largely excavated on the basis of arbitrary, horizontal dissection units, or 'absolute levels'. When the excavation was published long after the fieldwork was completed (Van Beek 1969), the stratigraphy was basically established through the seriation of pottery rather than well-defined physical layers (Blakely & Glanzman 1996, 3–4). The site's radiocarbon chronology, though elementary, filled a complete void of information and challenged datings based on subjective ideas of epigraphic style (see evaluations in Blakely 1985, 89; Edens & Wilkinson 1998, tab. viii). The HBH column in Figure 20.2 takes into account Glanzman's revision (1994, 538–547, tab. 2.6–2.8; see also Van Beek 1997) which, however, little modified the earlier part of the sequence. Within the scope of the present paper, a good agreement exists with the pottery correlation established by de Maigret (2010).

Thanks to the detailed work on ceramics, the HBH monograph has remained for decades a regional landmark. Understandably, other information was below standard. The recovery of ecofactual evidence including bones was extremely limited. The one-page faunal report from HBH (Van Beek 1969, 289–290) describes eight mammal bones and some shells and coral, which were identified by three American zoologists, including Smithsonian Institution's HW Setzer as a mammal specialist (cf Fedele 1992, 71). It appears that animal finds were only picked up in the field when unusual and ancient, and owing to this practice two camel bones individually ended up in the sample, having been found in the two lowermost strata of the site. They were a vertebra and an astragalus, both determined as 'cf *Camelus dromedarius*' (Setzer in Van Beek 1969, 289–290, pl. 56: b). On pages 15 and 16–7 of the monograph, respectively, their contexts were outlined as follows:

> The domestic character of this area [within Stratum S] is indicated by the presence of mortar, a clay spindle whorl, and a cervical vertebra of a camel (H3094).
>
> Significant nonceramic finds [from Stratum R] include a fragment of a steatite cooking bowl with an iron rivet, which was used to repair an ancient break, and represents the earliest evidence for the use of iron at the site; a single specimen of a mollusk shell [...]; a fragment of faviid coral, which was probably kept in Building R: 1 as a souvenir; and the astragalus of a camel (H2815).

'The finding of camel bones in the two earliest strata, S and R' – Van Beek remarks at the end of the faunal report – 'indicates that the camel, *either wild or domestic*, was used as food' (1969, 290, emphasis added). A fragment of a clay figurine representing a camel's head was found in Stratum O, attributed to the 8th century BC.

c. HR3, HT12, HBH: results

Hajar at-Tamra and Hajar Ibn Humayd, within the limits of their coarse-grained chronology, clearly hint at the existence of habitual human–camel relationships in Yemen in the 10th century and possibly as early as the 11th century BC. The main drawback concerning these sites is the impossibility of establishing whether the camel bones derived from wild populations or domesticated stock. The contribution of these sites in terms of chronology is thus – for the moment – considerably diminished by the indeterminacy of the human behaviour involved. At Hajar ar-Rayhānī the evidential basis is in a way the opposite: we may be dealing with a majority of animals which are osteologically domestic, according to the published measurements, but we lack information on their stratigraphic position and individual chronology. The collection was implicitly treated as corresponding to a homogeneous suite of domestic populations, on the assumption that the archaeological context suggests a domestic condition throughout the sequence. This is not the case, however, since a few bones would fall unequivocally within the range for the wild dromedary. Although the chronology of these bones is unreported, one can infer that during the lifespan of the city episodes of camel predation coexisted with camel husbandry. Similar observations from Barāqish will be discussed below.

Barāqish (Yathill)

Ancient Yathill, now known by its Arab name of Barāqish (de Maigret & Robin 1993; Robin & Brunner 1997), is one of the largest walled Iron Age cities of the Jawf Valley, in the northern predesertic belt of Yemen (Fig. 20.1). The walls are Minaean in origin (6th–3rd centuries BC; de Maigret 2004; de Maigret & Robin 1993; Fedele 2010), famously the best preserved in southern Arabia, and enclose an area of 4ha. Surface finds, mechanical corings, and the excavations currently in progress suggest that the Minaean city was preceded by an even more substantial Sabaean settlement. In particular, a stratigraphic testing outside the walls in 2005–6, excavation Area C, revealed a long sequence of occupations spanning the whole 1st millennium BC (Fedele 2010). Concurrently, a vertical sounding 8m deep within the city produced evidence of Sabaean occupation from as early as the 13th century BC, without reaching virgin soil or bedrock (excavation Area A, sounding A/S1, 2006; de Maigret 2010). These two stratigraphic columns are shown in Figure 20.2. Only Area C, indicated as 'C/S1', is of interest here.

In the Area C excavations (Fig. 20.7) a stratigraphic section was obtained by digging a radial, stepped trench down and away from the Minaean wall. The trench cut deeply into the talus surrounding the tell and expanded beyond the footslope into the adjacent plain, providing an essentially continuous cutting 52m long and 14m deep. In addition, an isolated test pit was sunk 30m away in the plain. At the upper end of the section, a sounding along the wall reached a depth of 8m and for the first time brought to light the base of the Minaean enclosure, revealing that it rested on Sabaean occupation deposits. The chronological control is assured by eleven AMS radiocarbon determinations, which indicate the 8th–7th centuries BC for the core of the Sabaean occupation (Fedele 2010, table iii and fig. 9). The deepest Sabaean levels from the lower part of the trench are yet undated, but on the basis of stratigraphic correlation can safely be attributed to the 9th–8th centuries BC (Figure 20.2).

Well-preserved animal remains have been systematically retrieved from both stratigraphic sequences, numbering several thousand pieces from Area C alone (de Maigret 2010; Fedele 2010). In addition, dung levels and coprolites characterise a number of Minaean layers in Area C. Only an interim account of the bone assemblages from this area can be given at this stage of study, because continuation of the faunal analysis undertaken in 2007, and work at the site in general, have been impossible. The results are summarised in Table 20.5, where the species composition for the Sabaean and Minaean periods is presented, accompanied by a qualitative indication of relative frequencies. Camel remains (Fig. 20.8) are fairly common and increase over time. Because of working-time constraints only a few bones could be measured, and a metrical (Table 20.6) and contextual evaluation will be made in this section. In spite of their paucity, as discussed below, the available data clearly bear on matters of early camel management.

Overall, the archaeofaunal picture from Sabaean Yathill closely matches Yalā's and Hajar ar-Rayhānī's. The economy relied heavily on domestic caprines – sheep and goat – and taurine cattle. The donkey was present. The high frequency of cattle points to a local oasis environment, sufficiently rich in pasture and probably still sparsely wooded, although the broader setting of the city was a semi-desert according to sediments. Additional similarity

Table 20.5. Barāqish Area C, over-wall excavations: preliminary species list for the Sabaean and Minaean phases (from Fedele 2010, table 8). The number of "×" is proportional to the frequency of finds; / = traces; 1 = a single find

Barāqish C/S1, 2005-06		
	Sabaean	*Minaean*
	kitchen/butchery refuse (often burnt or roasted)	*refuse from meals, some from trading camps? (cutmarks, trampling)*
DOMESTIC ANIMALS		
Sheep & goat (both)	×××××	×××× & coprolites; hornless sheep
Cattle, *Bos taurus*	×××××	×××××
Dromedary, domestic	××	××× & coprolites
Donkey	× & coprolites	×××
Dog, small	/	/
Domestic fowl, *?Gallus*	/	
? Mule		/ [horse x donkey hybrid?]
WILD ANIMALS		
Camelus sp. wild	1 [C.151.1, quadr. *E11*]	?
Gazelle, *Gazella* sp.	?	/
? Ibex	1 abraded talus [an artefact from *F1*]	/
Ostrich [bones]		× & 2 eggshell frags
Cypraeids, marine shells	/	1 [from sounding *F1*]
cf. *Arca*, marine shell		1 [from sounding *A*]

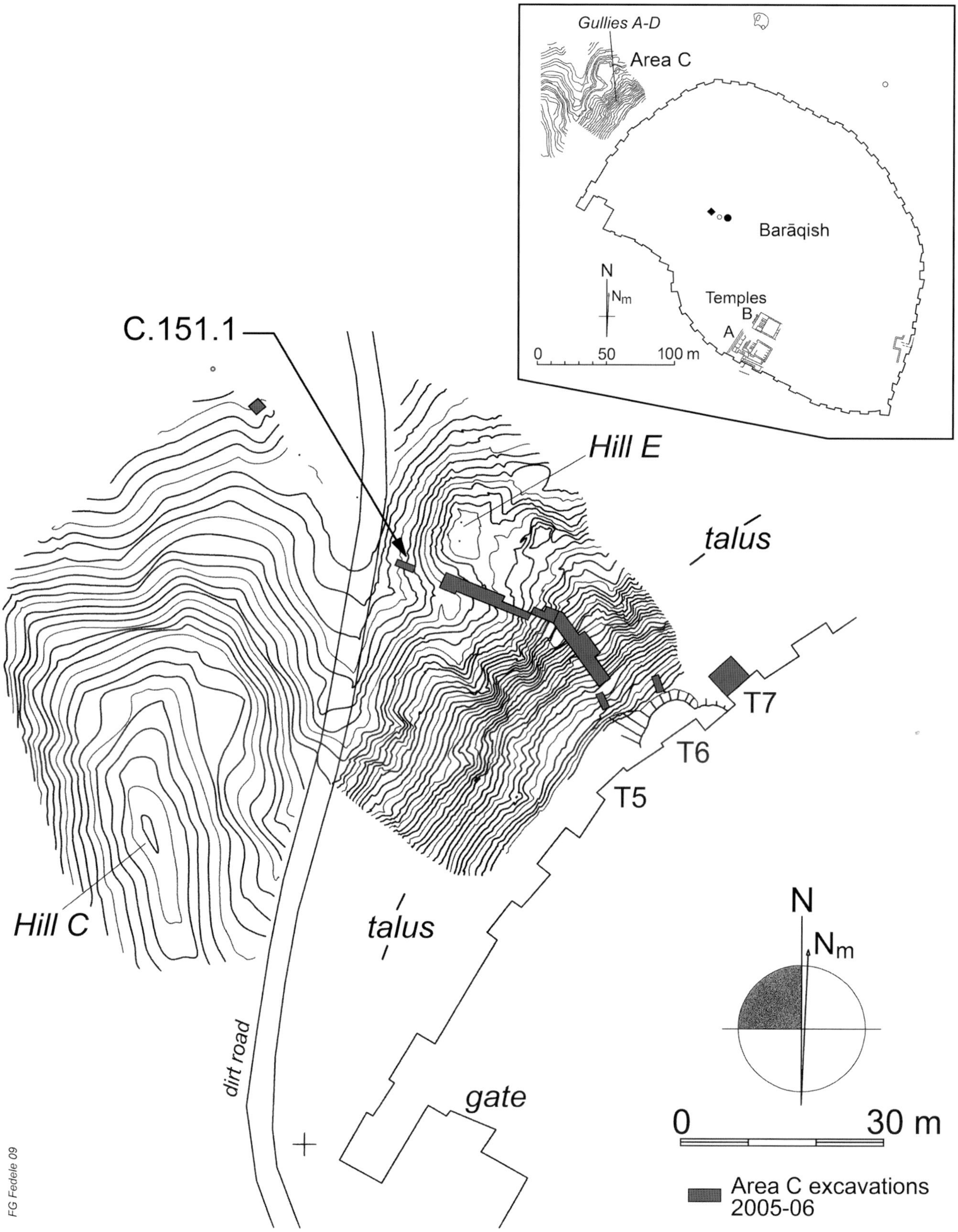

Fig. 20.7. Barāqish Area C, plan of the over-wall excavations 2005–6 (after Fedele 2010, with additions). The arrow marks the findspot of camel humerus C.151.1.

Table 20.6. Barāqish Area C, over-wall excavations: a selection of Camelus bone finds. Measurements in mm, with estimates in parentheses, as defined in von den Driesch (1976) and in Table 20.1 (accessory measurements). LSI is the averaged logarithmic size index

Barāqish: *Camelus dromedarius* and *Camelus* sp.

Sabaean 3 (7th century BC)									
Find no.	*Grid and layer*	*Skeletal element*	*Measurements*						*LSI*
			Bd	*BT*	*SD*	*HT*	*HTC*	*Dd*	
C.151.1	*E11*: H11, Locus 151	Humerus$_s$ di	101.6	89.3	(53)	68.3	46.0	(87)	+0.045
			GLV						
C.199:41	*A1–A31* E: SYH	Rib fr	52.2						
C.199:42	*A1–A31* E: SYH	Rib fr, head fusg	51.5						
*Minaean 1, counterfort fill (6th–?5th century BC)**									
			GL	*SD*	*Bp*	*BFp*	*Dp*	*Bd*	
C.195:1	*A1–A31*: T79J	Phalanx 1$_{3s}$ an	102.3	(19.5)	39.8	–	–	–	–0.015
C.195:2	*A1–A31*: T79J	Phalanx 2$_{3s}$ an	62.8	26.9	32.6	30.2	24.4	34.5	–0.045
C.190:45	*A1–A11 W*: T77 etc.	Phalanx 1 an, px	–	–	44.1	–	38.0	–	+0.034
			P_4: Lo	*P_4: Bo*	*M_1: Lo*	*M_1: Bo*	*M_2: La*	*M_2: Ba*	
C.195:2–3	*A1–A31*: T79J	Mandible$_s$ fr + rM	–	–	–	–	40.1	19.3	
								$H_{P/M}$	
C.194:5	*A1–A11*: T79L	Mandible$_d$ + P–M	22.5	9.8	39	(19.5)	42.2	(48)	
			Bp	*Dp*					
C.190:46	*A1–A11 W*: T77 etc.	Metatarsal$_d$ px	59.0	44.3					–0.015
Some additional finds			*Chronology, remarks*						
C.112:1	*E3*: SI3, PV3, ?V3	Phalanx 3	Sabaean 1: ca. 820–700 BC.						
C.102:1	*E2*: V3, SE3, SI1	Metacarpal$_s$ J	Sabaean 1–2: 8th century BC.						
C.39:1	*D5*: GK4	Phalanx 1, pathol.	Sabaean 3 (1): ca. 650 BC.						
C.151:02	*E11*: H11–H11$_0$	Humerus$_d$ px fusg	Sabaean 3: 7th century BC. Roasted.						
C.151:01	*E11*: H11–H11$_0$	Ulna$_s$ fr	Sabaean 3: 7th century BC. Roasted.						
C.198:4	*A1–A31*: AR–T7Z1	Humerus di fr	Minaean 1, foundation trench of the wall (6th–?5th century BC).*						
C.198:5	*A1–A31*: AR–T7Z1	Epistropheus fr	Minaean 1, foundation trench of the wall (6th–?5th century BC).*						
C.191:7	*A1–A31*: T77μ	Radius$_s$ px fr	Minaean 1, counterfort fill (6th–?5th century BC).*						
C.189:74	*A1–A31*: lower T7LP	Pubis$_s$ frs, fusg	Minaean 2: 5th–3rd century BC.						
C.188:4	*A1–A31*: upper T7LP	Humerus di fr	Minaean 2: 5th–3rd century BC.						
C.188:5	*A1–A31*: upper T7LP	Metacarpal di fr	Minaean 2: 5th–3rd century BC. Metal tool cutmarks.						
C.185:16	*A1–A11*: TSR–T76	Astragalus fr	Minaean 3: 2nd–1st century BC. Carbonised.						
C.185:18	*A1–A11*: TSR–T76	Mandible fr + rM	Minaean 3: 2nd–1st century BC.						
C.211:1	*F1*: F02 etc.	Humerus di fr	Minaean 3–4: 2nd–1st century BC. Sandblasted.						
C.184B:2	*A1*south: lower T75	Cervical vertebra	Minaean 4: 1st century BC. Dorsocaudal fr.						
C.183:1	*A1–A11*: T74K–T74η	Mandible + P–M	Post–Minaean interval: 1st–?2nd century AD						
C.184.2	*A* south: upper T75	Whole coprolite	Minaean 4: 1st century BC.						

* Most bones from the foundation trench and some from the fill of the so-called counterfort are reworked Sabaean 3 material (ca. 650–600 BC; see Fedele 2010).

Osteology codes, in alphabetic order: an = anterior; $_d/_s$ = right/left; di = distal; fr/frs = fragment/s; fusg = fusing; J = juvenile; px = proximal; rM = root of molar tooth; $_3$ = third or axial toe (phalanges)

between Yathill and other Sabaean cities along the desert fringe comes from the fact that evidence for the hunting or capture of wild animals is very scanty, and game species are virtually nonexistent at Yathill. However, a humerus from a wild camel, a unique find for the site, deserves special mention (see no. C.151.1, below)

A key information from Barāqish is the dromedary's earliest appearance in the Sabaean 1 phase of the over-

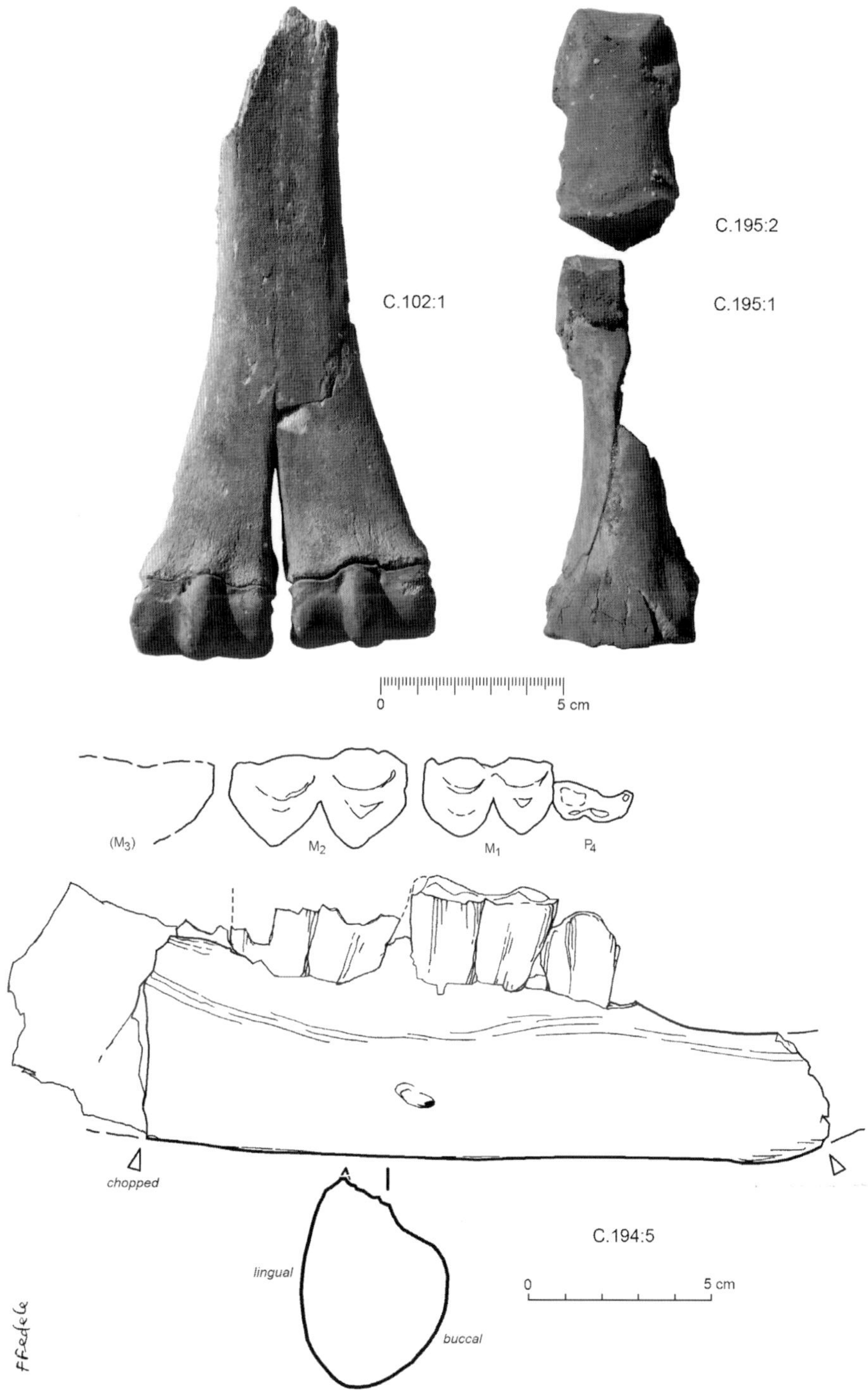

Fig. 20.8. Barāqish Area C, over-wall excavations. Camelus dromedarius: juvenile metacarpal C.102:1; 1st and 2nd anterior phalanges 195:1, 2; right mandible C.194:5.

wall sequence, from the turn of the 9th to 8th century BC (Stratum T in Figure 20.2; see Fig. 20.8 and Table 20.6 for bone examples, e.g. C.112:1, C.102:1). Equally interesting is the evidence that camels were commonly present throughout the sequence. More difficult to establish at this stage of research is the particular economic basis underscoring the occurrence of camels during each period. Clearly, the dromedary was butchered and consumed. And here again, like at Yalā and elsewhere, this species is presumably under-represented in terms of actual remains, therefore resulting in a defective picture. An animal kept for non-food purposes, especially one not frequently consumed, may leave comparatively fewer remains (Ermolova, NM 1970, cited in Compagnoni & Tosi 1978, 100; Horwitz & Rosen 2005, 128; Köhler 1984, 201). During at least the local Later Sabaean (7th century BC) and the subsequent Minaean, when the archaeological indications of trade become more evident, the greatest part of the dromedary

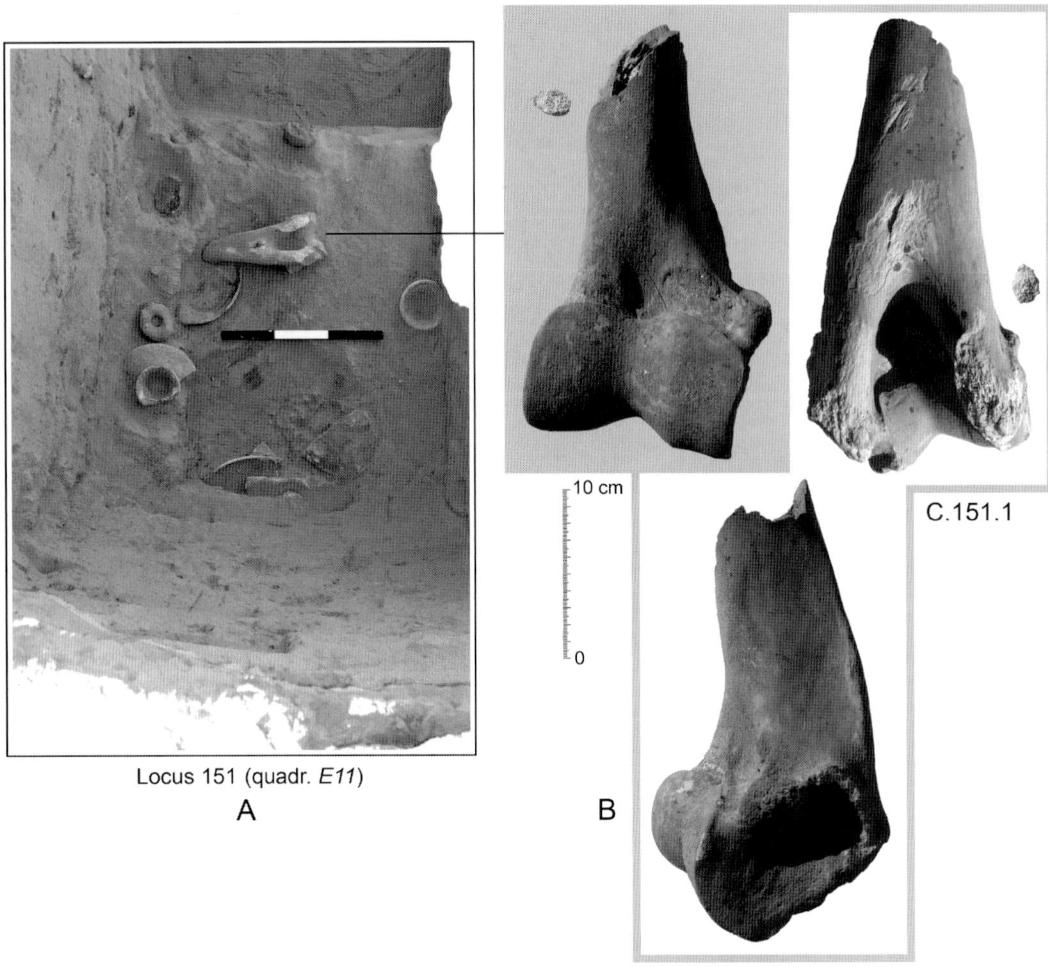

Fig. 20.9. Barāqish Area C, over-wall excavations (2005). A: Locus 151 in quadrangle E11, with Sabaean pottery and dromedary humerus, as exposed. B: Camelus sp. wild, humerus C.151.1.

bones at Barāqish probably derive from transport animals subsequently utilised as a food source (like at Tell Jemmeh, Israel, for instance; Wapnish 1984, 179). Herd assessment is not yet available, nor is – in particular – sex and age profiles for individual strata or periods. This information will certainly be forthcoming from both the city and the over-wall excavations at a future stage of study.

Logarithmic size indices were calculated for five bones, one belonging to the Later Sabaean and four to the fill of a 'counterfort' associated with the Minaean wall construction (Table 20.6); most bones from this fill, in fact, might be reworked Later Sabaean elements. The comparisons with the Arabian biometric records of Figure 20.5 show that three Minaean bones fall within the domestic range (first anterior phalanx and proximal metatarsal, both LSI = –0.015; second anterior phalanx LSI = –0.045). Taken together they provide an osteometric mean of –0.025, virtually identical to the value for Muweilah, Iron Age II (Uerpmann & Uerpmann 2002). On preliminary inspection most of the camel bones appear to be osteologically domestic throughout the Barāqish sequence.

Another first anterior phalanx (C.190:45) has instead an LSI of 0.034, which suggests a much heavier build. The plausible alternatives point to a wild or a feral animal, while the relatively early dating would rule out a camel hybrid. Whether feral camels roamed the Jawf during the Earlier Minaean period cannot be determined. It is safer to think of a wild animal, like at Hajar ar-Rayhānī (see above) or at Muweilah, where in a context generally similar to Barāqish' some bones fall in the range of the wild Bronze Age camels. As to the theoretical possibility of a hybrid, the view is generally accepted that the practice of crossbreeding did not begin earlier than the 2nd century BC, during the Parthian period (Bulliet 1990, 141–170; Köhler-Rollefson 1989; Potts 2004; 2005). Bactrian-dromedary hybrids were identified for the first time in Arabia at Mleiha (Fig. 20.1) in a graveyard probably dating from the 1st or 2nd century AD (Uerpmann 1999; see Mashkour 1997 for the graveyard context). Phalanx C.190:45 is slightly smaller than Mleiha grave 4 (Bp=44.8, Dp=38.5) and below the Bp range of the hybrids from Pella (46.0–51.3; Köhler-Rollefson 1989, tab. 16). Since the bone is incomplete, a comparison with Uerpmann's (1999, fig. 9) GL/Bp scattergram is not possible.

The case for a wild dromedary can be confidently argued for C.151.1, a distal humerus from a findspot on the outskirts of the Later Sabaean settlement, bordering the plain (Fig. 20.9). The same archaeological cluster, Locus 151, which as part of Stratum M can be dated within the 7th century BC, contained two more forelimb fragments of *Camelus*. In size and surface appearance the humerus is very different from all other camel material from the over-wall excavations. Its LSI of 0.045 (Fig. 20.5 & Table 20.6) leaves no doubt that this large bone belongs to a wild animal of dromedary morphology. The bone was chopped by heavy percussion and partially roasted, resulting in a reddish hue on its anterodistal part, while the orange pigmentation on the rest of the bone is somewhat unusual. The other bones from the locus are roasted as well.

The humerus, together perhaps with phalanx C.190:45 above, is of particular importance because it indicates that populations of wild dromedaries still existed in Yemen during the mid-1st millennium BC (cf Uerpmann & Uerpmann 2002, 257, speaking of the U.A.E.), when the domestication of the species had been under way for a long time in the Arabian Peninsula, including Yemen (this study), and was being perfected. Such populations may have survived much longer, since rock depictions of what have been interpreted as wild camels (e.g. Peters 2001; Spassov & Stoytchev 2004; Zarins 1989) continue to be found in southern and central Arabian contexts as late as the 2nd and possibly throughout the 1st millennium BC (see Khan 1998 and Bednarik & Khan 2005 for a redating and the concomitant demonstration that previous chronological claims were unfounded). The wild population inferred by humerus C.151.1 may be located in the wider area of Barāqish, but this is unsure. Although the bone may be a result of episodical hunting by residents of Yathill, it might, alternatively, derive from a chance encounter with a wild herd in the course of travel or caravan traffic, and thus have been brought in from greater distance for meat.

From this perspective it is significant that the general context of the find suggests a tent campsite outside the settlement, one in a long series of such traces stratified on the plain along the footslope of the talus (Fedele 2010, 125, 147–148). At Locus 151, the composition and condition of materials do not suggest 'normal' kitchen refuse. Rather than intensive occupation they indicate repeated episodes of activity in the open, possibly associated with light-shelter constructions. The archaeological association – cooking pots, tripods, pans, evidence of burning, 'torpedo-type' trading jars – combined with what looks like a patchy distribution of activity clusters, suggests a scenario of bivouacs or tents on the plain (cf Cribb 2004, Saidel 2009, and several contributions in Bar-Yosef & Khazanov 1992, for the relevant interpretive framework). Some unusual or exotic items may further suggest that foreign parties – merchant caravans? – were possibly involved. If trade, unsurprisingly, was a pivotal factor in Yathill's *raison d'être*, the over-wall excavations provide first-hand evidence for it as early as the lowermost Sabaean strata: 'torpedo'-jar handling, pack animals, exotic items, camping in tents. The subject has been developed elsewhere (Fedele 2014).

In this context, it may be of interest briefly to explore the theoretic possibility of similarities – if not outright exchanges – between the Jawf node in the trade routes and southeast Arabia. Muweilah, in particular, enjoyed a broadly symmetric position to Barāqish across the expanse of southern Arabia, relative to the desert margin (Fig. 20.1). With its *floruit* in the 9th century BC, Muweilah should be singled out because the growth and functioning of the city as reconstructed by Peter Magee (2004; 2007) may indeed suggest analogies with Earlier Sabaean Barāqish in spite of their cultural differences. For instance, Muweilah has revealed among its long distance connections some obvious contact with the Sabaean area (a pottery vessel with South Arabian script; Magee 1999). One can further note that the dromedaries from Muweilah, on present evidence, were slightly heavier than those of contemporary Tell Abraq on the coast (Magee 2007; Uerpmann & Uerpmann 2002). If a different camel type or ecotype was indeed circulating at Muweilah, this might hint to selection and/or participation in a different network. (On 'pack' types and other specialised camel variants see Wilson 1997.) The possibility that animals travelled along east-west or 'transverse' Arabian routes, together with information and goods, probably deserves consideration. Incidentally, any transverse contact may have helped spreading camel domestication at the beginning of the 1st millennium BC within central-southern Arabia itself (see next section). Future research is needed to test such propositions.

At Barāqish, like at Yalā, all the suitably preserved camel finds show a '*dromedarius*' morphology. There is certainly no reason to classify the remains from domestic animals other than the Arabian species. For instance, a mandibular body (C.194:5; Fig. 20.8) exhibits the relatively rounded cross-section in proximity of the diastema which characterises at least the older individuals of the dromedary (Uerpmann & Uerpmann 2008, 102). The specimen or specimens that in terms of the LSI lie in the commonly accepted range of the wild dromedary raise a nomenclature problem, due to this wild progenitor still lacking a name, and '*dromedarius*' being ruled out by the International Commission of Zoological Nomenclature (Gentry *et al.* 2004). Following Uerpmann and Uerpmann (2008, 102–3) the noncommittal label '*Camelus* sp.' is here employed.

Discussion and conclusions

Evidence from Yemen

Pending further work at Barāqish and hopefully elsewhere, the above information is fragmentary, but at the same time sufficient to signal early camel domestication from Yemen. Prima facie, Hajar Ibn Humayd and the Wādī al-Jūbah sites could be discounted for lack of anatomical and/or stratigraphic information. However, combined with circumstantial evidence, the relative frequency data reviewed above may well indicate a degree of camel management in Wādī al-Jūbah as early as the 10th century BC (Hajar at-Tamra). This is matched by the data from the main local centre, Hajar ar-Rayhānī, allowing for the domestic status of three camel bones found in Stratum VII, or the 1st and 2nd Occupational Phases (10th–8th centuries BC).

A firmer testimony to the presence of domesticated dromedaries is provided by the Early Sabaean settlements at Yalā and Barāqish. These sites suggest that between about 850 and 650 BC camels under cultural control had become common in a part of Yemen. In fact, the animals had already become sufficiently numerous by the late 9th century BC for their remains to be found consistently among food refuse in Sabaean cities along the semidesert belt. The two cities provide sufficient evidence to suggest that dromedaries were employed both on the main caravan traffic routes (Barāqish) and at a distance from them (Yalā). Furthermore, a 9th century BC date for the domesticated dromedary in Yemen compares well with its definite appearance in southeast Arabia ca. 900 BC (Magee 2007; Uerpmann & Uerpmann 2002).

A provisional 'first appearance datum' (FAD) is thus indicated for the domestic dromedary in Yemen. FADs are among the primary datum events by which biochronology on a regional scale is constructed. 'The FAD of any particular species in a certain geographic area usually is considered due to either the in situ evolution of the species from an ancestral form or the immigration of the form [i.e. the species] from another area' (Schoch 1989, 201). While important, FADs also require caution when used historically in zooarchaeological contexts. Two aspects are relevant. Once a species has evolved, it may take a certain amount of time to spread throughout what will eventually be its total geographic range (the 'prochoresis' process), and it may also take time to become sufficiently numerous in its range so that its remains will regularly appear in the record (Schoch 1989, 202). The second aspect has to do with site formation – what material remains will end up in the record and be preserved, and how. In addition to the usual biases (Chang & Koster 1986; Kent 1987; Schiffer 1996), specific cultural-functional biases may have acted in the case of the camel. For instance, reference was made above to its defective representation among food residue. But although sporadically mentioned, the issue is in fact unexplored. These selective and distorting factors may combine to produce an apparent, 'archaeological' FAD for the domestic dromedary that differs more or less significantly from a 'true' FAD in any given locale.

All Yemen finds come from urban contexts (Fig. 20.1). Like elsewhere in the arid parts of Western Asia, there has been a bias against locating such archaeological sites as ephemeral camps or trade-route stations, and that translates into a lack of bone finds from similar contexts, camel finds in particular (although current work by W Glanzman [2004b] may redress the balance; see Bar-Yosef & Khazanov 1992 and Szuchman 2009 for potential approaches). Thus a presumably large part of direct evidence for the early, specific handling of camels is missing. On the positive side, initial work at Barāqish/Yathill has begun to connect camel remains directly to Sabaean and subsequently Minaean archaeological trade contexts, from at least the 8th century BC. A city that played a pivotal role in caravan trade represents a privileged site. In addition, the Yemen evidence contributes to charting the distribution of wild dromedary populations in space and time, well into the 1st millennium BC. This is of particular interest if one considers that until very recently the wild ancestor of the Arabian camel was unknown in the Peninsula (compare, e.g., Köhler 1981, 129, with Uerpmann & Uerpmann 2002). The co-occurrence of wild and domesticated animals at such locales as Barāqish or Wādī al-Jūbah suggests repeated if not habitual interactions with wild camel populations. However, it is impossible to ascertain the frequency and modes of interaction, or the distances at which wild camel herds were found.

Optimally, the ascription of domestic status to camel finds must rely on a combination of body size and proportions (including use of index scaling techniques), faunal composition, and archaeological context, basically in this order. Data from Yemen that support a domestic-status interpretation have been reviewed with due qualifications concerning biometric or chronological aspects. On the other hand, the information is still defective in several areas, and the bone finds are insufficient to derive useful age profiles – a severe limitation. As an approximation, the presence of immature or even foetal animals was noted when possible. To mention another glaring deficiency, large bodies of individual metrical data are clearly needed in order to document the emergence and evolution (if any) of camel domestication. The example set by the Uerpmanns in southeast Arabia cannot presently be matched in the southwestern part of the Peninsula.

The problem of dromedary domestication

Knowledge on the domestication history of the dromedary is not only scanty but controversial. Several fundamental issues are unresolved, and, in particular, 'the date and location of camel domestication is still much debated' (Magee 2007, 94; cf Kitchen 1997, 135). A faunally well-studied area, the southern Levant, including the Negev Desert and the Sinai Peninsula, has often been used to monitor the emergence of domestication in Arabia or the adjoining regions. From this observatory an interpretation can be advanced that expands and in part supersedes more conservative reconstructions. Its strength derives from a particular convergence of bone data, imagery, and historical-political context, with due allowance for an uneven archaeological visibility of initial camel use.

This interpretation depends on the cultural status of the camel bones found in large quantity at some sites in the famous mining area of Timna', Negev (Rothenberg 1972, 105; 1997, 28; 1999; 2003; Rothenberg & Glass 1983, 122; see also Grigson 1998, 259; Zarins 1989, 148–149). The context is dated to 1300–1150 BC by the materials and structures associated with the Egyptian control of copper-ore exploitation, which climaxed during the reign of Ramesses III, 1184–1153 BC. The workers included Negevite pastoral groups and 'Midianites' from northwest Arabia, these latter representing a polity of the northern Hijāz roughly associated with the Qurayyah painted ware (de Maigret 1999; Edens & Bawden 1989; Knauf 1988; Levy 2009a, 255–257; Parr 1982; 1992a; 1992b; Rothenberg 1998; Sawyer & Clines 1983; Tebes 2007). Although donkeys were common, the Midianite miners 'already used many camels': 'supplies were actually brought over from the Hijaz on camel-back, [and] camels were also widely used as pack animals in the mines and smelters' (Rothenberg 2003, 14). Large camel-meat consumption might of course be related to large wild herds in the area – a cheap larder – rather than to a surplus of pack animals. However, if the above claims are substantiated, a thoroughly specialised use of camels will be demonstrated already in the 13th–12th centuries BC (Artzy 1994; Edens & Wilkinson 1998, 105–8; Finkelstein 1988; 1995, 103–126; Holladay 2001, 175; Knauf 1988, 14–15; Wapnish 1997; different views in Barako 2000; Retsö 1991). This interval corresponds to the Late Bronze II and Iron I sequence of the Levant and the Ramesside period in Egypt (for a discussion of Egypt's impact on regional or Arabian trade see Tebes 2008 and Wengrow 1996, with bibliography).

While the zooarchaeological details from Timna' (cf Rothenberg 2003, 14) are eagerly awaited, the contextual evidence looks compelling, as it joins other independent indicators that domesticated camels were known or used in that general context and period. Sparse but credible pictorial data come from northwest Arabia as well as Ramesside Egypt (Budka 2004; Ingraham *et al.* 1981, pl. 79: 14; Kitchen 1980; 1997, 135 and n. 22, *contra* Midant-Reynes & Braunstein-Silvestre 1977; Pusch 1996; Saber 1998; worth reading is the original statement by Petrie 1907, 23, on the water-bearing camel figurine from Dēr Rifeh). The Hijāz-Negev-Sinai region might even emerge as a potential centre of dromedary domestication. The marauding of the southern Levant by Midianite 'nomads', famously echoed in Judges 6:4–5 (see Barako 2000, 520, on this passage's chronology), is symptomatic of the initial consolidation of camel riding in Iron Age I, perhaps as a response to horseback warfare (the *Dromedarkriegertum*-hypothesis, Dostal 1967; cf Uerpmann & Uerpmann 2002). Furthermore, a scenario defined by rapid state emergence in southern Arabia and growing communication between this region and the Levant is apparent by 1200 BC, inevitably implying desert travel by use of dromedaries, as articulated by Kitchen (1997, with references). Half a century later Egypt was acquiring her supply of incense and other south Arabian aromatics by overland trade, no longer by sea, as did other expanding markets in the Near East. However, until about 1000 BC skilled camel use may have remained restricted to particular groups, areas or networks, giving the archaeological impression of 'low numbers and/or slow acceptance' of this animal (Horwitz & Rosen 2005, 126).

With the 10th century BC (Iron Age II) camel bones begin to be commonly found in archaeological sites of the southern Levant in contexts normally implying camel control, such as caravan-linked Tell Jemmeh or Beersheva (Hakker-Orion 1984; Köhler-Rollefson 1993; Wapnish 1983; 1984, supplemented by Hesse & Wapnish 1985, fig. 87; see a recent summary, Timna'excluded, in Horwitz & Rosen 2005, table 2). As the exploitation of wild camels has been shown to persist later than thought, an osteological verification on a case-by-case basis is essential. More dubious still is attributing domestic status to camels depicted in petroglyphs (e.g. Khan 1993; Younker 1997). That the 10th century was especially significant, as suggested by the bone evidence, is strongly supported by texts and imagery from the wider Near East. The domestic dromedary unambiguously appears in two famous reliefs of a camel and rider from Karkemish and Guzana (today Tell Halaf) in northern Syria, presently dated to the 900±30 BC interval (Genge 1979; Hawkins 1980; Winter 1983; see also Gerlach 2000). In addition to their chronology, a combination of ideology, economic-political context, and sheer distance from Arabia gives them special interest. Repetition makes the camel-and-rider a standard motif, suggesting that by the late 10th century the subject had acquired iconic value from both its novelty and visual impact.

Economically, these north Syrian polities were tied into a network which ostensibly had in the city of Khindānu on the middle Euphrates its main 'outlet' for Arabian products, hence Arabian innovations, as cogently argued by Liverani (1992; Holladay 2006; Kitchen 1997, 134–135; Na'aman 2007). According to cuneiform sources these goods included dromedaries, antimony, alabaster and myrrh; this last also traded northwards (Liverani 1992, fig. 1). Such commodities and network imply that by the mid-10th century caravan-cities were flourishing in Yemen and western Arabia, and further suggest that a direct crossing between this latter region and the middle Euphrates was perhaps normal (Liverani 1992; cf de Maigret 1997; 1999; Holladay 2006, fig. 2). From about 890 BC onwards the Arabian contributions are repeatedly confirmed by Assyrian texts (e.g. Grayson 1991, 175 [where the context supports the identification as 'dromedaries']; 1996; Cavigneaux & Ismail 1990, 346–357; Eph'al 1997; Frame 1995, 300; Retsö 2003, 121–129; Lanfranchi 2004; see also Bulliet 1990, 77 ff.). Here the dromedary enters 'history' as an animal to be ridden, an object of prestige increasingly involved in tributes and exchanges. Immediately afterwards – mid-9th century – it emerges as a uniquely important beast of burden, not only critical for the Arabian mastery of desert transport but useful in war (as first depicted on the Balawāt bronze gates; for the Assyrian stone reliefs see Mitchell 2000). As mentioned above, this is the very time frame in which the osteological evidence indicates a beginning of dromedary husbandry in southeast Arabia (Magee 2007; Uerpmann & Uerpmann 2002; 2008).

Several influential authors have argued for a long 'prehistory' of dromedary domestication, rooted in the familiarity of people and dromedaries in parts of southern or central Arabia resulting from habitual propinquity (e.g. Bulliet 1990, 58–65; Dostal 1958; Köhler 1981; 1993; Wapnish 1983, 104; Uerpmann & Uerpmann 2002; Zarins 1989, 148–149; 1992a; see also Borowski 1998, 112–121; Finkelstein 1988). In this perspective, herding would have been a final outcome in a limited number of suitable enclaves, presumably with marked regional differences (cf Dostal's [1958; 1967] distinction between southern and northern Arabia, or Kitchen's [1997, 127] justified contention that there are 'two Arabias, culturally and historically – East and West'). Specialised exploitation of the dromedary for long-distance transport may have been a closely related development, favoured if not ignited by a new international scene as discussed above. Suitable body types would have been selected – and equipment created – for particular tasks, such as carrying goods, or as a war mount on the northern Arabian fringe (Köhler-Rollefson 1993; Sweet 1965a; 1965b). In all probability, the core novelty from a cultural viewpoint was a broader process: the mastery of the desert, the unprecedented human use of its space and biome. This, I would suggest, is the backcloth against which the still modest faunal data from Yemen have to be set.

In terms of both ecology and geographic position the Yemeni Sayhad, at the broad interface of foothills and desert, would be ideally located for nurturing a large biomass of wild *Camelus* as well as proximity of camels and people. It is not a minor factor that camels need salty desert or semidesert shrubs (e.g. Khazanov 2009, 123; Köhler-Rollefson 1991; Wilson 1984), such as abound in the Sayhad (Wood 1997). The association of the Sayhad with interregional dromedary traffic, and the trade of aromatics in particular, may thus have emerged as a local development, based on a longer preparation than normally acknowledged. Here the highly organised kingdom of Saba' was best placed to exploit this phenomenon, shortly heading a federation of caravan-cities (early 10th century BC minimum on a ca. 680 BC date for Karib'īl Watār I; cf Kitchen 1997, 151); T. E. Levy's (2009b) model for the emergence of a 'chiefly confederacy' in a similar context may be relevant here. If there was in the 13th–12th centuries BC a domestication enclave in the 'Midian'-Negev area, linked to desert travel, one could venture the idea that camel use rapidly developed as a desert-border specialisation along the whole belt of western Arabia, with Yemen as its main southern focus at the same date (cf Zarins 1992b). This in principle would justify theories of Yemen's important role in camel domestication. Camel bone finds from exploratory surveys in interior Saudi Arabia (the southern Nejd; Zarins 1989, 148, with references) have a bearing on this possibility and suggest great potential for future research.

Concerning Yemen, only further archaeological work on both urban and non-urban sites of the Sayhad, supported by archaeofaunal analysis, will allow this set of hypotheses to be tested. At present, in general, any attempt to chart the spread of the domesticated dromedary in southern Arabia is hampered not only by the dearth of zooarchaeological reports, especially for the earlier periods, but by chronological deficiencies. If, predictably, a great deal still awaits to be discovered about Yemen's record of early camel use, from the glimpse already obtained Barāqish might be a choice place to look for it. Zooarchaeological results from deeper levels in the city, reaching back to perhaps the 13th century BC, will thus be of considerable interest.

Acknowledgements

This paper is the consequence of considerable encouragement from Marjan Mashkour, to whom I am particularly grateful; she also advised on the evaluation of bone size and kindly read my paper at the Al-Ain conference. My work in Yemen was undertaken in 1984 at the invitation of Alessandro de Maigret and has since been supported by funds from the Italian Archaeological Mission; I am equally indebted to de Maigret for background information on Yalā as well as introducing me to Barāqish. All faunal collections from Yalā and Barāqish are housed and were studied in Yemen. Jeffrey Blakely, Caroline Grigson, Brian Hesse, Ilse Köhler, Richard Meadow, already in the 1980s, and much more recently Alessandra Avanzini, Mark Beech, Nicole Boivin, William Glanzman, Peter Magee, Jöris Peters, Daniel Potts and Trevor Wilson provided relevant information or literature, while Gennaro Di Rosa (PhD candidate, Naples University 'L'Orientale') made available information on Assyrian records. Jill Morris critically improved the language of the paper. Needless to say, any errors in fact or formulation are my own.

References

Artzy, M. 1994, 'Incense, camels and collared rim jars: desert trade routes and maritime outlets in the second millennium'. *Oxford Journal of Archaeology*, 13(2), 121–147.

Barako, T. J. 2000, 'The Philistine settlement as mercantile phenomenon?'. *American Journal of Archaeology* 104(3), 513–30.

Bar-Yosef, O. & Khazanov, A. (eds). 1992, *Pastoralism in the Levant. Archaeological Materials in Anthropological Perspectives*. Prehistory Press, Madison.

Bednarik, R. G. & Khan, M. 2005, 'Scientific studies of Saudi Arabian rock art'. *Rock Art Research* 22, 49–81.

Beech, M., Mashkour, M., Huels, M. & Zazzo, A. 2009, 'Prehistoric camels in south-eastern Arabia: the discovery of a new site in Abu Dhabi's Western Region, United Arab Emirates'. *Proceedings of the Seminar for Arabian Studies* 39, 17–30.

Blakely, J. A. 1985, 'The stratigraphic probe at Hajar at-Tamrah'.

in J. A. Blakely, J. A. Sauer & M. R. Toplyn, *The Wadi al-Jubah Archaeological Project*, 2: *Site reconnaissance in North Yemen, 1983*, 55–145. American Foundation for the Study of Man, Washington, DC.

Blakely, J. A. 1996, 'Introduction', in M. J. Grolier, R. Brinkmann & J. A. Blakely, *The Wadi al-Jubah Archaeological Project*, 5: *Environmental research in support of archaeological investigations in the Yemen Arab Republic, 1982–1987*, 1–5. American Foundation for the Study of Man, Washington, DC.

Blakely, J. A. & Glanzman, W. D. 1996, 'Fron clay to stone? Cooking bowls of South Arabia', in J. D. Seger (ed.), *Retrieving the Past. Essays on Archaeological Research and Methodology in Honor of Gus W. Van Beek*, 3–26. Eisenbrauns (for Cobb Institute of Archaeology, Mississippi State University), Winona Lake, Indiana.

Borowski, O. 1998, *Every Living Thing. Daily Use of Animals in Ancient Israel*, AltaMira Press, Walnut Creek, CA.

Budka, J. 2004, 'Das einhöckrige Kamel (Camelus dromedarius) – Lastschiff der Wüste'. *Kemet* 13(4), 39–42.

Bulliet, R. W. 1990, *The Camel and the Wheel* (2nd edn). Morningside Books series, Columbia University Press, New York.

Cavigneaux, A. & Ismail, B. K. 1990, 'Die Statthalter von Suẖu und Mari im 8. Jh. v.Chr. anhand neuer Texte aus den irakischen Grabungen im Staugebiet des Qadissiya-Damms'. *Baghdader Mitteilungen* 21, 321–456.

Chang, C. & Koster, H. A. 1986, 'Beyond bones: toward an archaeology of pastoralism', in M. B. Schiffer (ed.), *Advances in Archaeological Method and Theory* 9, 97–148. Academic Press (Harcourt Brace Jovanovich, Publishers), Orlando.

Compagnoni, B. & Tosi, M. 1978, 'The camel: its distribution and state of domestication in the Middle East during the third millennium B.C. in light of finds from Shahr-i Sokhta', in R. H. Meadow & M. A. Zeder (eds), *Approaches to Faunal Analysis in the Middle East*, 91–103. Peabody Museum of Archaeology and Ethnology, Harvard University, Cambridge, MA.

Cribb, R. 2004, *Nomads in Archaeology* (2nd edn). Cambridge University Press, Cambridge .

de Maigret, A. (ed.) 1988, *The New Sabaean Archaeological Complex in the Wādī Yalā (Eastern Ḫawlān at-Tiyāl, Yemen Arab Republic). A Preliminary Report*. Istituto Italiano per il Medio ed Estremo Oriente, Rome.

de Maigret, A. 1997, 'The frankincense road from Najrān to Ma'ān: a hypothetical itinerary', in A. Avanzini (ed.), *Profumi d'Arabia. Atti del convegno*, 315–331. 'L'Erma' di Bretschneider, Rome.

de Maigret, A. 1999, 'The Arab nomadic people and the cultural interface between the "Fertile Crescent" and "Arabia Felix"'. *Arabian Archaeology and Epigraphy* 10, 220–224.

de Maigret, A. 2002, *Arabia Felix: an exploration of the archaeological history of Yemen*. Stacey International, London.

de Maigret, A. 2003, 'La ceramica sabea. Specificità e sviluppi da uno studio delle forme'. *Arabia* 1, 89–96, 217–44.

de Maigret, A. 2004, *Barāqish, Minaean Yathill. Excavation and restoration of the temple of Nakrah*. YICAR Papers 1, Italian Archaeological Mission in the Republic of Yemen, San'ā [*sic*].

de Maigret, A. 2010, 'A "Sabaean" stratigraphy from Barāqish' *Arabia* 4, 67–95, 205–240.

de Maigret, A. & Robin, C. 1989, 'Les fouilles italiennes de Yalā (Yémen du Nord): nouvelles données sur la chronologie de l'Arabie du Sud préislamique'. *Académie des Inscriptions & Belles-Lettres, Comptes rendus* 1989, 255–91.

de Maigret, A. & Robin, C. 1993, 'Le temple de Nakrah à Yathill (aujourd'hui Barāqish), Yémen, résultats des deux premières campagnes de fouilles de la Mission italienne'. *Académie des Inscriptions & Belles-Lettres, Comptes rendus*, 1993(2), 427–496.

Dostal, W. 1958, 'Zur Frage der Entwicklung des Beduinentums'. *Archiv für Völkerkunde* 13, 1–14.

Dostal, W. 1967, *Die Beduinen in Südarabien. Eine ethnologische Studie zur Entwicklung der Kamelhirtenkultur in Arabien.* Wiener Beiträge zur Kulturgeschichte und Linguistik 16, Ferdinand Berger & Sonne/Institut für Völkerkunde, Horn & Vienna.

Edens, C. & Bawden, G. 1989, 'History of Taymā' and Hejazi trade during the first millennium B.C.'. *Journal of the Economic and Social History of the Orient* 32, 48–103.

Edens, C. & Wilkinson, T. J. 1998, 'Southwest Arabia during the Holocene: recent archaeological developments'. *Journal of World Prehistory* 12(1), 55–119.

Eph'al, I. 1997, *The Ancient Arabs. Nomads on the Borders of the Fertile Crescent 9th–5th centuries B.C.* (2nd edn). The Hebrew University Magnes Press, Jerusalem.

Ermolova, N. M. 1970, 'Novye materialy po izučeniju ostatkov mlekopitajuščikh iz drevnikh poselenij Turkmenii'. *Karakumskie Drevnosti* 3, 205–232 (quoted in Compagnoni & Tosi 1978, 102).

Fedele, F. G. 1992, 'Zooarchaeology in Mesopotamia and Yemen: a comparative history'. *Origini. Preistoria e protostoria delle civiltà antiche* 16, 49–93.

Fedele, F. G. 2009, 'Sabaean animal economy and household consumption at Yalā, eastern Khawlān al-Tiyāl, Yemen'. *Proceedings of the Seminar for Arabian Studies* 39, 135–154.

Fedele, F. G. 2010, 'Barāqish, over-wall excavations 2005–2006: stratigraphy, environment and economy of the Sabaean-Islamic sequence' *Arabia* 4, 97–161, 241–257.

Fedele, F. G. 2014, 'Camels, donkeys and caravan trade: an emerging context from Barāqish , ancient Yathill (Wādī al-Jawf, Yemen). *Anthropozoologica* 49/2, 177–194.

Finkelstein, I. 1988, 'Arabian trade and socio-political conditions in the Negev in the twelfth-eleventh centuries B.C.E.'. *Journal of Near Eastern Studies* 47(4), 241–252.

Finkelstein, I. 1995, *Living on the Fringe. The Archaeology and History of the Negev, Sinai and Neighbouring Regions in the Bronze and Iron Ages.* Sheffield Academic Press, Sheffield.

Frame, G. 1995, *Rulers of Babylonia from the Second Dynasty of Isin to the end of Assyrian Domination (1157–612 BC).* The Royal Inscriptions of Mesopotamia, Babylonian Periods 2, University of Toronto Press, Toronto, Buffalo & London.

Gauthier-Pilters, H. & Dagg, A. I. 1981, *The Camel. Its Evolution, Ecology, Behavior, and Relationship to Man.* The University of Chicago Press, Chicago & London.

Genge, H. 1979, *Nordsyrisch-südanatolische Reliefs. Eine archäologisch-historische Untersuchung – Datierung und Bestimmung.* Historisk-filosofiske meddelelser 49(1–2), Det Kongelige Danske Videnskabernes Selskab, Copenhagen.

Gentry, A., Clutton-Brock, J. & Groves, C. 2004, 'The naming of wild animal species and their domestic derivates', *Journal of Archaeological Science* 31, 645–651.

Gerlach, I. 2000, 'Tradition – Adaptation – Innovation. Zur Reliefkunst Nordsyriens/Südostanatoliens in neuassyrischer Zeit', in G Bunnens (ed.), *Essays on Syria in the Iron Age*, 235–257. Ancient Near Eastern Studies Supplement 7, Peeters Press, Louvain.

Glanzman, W. D. 1987a, 'Radiocarbon analysis', in W. D. Glanzman & A. O. Ghaleb, *The Wadi al-Jubah Archaeological Project*, 3: *Site reconnaissance in the Yemen Arab Republic, 1984: the stratigraphic probe at Hajar ar-Rayhani*, 65–6. American Foundation for the Study of Man, Washington, DC.

Glanzman, W. D. 1987b, 'Conclusions', in W. D. Glanzman & A. O. Ghaleb, *The Wadi al-Jubah Archaeological Project*, 3: *Site reconnaissance in the Yemen Arab Republic, 1984: the stratigraphic probe at Hajar ar-Rayhani*, 207–8. American Foundation for the Study of Man, Washington, DC.

Glanzman, W. D. 1994, 'Toward a classification and chronology of pottery from HR3 (Hajar ar-Rayhani), Wadi al-Jubah, Republic of Yemen'. PhD dissertation, University of Pennsylvania, No. 9521036, ProQuest LLC, Ann Arbor, Michigan.

Glanzman, W. D. 2004a, 'Beyond their borders: a common potting tradition and ceramic horizon within South Arabia during the later first millennium BC through the early first millennium AD'. *Proceedings of the Seminar for Arabian Studies* 34, 121–138.

Glanzman, W. D. 2004b, 'Ancient South Arabia's camel caravans: a reassessment of the evidence, and the road ahead'. Unpublished paper presented at the Seminar for Arabian Studies, London, 22 July, 2004.

Glanzman, W. D. 2005, '"Is your cup half empty, or half full?". The function and significance of the Wavy Rim Bowl in pre-Islamic South Arabia'. *Bulletin of the Society for Arabian Studies* 10, 15–19, 37.

Glanzman, W. D & Ghaleb, A. O. 1987, 'The stratigraphic probe at Hajar ar-Rayhani'. in W. D. Glanzman & A. O. Ghaleb, *The Wadi al-Jubah Archaeological Project*, 3: *Site reconnaissance in the Yemen Arab Republic, 1984: the stratigraphic probe at Hajar ar-Rayhani*, 5–63. American Foundation for the Study of Man, Washington, DC.

Grayson, A. K. 1991, *Assyrian Rulers of the Early First Millennium BC I (1114–859 BC)*. The Royal Inscriptions of Mesopotamia, Assyrian Periods 2, University of Toronto Press, Toronto, Buffalo & London.

Grayson, A. K. 1996, *Assyrian Rulers of the Early First Millennium BC II (858–745 BC)*. The Royal Inscriptions of Mesopotamia, Assyrian Periods 3, University of Toronto Press, Toronto, Buffalo & London.

Grigson, C. 1998, 'Plough and pasture in the early economy of the southern Levant', in T. E. Levy (ed.), *The Archaeology of Society in the Holy Land*, 245–268, 573–576 (2nd edn). Leicester University Press, London & Washington.

Hakker-Orion, D. 1984, 'The rôle of the camel in Israel's early history', in J. Clutton-Brock & C. Grigson (eds), *Animals and Archaeology*, 3: *Early Herders and their Flocks*, 207–12. British Archaeological Report S202. British Archaeological Reports, Oxford.

Hawkins, J. D. 1980, 'Karkamiš', in D. O. Edzard (ed.), *Reallexikon der Assyriologie und vorderasiatischen Archäologie*, V(5–6), 426–446. Walter de Gruyter, Berlin & New York.

Hesse, B. 1996a, 'The Hajar ar-Rayhani fauna: a first look at Yemen's Iron Age pastoral economy', in J. D. Seger (ed.), *Retrieving the Past. Essays on Archaeological Research and Methodology in Honor of Gus W. Van Beek*, 103–122. Eisenbrauns (for Cobb Institute of Archaeology, Mississippi State University), Winona Lake, Indiana.

Hesse, B. 1996b, 'The Hajar ar-Rayhani fauna: variation in the animal economy in Iron-Age Yemen', in M. J. Grolier, R. Brinkmann & J. A. Blakely, *The Wadi al-Jubah Archaeological Project*, 5: *Environmental Research in Support of Archaeological Investigations in the Yemen Arab Republic, 1982–1987*, 263–292. American Foundation for the Study of Man, Washington, DC.

Hesse, B. & Wapnish, P. 1985, *Animal Bone Archaeology*, Manuals in Archaeology 5, Taraxacum, Washington, DC.

Holladay, J. S. Jr. 2001, 'Toward a new paradigmatic understanding of long-distance trade in the ancient Near East: from the Middle Bronze II to Early Iron II – a sketch', in P. M. M. Daviau, J. W. Wevers & M. Weigl (eds), *The World of the Aramaeans II. Studies in History and Archaeology in Honour of Paul-Eugène Dion*, 136–198. JSOT Supplement Series 325, Sheffield Academic Press, Sheffield.

Holladay, J. S. Jr. 2006, 'Hezekiah's tribute, long-distance trade, and the wealth of nations ca. 1000–600 BC: a new perspective', in S. Gitin, J. E. Wright & J. P. Dessel (eds), *Confronting the Past. Archaeological and Historical Essays on Ancient Israel in honor of William G. Dever*, 309–331. Eisenbrauns, Winona Lake, Indiana .

Horwitz, L. K. & Rosen, B. 2005, 'A review of camel milking in the Southern Levant', in J. Mulville & A. K. Outram (eds), *The Zooarchaeology of Fats, Oils, Milk and Dairying. Proceedings of the 9th Conference of the International Council of Archaeozoology, Durham, August 2002*, 121–131. Oxbow Books, Oxford.

Ingraham, M. L., Johnson, T. D., Rihani, B. & Shatla, I. 1981, 'Saudi Arabian Comprehensive Survey Program: preliminary report on a reconnaissance survey of the Northwestern Province (with a note on a brief survey of the Northern Province)'. *Atlal. The Journal of Saudi Arabian Archaeology* 5, 59–84.

Japp, S. 2008, 'New research in the capital of Saba', Mārib'. Unpublished paper presented at the XII Rencontres Sabéennes, held at IsIAO, Rome, June 2008.

Kent, S. 1987, *Method and Theory for Activity Area Research. An Ethnoarchaeological Approach*, Columbia University Press, New York.

Khan, M. 1993, *Prehistoric Rock Art of Northern Saudi Arabia*. Ministry of Education, Department of Antiquities and Museums, Riyadh.

Khan, M. 1998, 'A critical review of rock art studies in Saudi Arabia', *East and West* 48(3–4), 427–437.

Khazanov, A. M. 2009, 'Specific characteristics of Chalcolithic and Bronze Age pastoralism in the Near East', in J. Szuchman (ed.), *Nomads, Tribes, and the State in the Ancient Near East. Cross-disciplinary Perspectives*, 119–127. Oriental Institute Seminars 5, The Oriental Institute of the University of Chicago, Chicago.

Kingdon, J. 1988, *East African Mammals. An Atlas of Evolution in Africa* 3, *Part B: Large mammals*. University of Chicago Press, Chicago.

Kitchen, K. A. 1980, 'Camel', in J. D. Douglas (ed.), *The Illustrated Bible Dictionary* 1, 228–230. Inter-Varsity Press, Leicester/Tyndale House, Wheaton, Illinois.

Kitchen, K. A. 1997, 'Sheba and Arabia', in L. K. Handy (ed.), *The Age of Solomon. Scholarship at the Turn of the Millennium*, 126–153. Brill, Leiden, New York & Köln.

Knauf, E. A. 1988, *Midian. Untersuchungen zur Geschichte Palästinas und Nordarabiens am Ende des 2. Jahrtausends v.Chr.*. Abhandlungen des Deutschen Palästinavereins, Harrassowitz, Wiesbaden.

Köhler, I. 1981, 'Zur Domestikation des Kamels'. Inaugural-

Dissertation, Institut für Zoologie der Tierärztlichen Hochschule Hannover, Hannover.

Köhler, I. 1984, 'The dromedary in modern pastoral societies and implications for its process of domestication', in J. Clutton-Brock & C. Grigson (eds), *Animals and Archaeology* 3: *Early Herders and their Flocks*, 201–206. British Archaeological Report S202. British Archaeological Reports, Oxford.

Köhler-Rollefson, I. 1989, 'Zoological analysis of camel skeletons', in R. H. Smith & L. P. Day (eds), *Pella of the Decapolis* 2: *Final report on The College of Wooster excavations in Area IX, the Civic Complex, 1979–1985*, 142–164. The College of Wooster, Wooster, Ohio.

Köhler-Rollefson, I. 1991, 'Camelus dromedarius'. *Mammalian Species* 375, 1–8.

Köhler-Rollefson, I. 1993, 'Camels and camel pastoralism in Arabia'. *The Biblical Archaeologist* 56(4), 180–188.

Köhler-Rollefson, I. 1996, 'The one-humped camel in Asia: origin, utilization and mechanisms of dispersal', in D. R. Harris (ed.), *The Origins and Spread of Agriculture and Pastoralism in Eurasia*, 282–294. UCL Press, London/ Smithsonian Institution Press, Washington, DC.

Lanfranchi, G. B. 2004, 'Mesopotamia e Arabia nelle iscrizioni reali assire', in F. M. Fales & D. Morandi Bonacossi (eds), *Mesopotamia e Arabia. Scavi archeologici e studi territoriali delle Università trivenete (1994–1998)*, 217–253. Istituto Veneto di Scienze, Lettere ed Arti, Venice.

Lesur, J. 2007, *Chasse et élevage dans la Corne de l'Afrique entre le Néolithique et les temps historiques*. Cambridge Monographs in African Archaeology 68/British Archaeological Report S1602, Archaeopress, Oxford.

Levy, T. E. 2009a, 'Ethnic identity in Biblical Edom, Israel, and Midian: some insights from mortuary contexts in the lowlands of Edom', in J. D. Schloen (ed.), *Exploring the Longue Durée. Essays in honor of Lawrence E. Stager*, 251–261. Eisenbrauns, Winona Lake, Indiana.

Levy, T. E. 2009b, 'Pastoral nomads and Iron Age metal production in ancient Edom', in J. Szuchman (ed.), *Nomads, Tribes, and the State in the Ancient Near East. Cross-disciplinary Perspectives*, 147–177. Oriental Institute Seminars 5, The Oriental Institute of the University of Chicago, Chicago.

Liverani, M. 1992, 'Early caravan trade between South-Arabia and Mesopotamia'. *Yemen. Studi Archeologici, Storici e Filologici sull'Arabia Meridionale* 1, 111–115.

Loreto, R. 2009, 'House and household: a contextual approach to the study of South Arabian domestic architecture. A case study from seventh- to sixth-century BC Yalā/ad-Durayb'. *Proceedings of the Seminar for Arabian Studies* 39, 255–269.

Magee, P. 1999, 'Writing in the Iron Age: the earliest South Arabian inscription from southeastern Arabia'. *Arabian Archaeology and Epigraphy* 10, 43–50.

Magee, P. 2004, 'The impact of southeast Arabian intra-regional trade on settlement location and organisation during the Iron Age II period'. *Arabian Archaeology and Epigraphy* 15, 24–42.

Magee, P. 2007, 'Beyond the desert and the sown: settlement intensification in late prehistoric southeastern Arabia'. *Bulletin of the American Schools of Oriental Research* 347, 83–105.

Manzo, A. 2005, 'The stratigraphic sequence at Yalā (Yemen): a statistical evaluation' *Radiocarbon* 47(1), 147–158.

Mashkour, M. 1997, 'The funeral rites at Mleiha (Sharjah-U.A.E.); the camelid graves'. *Anthropozoologica* 25–26, 725–736.

Midant-Reynes, B. & Braunstein-Silvestre, F. 1977, 'Le chameau en Égypte'. *Orientalia* n.s. 46(3), 337–362.

Mitchell, T. C. 2000, 'Camels in the Assyrian bas-reliefs' *Iraq*, 62, 187–194.

Na'aman, N. 2007, 'The contribution of the Suḫu inscriptions to the historical research of the kingdoms of Israel and Judah'. *Journal of Near Eastern Studies* 66(2), 107–122.

Parr, P. J. 1982, 'Contacts between Northwest Arabia and Jordan in the Late Bronze and Iron Ages', in A. Hadidi (ed.), *Studies in the History and Archaeology of Jordan* 1, 127–34. Department of Antiquities of Jordan, Amman.

Parr, P. J. 1992a, 'Edom and the Hejaz', in P. Bienkowski (ed.), *Early Edom and Moab. The Beginning of the Iron Age in Southern Jordan*, 41–46. Sheffield Archaeological Monographs 7, J. R. Collis Publishers, Sheffield.

Parr, P. J. 1992b, 'Qurayya', in D. N. Freedman (ed.), *The Anchor Bible Dictionary* 5, 594–596. Doubleday and Co, New York.

Payne, W. J. A. & Wilson R. T. 1999, 'Old World camels', in W. J. A. Payne & R. T. Wilson (eds), *An Introduction to Animal Husbandry in the Tropics* (5th edn), 485–520. Wiley-Blackwell (Blackwell Science), Chichester.

Peters, J. 2001, 'Aspekte der Domestikations- und Kulturgeschichte des Dromedars (*Camelus dromedarius*)', in H. Buitenhuis & W. Prummel (eds), *Animals and Man in the Past. Essays in Honour of Dr. A.T. Clason*, 330–341. ARC-Publicatie 41, ARC/Archaeological Research and Consultancy, Groningen.

Petrie, W. M. F. 1907, *Gizeh and Rifeh*. School of Archaeology in Egypt, University College/Bernard Quaritch, London.

Potts, D. T. 1991, *Further Excavations at Tell Abraq: the 1990 season*. Munksgaard, Copenhagen.

Potts, D. T. 1993, 'Four seasons of excavation at Tell Abraq (1989–1993)'. *Proceedings of the Seminar for Arabian Studies* 23, 117–126.

Potts, D. T. 2004, 'Camel hybridization and the role of *Camelus bactrianus* in the ancient Near East'. *Journal of the Economic and Social History of the Orient* 47, 143–165.

Potts, D. T. 2005, 'Bactrian camels and Bactrian-dromedary hybrids'. *The Silk Road* 3(1), 49–58. Online, viewed 20 October 2009, <http://www.silk-road.com/newsletter/vol3num1/7_bactrian.php>.

Pusch, E. B. 1996, 'Ein Dromedar aus der Ramses-Stadt'. *Ägypten und Levante/Egypt and the Levant* 6, 107–118.

Retsö, J. 1991, 'The domestication of the camel and the establishment of the Frankincense Road from south Arabia'. *Orientalia Suecana* 40, 187–219.

Retsö, J. 2003, *The Arabs in Antiquity. Their History from the Assyrians to the Umayyads*. Routledge-Curzon, Abingdon & New York.

Robin, C. 2002, 'Saba' and the Sabaeans', in St J. Simpson (ed.), *Queen of Sheba. Treasures from Ancient Yemen*, 51–8. The British Museum Press, London.

Robin, C. & Brunner, U. 1997, *Map of Ancient Yemen/Carte du Yémen Antique 1:1,000,000*. Staatliches Museum für Völkerkunde, Munich.

Rothenberg, B. 1972, *Timna. Valley of the Biblical Copper Mines*, Thames and Hudson, London.

Rothenberg, B. 1997, 'Researches in the southern Arabah 1959–1990. Summary of thirty years of archaeo-metallurgical field work in the Timna Valley, the Wadi Amram and the southern Arabah (Israel)'. *Arx* 2–3, 5–42.

Rothenberg, B. 1998, 'Who were the "Midianite" copper miners of the Arabah? About the "Midianite enigma"', in T. Rehren,

A. Hauptmann & J. D. Muhly (eds), *Metallurgica Antiqua. In Honour of Hans-Gert Bachmann and Robert Maddin*, 197–212. *Der Anschnitt* Beiheft 8, Deutsches Bergbau-Museum, Bochum.

Rothenberg, B. 1999, 'Archaeo-metallurgical researches in the southern Arabah 1959–1990. Part 2: Egyptian New Kingdom (Ramesside) to Early Islam'. *Palestine Exploration Quarterly* 131, 149–175.

Rothenberg, B. 2003, 'Egyptian chariots, Midianites from Hijaz/Midian (Northwest Arabia) and Amalekites from the Negev in the Timna Mines. Rock drawings in the Ancient Copper Mines of the Arabah – new aspects of the region's history II'. *iams – Institute for Archaeo-Metallurgical Studies* 23, 9–14.

Rothenberg, B. & Glass, J. 1983, 'The Midianite pottery', in J. F. A. Sawyer & D. J. A. Clines (eds), *Midian, Moab and Edom. The history and archaeology of Late Bronze and Iron Age Jordan and North-west Arabia*, 65–124. *Journal for the Study of the Old Testament* Supplement series 24, JSOT Press, Sheffield.

Saber, A. S. 1998, 'The camel in ancient Egypt', in Gh. Alhadrami & M. A. Ayoub (eds), *Proceedings of the Third Annual Meeting for Animal Production Under Arid Conditions: 'The International Conference on Camel Production and Future Perspectives'* 1, 208–215. United Arab Emirates University, Al Ain.

Saidel, B. A. 2009, 'Pitching camp: ethnoarchaeological investigations of inhabited tent camps in the Wadi Hisma, Jordan', in J. Szuchman (ed.), *Nomads, Tribes, and the State in the Ancient Near East. Cross-disciplinary Perspectives*, Oriental Institute Seminars 5, 87–104. The Oriental Institute of the University of Chicago, Chicago.

Sawyer, J. F. A. & Clines, D. J. A. (eds) 1983, *Midian, Moab and Edom. The History and Archaeology of Late Bronze and Iron Age Jordan and North-west Arabia. Journal for the Study of the Old Testament* Supplement series 24, JSOT Press, Sheffield.

Schiffer, M. B. 1996, *Formation Processes of the Archaeological Record.* University of Utah Press, Salt Lake City.

Schoch, R. M. 1989, *Stratigraphy. Principles and Methods.* Van Nostrand Reinhold, New York.

Spassov, N. & Stoytchev, T. 2004, 'The dromedary domestication problem: 3000 BC rock art evidence for the existence of wild one-humped camel in Central Arabia'. *Historia Naturalis Bulgarica* 16, 151–158.

Steiger, C. 1990, 'Vergleichend morphologische Untersuchungen an Einzelknochen des postkranialen Skeletts der Altweltkamele'. Inaugural-Dissertation der Tierärztlichen Fakultät der Universität München, Institüt für Palaeoanatomie, Domestikationsforschung und Geschichte der Tiermedizin, München.

Sweet, L. E. 1965a, 'Camel raiding of North Arabian Bedouin: a mechanism of ecological adaptation'. *American Anthropologist* 67, 1132–1150.

Sweet, L. E. 1965b, 'Camel pastoralism in north Arabia and the minimal camping unit', in A. Leeds & A. P. Vayda (eds), *Man, Culture, and Animals. The Role of Animals in Human Ecological Adjustments*, 129–152. AAAS Publication 78, American Association for the Advancement of Science, Washington, DC.

Szuchman, J. (ed.) 2009, *Nomads, Tribes, and the State in the Ancient Near East. Cross-disciplinary Perspectives*, Oriental Institute Seminars 5, The Oriental Institute of the University of Chicago, Chicago.

Tebes, J. M. 2007, 'Pottery makers and pre-modern exchange in the fringes of Egypt: an approximation to the distribution of Iron Age Midianite pottery'. *Buried History* 43, 11–26.

Tebes, J. M. 2008, *Centro y periferia en el mundo antiguo. El Negev y sus interacciones con Egipto, Asiria, y el Levante en la Edad del Hierro (1200–586 a.C.)* (2nd edn). Society of Biblical Literature – Centro de Estudios de Historia del Antiguo Oriente (Universidad Católica Argentina), Buenos Aires.

Tiede, L. J. & Overstreet, W. C. 1988, 'Gazetteer of place names with explanatory notes on transliteration', in W. C. Overstreet, M. J. Grolier & M. R. Toplyn, *The Wadi al-Jubah Archaeological Project*, 4: *Geological and archaeological reconnaissance in the Yemen Arab Republic, 1985*, xxxiii–xlii. American Foundation for the Study of Man, Washington, DC.

Toplyn, M. R. 1985, 'Faunal evidence', in J. A. Blakely, J. A. Sauer & M. R. Toplyn, *The Wadi al-Jubah Archaeological Project*, 2: *Site reconnaissance in North Yemen, 1983*, 127–128. American Foundation for the Study of Man, Washington, DC.

Toplyn, M. R. 1988, 'The 1984 geological probes GP1–GP3 in Wadi al-Jubah, Yemen Arab Republic', in W. C. Overstreet, M. J. Grolier & M. R. Toplyn, *The Wadi al-Jubah Archaeological Project*, 4: *Geological and archaeological reconnaissance in the Yemen Arab Republic, 1985*, 85–104. American Foundation for the Study of Man, Washington DC.

Uerpmann, H.-P. 1999, 'Camel and horse skeletons from protohistoric graves at Mleiha in the Emirate of Sharjah (U.A.E.)'. *Arabian Archaeology and Epigraphy* 10, 102–118.

Uerpmann, H.-P. & Uerpmann, M. 2002, 'The appearance of the domestic camel in south-east Arabia'. *The Journal of Oman Studies* 12, 235–260.

Uerpmann, M. & Uerpmann, H.-P. 2008, 'Neolithic faunal remains from al-Buhais 18 (Sharjah, UAE)', in H.-P. Uerpmann, M. Uerpmann & S. A. Jasim (eds), *The Archaeology of Jebel al-Buhais* 2: *The natural environment of Jebel al-Buhais: past and present*, 97–132. Kerns Verlag, Tübingen.

Van Beek, G. 1969, *Hajar Bin Humeid. Investigations at a Pre-Islamic site in South Arabia.* Publications of the American Foundation for the Study of Man 5, The Johns Hopkins Press, Baltimore.

Van Beek, G. 1997, 'Hajar Bin Humeid', in E. M. Meyers (ed.), *The Oxford Encyclopedia of Archaeology in the Near East* 2, 457–458. Oxford University Press, Oxford & New York.

Vigne, J.-D., Helmer, D. & Peters, J. 2005, 'New archaeozoological approaches to trace the first steps of animal domestication: general presentation, reflections and proposals', in J.-D. Vigne, J. Peters & D. Helmer (eds), *First Steps of Animal Domestication. New Archaeozoological Approaches. Proceedings of the 9th Conference of the International Council of Archaeozoology, Durham, August 2002*, 1–16. Oxbow Books, Oxford.

von den Driesch, A. 1976, *A Guide to the Measurement of Animal Bones from Archaeological Sites*. Peabody Museum of Archaeology and Ethnology, Harvard University, Cambridge, Massachusetts.

Wapnish, P. 1983, 'Camel caravans and camel pastoralists at Tell Jemmeh'. *Journal of the Ancient Near Eastern Society of Columbia University* 13 (1981), 101–121.

Wapnish, P. 1984, 'The dromedary and Bactrian camel in Levantine historical settings: the evidence from Tell Jemmeh', in J. Clutton-Brock & C. Grigson (eds), *Animals and Archaeology* 3: *Early Herders and their Flocks*, 171–200.

British Archaeological Report S202. British Archaeological Reports, Oxford.

Wapnish, P. 1997, 'Camels', in E. M. Meyers (ed.), *The Oxford Encyclopedia of Archaeology in the Near East* 1, 407–408. Oxford University Press, Oxford & New York.

Wengrow, D. 1996, 'Egyptian taskmasters and heavy burdens: highland exploitation and the collared-rim pithos of the Bronze/Iron Age Levant'. *Oxford Journal of Archaeology* 15(3), 307–326.

Wilson, R. T. 1984, *The Camel.* Longman, London & New York.

Wilson, R. T. 1997, 'Types and breeds of the one-humped camel'. *Journal of Camel Practice and Research* 4(2), 111–117.

Winter, I. J. 1983, 'Carchemish *ša kišad puratti*'. *Anatolian Studies* 33, Special number in honour of the seventy-fifth birthday of Dr Richard Barnett, 177–197.

Wood, J. R. I. 1997, *A Handbook of the Yemen Flora.* Royal Botanic Gardens, Kew.

Younker, R. W. 1997, 'Late Bronze Age camel petroglyphs in the Wadi Nasib, Sinai'. *Near East Archaeological Society Bulletin* 42, 47–54.

Zarins, J. 1989, 'Pastoralism in southwest Asia: the second millennium BC', in J Clutton-Brock (ed.), *The Walking Larder. Patterns of Domestication, Pastoralism, and Predation*, 127–155. Unwin Hyman, London.

Zarins, J. 1992a, 'Camel', in D. N. Freedman (ed.), *The Anchor Bible Dictionary* 1, 824–836. Doubleday and Co, New York.

Zarins, J. 1992b, Review of 'The Arabian Gulf in Antiquity 1: From Prehistory to the Fall of the Achaemenid Empire', by D. T. Potts, 1990. *The Biblical Archaeologist* 55(4), 234.

Update April 2016

During the six years since this article was completed (December 2009) and accepted for publication (January 2010) a substantial amount of literature on subjects discussed in its final section has appeared: here is a highly selective account. A masterful treatment of dromedary domestication from an Arabian perspective, in the context of the Near and Middle East, has been presented by P. Magee: *The Archaeology of Prehistoric Arabia*, Cambridge University Press, 2014, 197–213 (ch. 7), 258–74; and 'When was the dromedary domesticated in the ancient Near East?', *Zeitschrift für Orient-Archäologie* 8 (2015), 252–77. Magee would argue for a pre-eminent role of southeastern Arabia in developing the dromedary for desert transport, possibly in emulation of Bactrian camel use across the Gulf, although acknowledging my hypothesis of such a role for southwestern Arabia. Equally comprehensive, from a predominantly 'Near Eastern' perspective, is M. Heide's 'The domestication of the camel. Biological, archaeological and inscriptional evidence from Mesopotamia, Egypt, Israel and Arabia, and literary evidence from the Hebrew Bible', *Ugarit-Forschungen* 42 (2010), 331–82; soon to be followed by a major book, *The Camel in the Biblical World* (which I was kindly allowed to read before publication). Two multidisciplinary collections include reports on ancient dromedaries in Arabia: E.-M. Knoll & P. Burger (eds), *Camels in Asia and North Africa: Interdisciplinary Perspectives and their Past and Present Significance*, Wien, Österreichische Akademie der Wissenschaften, 2012, with contributions by R. W. Bulliet, M. and H.-P. Uerpmann, D. M. Varisco, and the late W. Dostal; and M. Mashkour & M. Beech (eds), *Ancient Camelids in the Old World – Between Arabia and Europe*, *Anthropozoologica* 49/2 (2014), particularly with articles by C. Grigson and H. Monchot; my own contribution can be read in conjunction with 'The wall and talus at Barāqish, ancient Yathill (al-Jawf, Yemen): a Minaean stratigraphy', *Proceedings of the Seminar for Arabian Studies* 41 (2011), 101–20. On particular topics or areas, the most significant update concerns the mining districts of the southern Levant, Timna' and Faynān. Not only have the old camel finds been published (C. Grigson, 'Camels, copper and donkeys in the early Iron Age of the southern Levant: Timna revisited', *Levant* 44/1 (2012), 82–100), but new research has led to a redating of dromedary utilization in the region to not earlier than ca. 930 BC (Iron Age II): Ben-Yosef et al., 'A new chronological framework for Iron Age copper production in Timna (Israel)', *Bulletin of the American Schools of Oriental Research* 366 (2012), 1–41; L. Sapir-Hen & E. Ben-Yosef, 'The introduction of domestic camels to the southern Levant: evidence from the Aravah Valley', *Tel Aviv* 40 (2013), 277–85 (see also 'The socioeconomic status of Iron Age metalworkers: animal economy in the "Slaves' Hill", Timna, Israel', *Antiquity* 88/341 (2014), 775–90). This pattern is repeated in Syria (C. Becker, 'Die Tierknochenfunde', in H. Kühne et al. (eds), *Berichte der Ausgrabung aus Tall Shēkh Hamad/Dūr-Katlimmu 8*, 61–131, Wiesbaden, Harrassowitz, 2008), apart from a report of 'domestic' dromedaries in Late Bronze II: C. Grigson, 'The fauna of Tell Nebi Mend (Syria) in the Bronze and Iron Age—a diachronic overview. Part 1', *Levant* 47/1 (2015), 5–29. As to Barāqish, the distribution and status of the dromedary finds obtained from the crucially important 13th–8th century BC levels in sounding A/S1 remain unknown. A return to the site proved impossible after 2007, both due to A. de Maigret's death (2011) and warfare in Yemen; in 2015 the Italian Mission base was razed to the ground.

21. The Iron Age site of Muweilah (Sharjah, UAE) and the problems of dromedary domestication

Margarethe Uerpmann and Hans-Peter Uerpmann

Muweilah is one of the few archaeological sites in southeast Arabia where animal remains are preserved in larger quantities. It is a small fortified town of the Iron Age II period situated in the outskirts of Sharjah City. Of some 34,000 identified animal bones almost half are from fish. Among the 16,500 bones of domesticates there are more than 700 bones of dromedaries. In terms of find numbers only sheep and goat are more frequent. According to bone weights dromedaries are second in importance after sheep. Cattle amount to one-third of the camel bones in terms of numbers and to only one-fifth in terms of weight. They are the only other domesticates of any importance for Muweilah's subsistence. Assuming that dromedaries were not only kept for meat, but also for labour and milk, they were probably the most important living resource for the inhabitants of the township.

There is no commonly accepted evidence for the occurrence of the domestic dromedary prior to the Iron Age II period. Thus, the observation on the economic importance of this animal at Muweilah raises interesting questions with regard to the origins of the domestic dromedary and its early history. The wild dromedary, well known from several Bronze Age and Neolithic sites in southeast Arabia, is also represented at Muweilah together with a number of other wild animals

Keywords Dromedary, domestication, UAE, Iron Age

Introduction

The archaeological site of Muweilah is situated in the eastern outskirts of Sharjah (the Emirate of Sharjah), just a few kilometres south of the International Airport of the city. It was discovered during archaeological reconnaissance carried out by a French mission to the Emirate in the 1980s. Its archaeological importance was realised after the first test excavations made by Peter Magee in 1994 (Magee 1996). Continued excavations since that time have meanwhile uncovered large areas (Fig. 21.1) of a complex Iron Age (IA) settlement surrounded by a ditch and a wall, and with diverse buildings, mostly constructed of mud-brick with roofs probably made of organic materials. Some of the buildings seem to have had special functions which cannot be described here in any detail (Magee *et al.* 2002). Of particular importance for the topic of the present paper is the fact that the town was apparently founded early in the 1st millennium BC and destroyed by fire during the late 8th or early 7th century BC and that it was not inhabited again afterwards. It seems that the remains of the mudbrick walls, decaying after the destructive event, embedded the floor horizons of the former town. This cover preserved a multitude of archaeological finds and among them many animal bones – which is a most fortunate circumstance in a geographic area where the preservation of bones at archaeological sites is usually precarious to non-existent.

Muweilah faunal remains

Up to now some 200,000 bone fragments have been analysed with a total weight of more than 100kg.

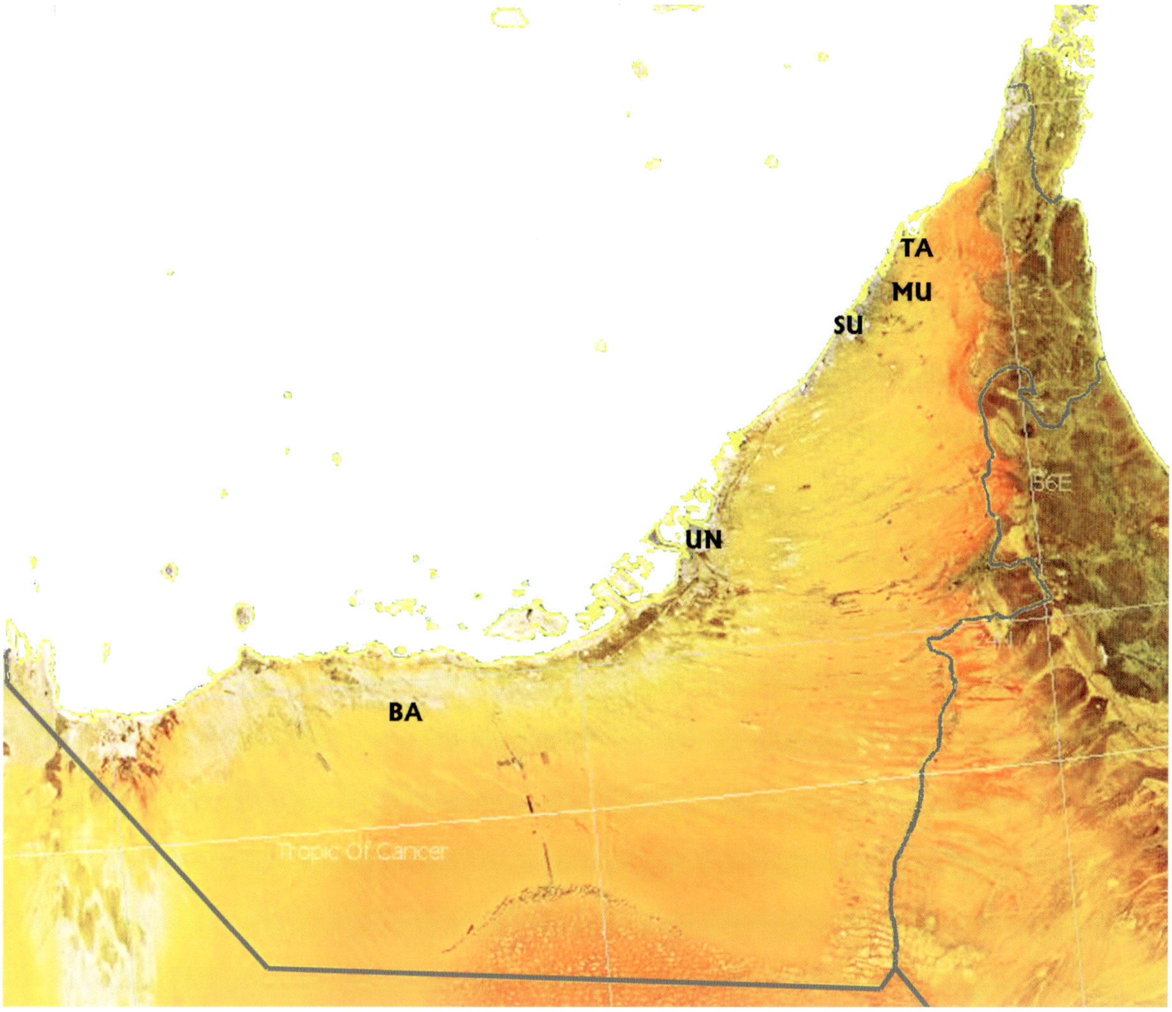

Fig. 21.1. Positions of the camel-sites mentioned in the text (TA=Tell Abraq, MU=Muweilah, SU=al-Sufou2, UN=Umm an-Nar, BA= Baynunah).

However, only 37,676 of these fragments with a total weight of 71.5kg could be identified to some extent. The low degree of identification – 17% by number and 59% by weight – is indicative of the fact that systematic sieving of excavated sediments yielded large amounts of small bone fragments. These were classified according to animal size, mainly based on cortical thickness, in order to control potential biases caused by the low degree of identification. The obtained size-classification of the unidentified fragments was then compared to the same classes of animal sizes as observed in the identified bones. Since the distribution of size classes in both assemblages is quite similar, one may assume that the identified portion of the bone finds from the site is representative for all of the studied bone finds. Table 21.1 gives an overview of the so far identified faunal remains from Muweilah. The role of the domestic dromedary is highlighted in the summary of Table 21.1.

It is obvious from the bone weights in Table 21.1 that the contribution of the dromedary to the meat consumed by the inhabitants of Muweilah was quite high. It was much more important than cattle, and when separating the contributions of sheep and goat according to the weight-relation of their identified bones, the camel was even more important than the goat. Only sheep contributed more meat to the subsistence economy of Muweilah than the dromedary. However, in order to assess the contribution of a particular species to human nutrition, milk as a potential life-time product has to be taken into consideration as well. The importance of the goat and perhaps also of cattle was probably much higher at Muweilah than indicated by bone-weights. This cannot be discussed here in any detail for cattle, sheep and goat, because this would require detailed insight into the age-structures revealed by their bone-finds. Only the slaughtering ages of the Muweilah dromedaries can be considered in this respect. As indicated by Figure 21.5 (below) they were probably exploited for milk as well. Thus, the importance of this animal for the subsistence economy of Muweilah would probably not have changed much if – in addition to meat – milk as a lifetime-product is also taken into account. We may in any case conclude that the domestic dromedary had attained very high economic

Table 21.1. Animal remains from Muweilah (as identified up to 2008)

TAXA:	*NIS**	*NIS%*	*WIS***	*WIS%*
Domestic cattle, *BOS*	239	1.32	2660.1	4.80
Domestic sheep, *OVIS*	1125	6.19	6153.7	11.09
Domestic goat, *CAPRA*	484	2.66	3062.4	5.52
Domestic sheep or goat *(OVIS/CAPRA)*	15,427	84.89	27,063.4	48.79
Domestic sheep *and* goat *(OVIS+CAPRA)*	17,036	93.74	36,279.5	65.41
Unidentified equid	2	0.01	37	0.07
Dromedary, *DROMEDARIUS*	891	4.90	16,473.3	29.70
Dog, *CANIS*	5	0.03	18.2	0.03
Domestic animals: total	*18,173*	*100.00*	*55,468.1*	*100.00*
Small whales or dolphins, CETACAE	3	0.46	7.8	0.15
Hare, *Lepus* sp.	58	8.94	39.2	0.77
Wolf, *Canis lupus*	7	1.08	85.9	1.70
Red fox, *Vulpes vulpes*	17	2.62	18.4	0.36
Rüpell's fox, *Vulpes rueppelli*	2	0.31	4.8	0.09
Striped Hyaena, *Hyaena hyaena*	4	0.62	16.1	0.32
Wild cat, *Felis silvestris*	8	1.23	9.2	0.18
Caracal, *Lynx caracal*	1	0.15	0.8	0.02
Leopard, *Panthera pardus*	1	0.15	8.8	0.17
Carnivora indet., small	2	0.31	2.8	0.06
Carnivora indet., medium sized	2	0.31	3.2	0.06
Carnivora indet., large	1	0.15	3.3	0.07
Seacow, *Dugong dugon*	1	0.15	25	0.49
Wild dromedary, *Camelus* spec.	9	1.39	406.5	8.03
White Oryx, *Oryx leucoryx*	316	48.69	3508.2	69.28
Sand gazelle, *Gazella subgut. marica*	4	0.62	23.7	0.47
Gazelle, *Gazella* spec.	213	32.82	900	17.77
Wild mammals: total	*649*	*100.00*	*5063.7*	*100.00*
Cormorant, *Phalacrocorax nigrogularis*	662	89.22	806.6	90.29
unidentified birds	80	10.78	86.7	9.71
Wild birds - Total	*742*	*100.00*	*893.3*	*100.00*
unident. sea turtles, Chelonidae indet.	12	50.00	58.4	94.19
other unident. reptiles	12	50.00	3.6	5.81
Reptiles - Total	*24*	*100.00*	*62*	*100.00*
Sharks, CARCHARHINIDAE	23	0.13	15.3	0.15
Hammerhead sharks, SPHYRNIDAE	10	0.06	3.5	0.03
Sawfishes, PRISTIDAE	8	0.04	10.4	0.10
Guitarfishes, RHINOBATIDAE	114	0.63	17.9	0.18
Giant guitarfishes, RHYNCHOBATIDAE	32	0.18	26.9	0.27
Eaglerays, MYLIOBATIDIDAE	7	0.04	1.3	0.01
Unidentifiable CHONDRICHTHYES	106	0.59	29.5	0.29
Herrings, CLUPEIDAE	2	0.01	1.1	0.01
Sea catfishes, ARIIDAE	17	0.09	11.1	0.11
Needlefishes, BELONIDAE	43	0.24	9.5	0.09
Flatheads, PLATYCEPHALIDAE	4	0.02	1.1	0.01
Groupers, SERRANIDAE	663	3.67	1358.9	13.55
Remoras, ECHENEIDIDAE	3	0.02	1	0.01
Jacks, CARANGIDAE	357	1.97	482.6	4.81
Snappers, LUTJANIDAE	5	0.03	1.1	0.01
Grunts, HAEMULIDAE/ POMADASIDAE	9	0.05	6	0.06
Emperors, LETHRINIDAE	120	0.66	65.1	0.65
Seabreams, SPARIDAE	3405	18.82	1747.7	17.42
Butterflyfishes, CHAETODONTIDAE	1	0.01	1	0.01
Mullets, MUGILIDAE	456	2.52	142.7	1.42
Baracudas, SPHYRAENIDAE	77	0.43	40.3	0.40
Parrotfishes, SCARIDAE	32	0.18	7.5	0.07
Rabbitfishes, SIGANIDAE	4	0.02	0.5	0.00
Tunas, Mackerels, SCOMBRIDAE	880	4.87	880.2	8.78
Unidentifiable OSTEICHTHYES	11,710	64.74	5167.8	51.52
Fish: total	*18,088*	*100.00*	*10,030.0*	*100.00*
identified animal remains	37,676	16.72	71,517.1	59.02
unidentified animal remains	187,675	83.28	49,654.9	40.98
Animal remains: total	*225,351*	*100.00*	*121,172.0*	*100.00*
Summary totals				
Domestic animals	18,173	48.23	55,468.1	77.56
Wild mammals	649	1.72	5063.7	7.08
Wild birds	742	1.97	893.3	1.25
Reptiles	24	0.06	62	0.09
Fish	18,088	48.01	10,030.0	14.02
Identified Animal Remains	37,676	100.00	71,517.1	100.00
Dromedary, DROMEDARIUS	*891*	*2.36*	*16,473.3*	*23.03*

* NIS = Number of identified bone specimens;
** WIS = Weight of identified bone specimens

importance for the community living at Muweilah during the 1st quarter of the 1st millennium BC.

Not far from Muweilah, ca. 25km to the southwest and now in the centre of the city of Dubai, there is another new site where camel remains form the bulk of the excavated faunal remains: the so-called camel butchering site of Al-Sufouh 2 yielded many more camel bones than Muweilah. These bone finds are from a context of the Middle and Late Bronze Age (MBA/LBA) Wadi-Suq-Culture representing the time period from around 1700–1200 BC. Thus, the finds are around 500 years older than those from Muweilah. They are considered to be from wild camels (von den Driesch & Obermeier 2007), and there is indeed nothing to indicate the presence of domestic dromedaries at al Sufouh 2. The same is true for the Early Bronze Age (EBA) site of Umm an-Nar, where camel bones are also an important constituent of the faunal remains excavated from this settlement dating to the middle of the 3rd millennium BC and situated near Abu Dhabi some 120km farther southwest along the coast of the Gulf (Uerpmann & Uerpmann 2008). Another new sample of the early Holocene period from Baynunah in the Western Region of the Emirate of Abu Dhabi was recently evaluated by Beech *et al.* (2009 and this volume) and adds new data for the Holocene wild dromedary population in SE-Arabia.

Up to now, assumptions about early dromedary domestication in this general area rested mainly upon the faunal remains from Tell Abraq – one of the few stratified sites of the late prehistoric periods in southeast Arabia (Fig. 21.2). It is situated only about 25km north of Muweilah. Its stratigraphy begins with layers of the late Umm an-Nar culture (EBA), superimposed by four strata of the Wadi Suq culture (MBA and LBA) and three Iron Age layers.

The general development of animal economy at Tell Abraq was described by M. Uerpmann (2001).

Wild dromedaries were the major terrestrial animals hunted by the inhabitants of Tell Abraq during all phases of the Bronze Age, leading to a marked decrease in their numbers among the faunal remains of the successive layers from the Umm an-Nar period (EBA) and into the beginning of the IA, when camel remains disappeared completely from the assemblage. Camel bones re-appear at Tell Abraq in the assemblage of the Iron Age II, the phase to which Muweilah also belongs. They are absent again in the uppermost layer of the Iron Age (IA III) and in the ed-Dur phase, but the faunal assemblages from these contexts are too small to be conclusive in regard to the non-occurrence of rare faunal elements. As described in detail by the present authors (Uerpmann & Uerpmann 2002), there is convincing evidence that most of the camel remains from the Iron Age II at Tell Abraq are from domestic dromedaries, while all the older finds are from wild animals.

The identification of the IA II camels from Tell Abraq as domestic dromedaries was based on a marked size difference between the camel bones from the Bronze and the Iron Age levels. While we still consider this evidence as convincing, there is an ongoing debate among archaeozoologists whether size differences should at all be used in attempts to identify early domestication. It would go to far to enter into all details in this discussion, but it should be made clear that there is a fundamental difference between identifying domestic animals and identifying the process of domestication. A domestic animal is – by definition – different from its wild ancestor; with size being one of the few characteristics of both wild and domestic animals which can objectively be measured in skeletal remains and then compared between the different bone assemblages. If there are differences between two populations and if there are archaeological indications that a wild versus a domestic status might be responsible for these differences, it is legitimate to consider them as evidence for the genetic separation between wild and domestic members of the respective species which is the result of the domestication process. One could also think about archaeological traces of this process itself, which might be visible in an archaeological context. However, in the desert areas of Arabia, the conditions of archaeological preservation are too poor to hope for such direct traces of human–animal interaction. Through diachronic approach it is in any case possible to identify the wild ancestors on the one side and their altered domestic offspring on the other, as soon as the observable characteristics are recognizably different. This implies that the animals under human control continue to breed in well maintained genetic isolation from their wild relatives. The time required for the process of differentiation does not matter in terms of its final result, which is the co-existence of two populations of the same species, one of them living wild and the other under human control.

With regard to changes in body size as potential indicators of domestication, this means that these changes in body size must have been restricted to the domestic population, while the wild population remained unchanged at the same time. As clearly shown in recent research on isolated animal populations, body size is a highly sensitive indicator for life circumstances and their temporal changes (Ozgul *et al.* 2009). Many reasons for a diminution of body size in early domestication have already been discussed (e.g. Uerpmann 2008). Depending on the particular circumstances, all of them may have contributed to the observed size changes in early domesticates. It is of course necessary to make sure that size changes within such a population are 'real' in a statistical sense, and not – for instance – only caused by a numerical increase of smaller animals (e.g. females) within the measured sample. The new complexes of camel remains from Baynunah and al-Sufouh 2 provide an opportunity to verify earlier observations on the size of wild dromedaries in southeast Arabia and also to examine changes in population structures as indicators of particular forms of exploitation of the respective populations.

Figure 21.3 represents the distribution of LSI values for camel remains of all collections from southeast Arabia presently available to the authors. LSI calculations are based on the same standard individual (skeleton CA4 in the Tübingen collection) as used in earlier publications (e.g. Uerpmann & Uerpmann 2002). The graphs clearly demonstrate that size variability is quite large at al-Sufouh 2 with regard to both maxima and minima[1] of the total size range. This is probably a function of the large sample size. The occasional occurrence of very large or very small animals in a wild population usually remains hidden in smaller samples.

The statistical indicators of variability – inter-quartiles[2] and standard deviation[3] – are not wider in the al-Sufouh 2 sample than in the much smaller sample of Bronze Age camel remains from Tell Abraq. At Umm an-Nar there is an apparent lack of very large animal remains, which reduces the statistical variability of the camel finds from this island site. Nevertheless, the mean values are identical to those of the al-Sufouh 2 sample. The small measurable sample from the Bronze Age (BA) layers of Tell Abraq on the other hand is apparently poor in remains of female species, which increases the mean values of the respective LSI graph. The sample from Baynunah, directly transferred from figure 9 in Beech *et al.* (2009, 26), is interesting insofar as smaller animals dominate. This indicates that important differences might exist between anthropogenic and natural accumulations of camel remains. The Baynunah finds most probably represent a natural death assemblage of Early Holocene wild dromedaries in a low lying desert area where water accumulated in temporary pools – sometimes leaving animals to dye of thirst if there was no rain to refill the ponds. This would have affected all members of a wild herd, perhaps with a slight advantage for large and strong individuals which might have been able to get away from the area before it was too late. In general, however, the size-structure of the accumulation is

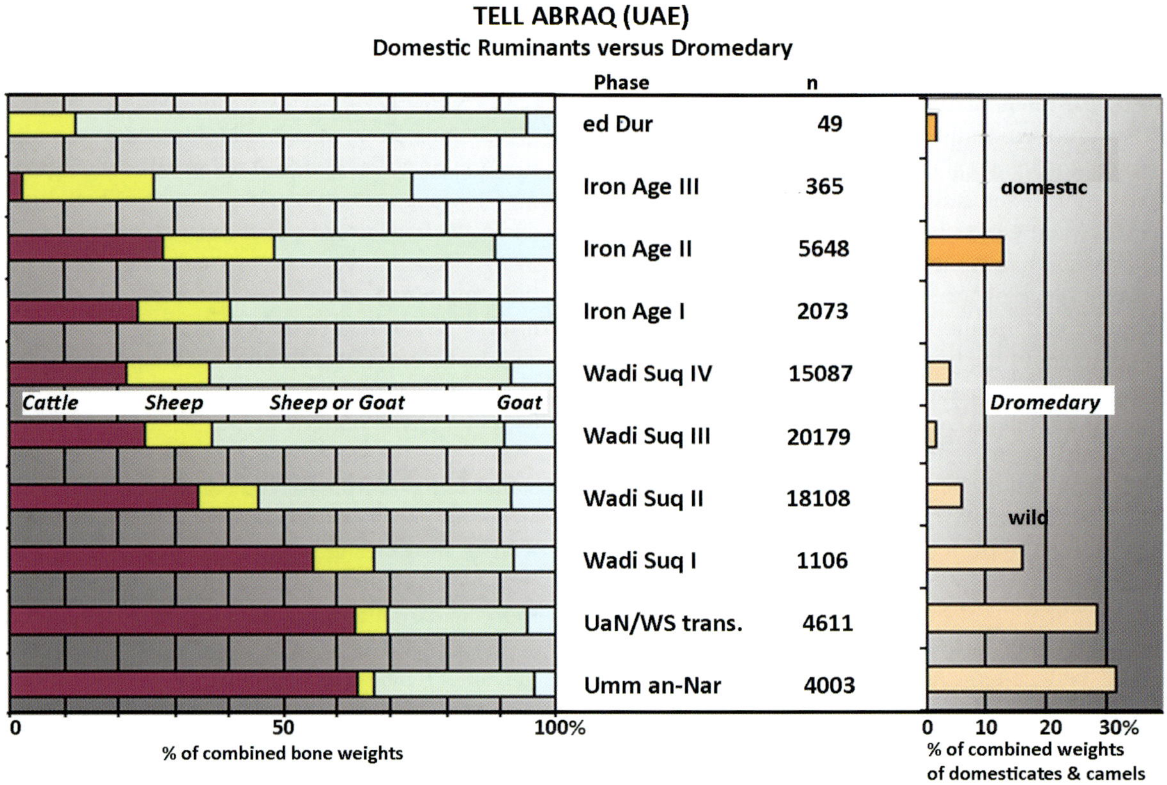

Fig. 21.2. Frequencies of domestic ruminants and camels in the stratigraphy of Tell Abraq (from Uerpmann 2001).

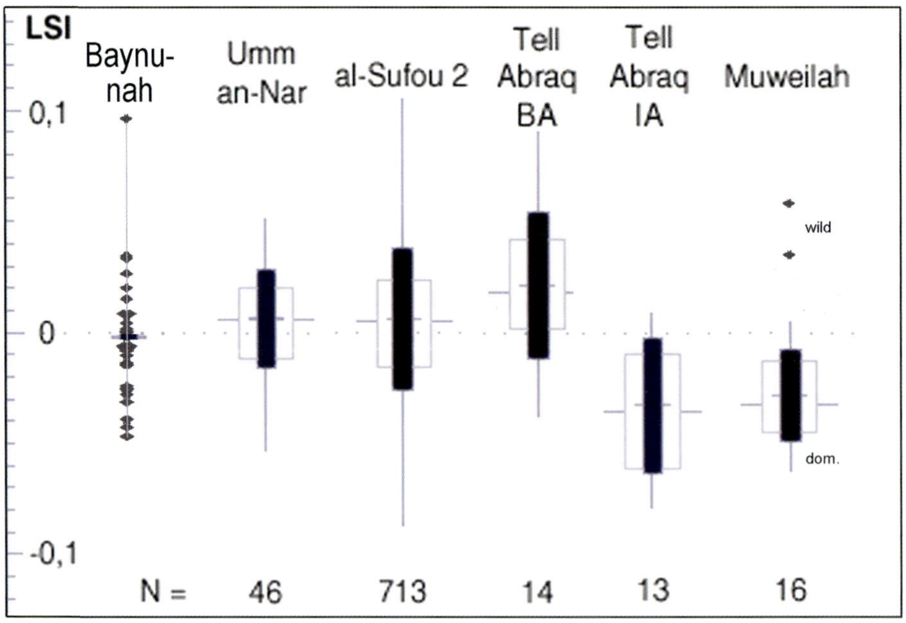

Fig. 21.3. LSI distributions of dromedary finds from sites in southeast Arabia.

likely to represent the living population – probably more so than assemblages accumulated by human hunters. Hunting might to some extent distort size-distributions, particularly in large and well-defended animals, where males tend to be less fugitive than females. Males may therefore be more or less overrepresented in assemblages accumulated by hunters. Nonetheless, the mean LSI of the camel bones from Baynunah is not much below that observed at Umm an-Nar and al-Sufouh 2 (Fig. 21.3). A very large male found at Baynunah corroborates the maximal LSI values for al-Sufouh 2.

At Muweilah two finds of clearly wild camels were

recorded. As far as represented by the few measurable bones, the domestic dromedaries from Muweilah are similar in size to the Iron Age dromedaries of Tell Abraq. Unfortunately the number of measurable finds is quite low at Muweilah due to the bad state of preservation of the bones. On the whole, the finds from Muweilah corroborate the earlier observations on a shift in the average sizes of dromedaries in southeast Arabia at the onset of the Iron Age, which most probably indicates domestication of this species around this time. Contrary to Tell Abraq, the domestic dromedary acquired real economic importance at Muweilah during the Iron Age II as indicated by the quantity of its bone remains in relation to the rest of the fauna.

The presence of domestic dromedaries at Muweilah is also indicated by several finds of statuettes, which represent one-humped camels with a load or a saddle on their hump (Fig. 21.4). Such statuettes are clear archaeological evidence for the existence of domestic dromedaries during the Iron Age II in the area. These finds from Muweilah are a cultural expression of the importance of this animal for the inhabitants of this settlement.

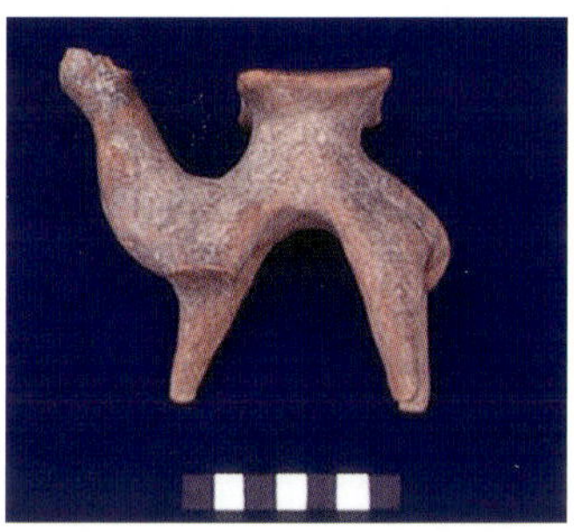

Fig. 21.4. Dromedary statuette from Muweilah (photo: P. Magee).

As a third line of evidence the amount of camel finds from Muweilah provides the opportunity to evaluate the kill-off patterns of dromedaries at this site. Kill-off patterns indicate how people exploited a particular animal species. Hunting them at random should produce another kill-off pattern than breeding and managing them for particular purposes. As tooth-bearing jaws of camels preserve badly – even worse than postcranial bones – kill-off patterns for the dromedaries of Muweilah had to be reconstructed from the evaluation of epiphyseal fusion. Although the exact timing of epiphyseal fusion in camels is badly known, estimates can be based on skeletal material in comparative collections and on the respective data for other artiodactyls. For the graphs in Figure 21.5 we have grouped epiphyses into an early, middle and late state of fusion, with the respective epiphyses fusing at roughly 12, 30 and 48 months. The early stage (≈12 months) includes the distal epiphysis of the humerus and the proximal epiphyses of radius and first and second phalanges.

For the middle stage (≈30 months) we used the distal tibia and for the late stage (≈48 months) we used the proximal epiphyses of humerus, ulna, femur and tibia and the distal epiphyses of radius and femur as well as the apophyses of vertebra. In addition to the counts of fused versus unfused epiphyses, the occurrence of some 15% of infantile bones at Muweilah was used as indication for slaughtering of young animals less than about 6 months old. Although the results represented in Figure 21.5 are liable to alteration once more detailed information on epiphyseal fusion in dromedaries becomes available in the future, the relative differences observed in comparison to the camel remains from al-Sufouh 2 will remain valid.

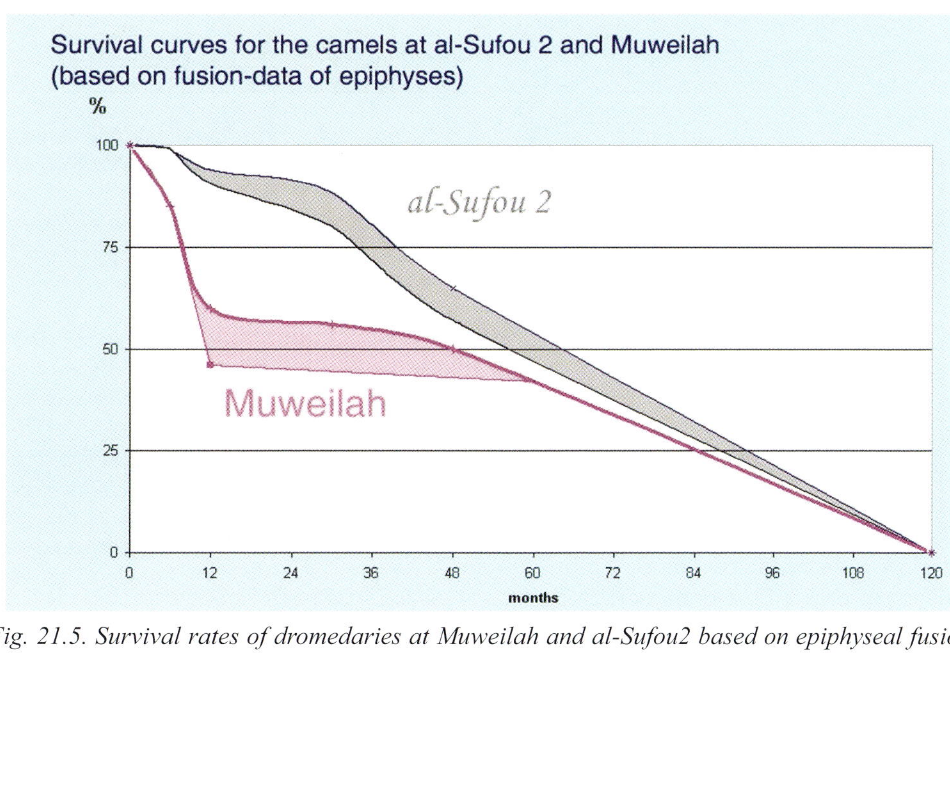

Fig. 21.5. Survival rates of dromedaries at Muweilah and al-Sufou2 based on epiphyseal fusion.

The counts for the early stage of epiphyseal fusion (12 months) for Muweilah vary between the different skeletal elements. This is reflected by a range of insecurities as indicated in Figure 21.5 by the pink area below the most probable survival curve for the Muweilah dromedaries. The kill-off pattern of dromedaries at al-Sufouh 2 as documented by von den Driesch and Obermaier (2007) is marked in grey with its respective range of insecurity. The differences between the two sites are obvious: At Sufouh 2 less than 10% of the animals were killed at an age under 12 months, while at Muweilah more than one-third of the animals did not survive this age. After this kill-off of juveniles, the survival curve for Muweilah flattens, probably indicating the exploitation of life-time products – labour and possibly milk – of the adults reaching ages of 4 years and beyond. It is obvious that the pattern of dromedary exploitation changed differed from al-Sufouh 2, where a wild dromedary population was extensively hunted, to Muweilah, where domestic dromedaries were utilised for their respective products together with domestic sheep, goats and cattle.

The clear differences – both in body -size and in the pattern of exploitation – between the Iron Age and the Bronze Age and older dromedaries in the area, which is now the UAE, indicates that either the domestication of this species happened locally in the badly known transitional phase from the Late Bronze to the Early Iron Age or that the domestic dromedary was introduced to the area from an hitherto unknown centre of camel domestication outside southeast Arabia. More evidence will probably come to light when other important archaeological sites in the United Arab Emirates and Oman will have been evaluated with regard to their animal remains. Large areas of the Arabian Peninsula, where the wild dromedary presumably occurred during the Early Holocene and where its domestication must have happened, are also completely unexplored in respect to their archaeozoology. An exception are sites in the Yemen, where thanks to the work of Francesco Fedele (2009 and this volume) we know about the presence of the domestic dromedary in the Sabaean town of Yalā at roughly the same time as the Iron Age II in the Emirates.

The introduction of the domestic dromedary into an area, where its wild ancestor did *not* exist beforehand, was recently published by Cornelia Becker for Tell Sheikh-Hamad on the Khabur River in northeast Syria (Becker 2008). Remains of dromedaries were discovered at Sheikh Hamad in layers of the Neoassyrian Phase, the beginnings of which correspond to the Iron Age in southeast Arabia. The domestic dromedary was already well known from the palace reliefs of this period in Mesopotamia. Textual evidence reports its contemporaneous presence in the Levant. It will be interesting to critically review the dating evidence for the appearance of this animal in the coastal areas of the Eastern Mediterranean, which also were outside the natural range of the wild dromedary.

Even more interesting than the dromedary finds from Tell Sheikh Hamad are the other findings of Cornelia Becker at this important place: There are remains of the two-humped or Bactrian camel at this stratified site from layers below those which yielded the dromedary finds. As the wild ancestor of the Bactrian camel is not known to have occurred in the Djazirah area of Syria or in Lower Mesopotamia, the respective finds also indicate an early introduction: In this case, of the domestic two-humped camel from its hitherto unknown area of original domestication. This area may well have been the dry steppes and deserts of the Iranian Highland, where the wild two-humped camel may have existed (Uerpmann 1987) and where it may have been domesticated during the Bronze Age – as possibly indicated by the early findings at Shar-e-Sokhta (Compagnoni & Tosi 1977). If domestic Bactrian camels existed in South-eastern Iran at the end of the 2nd millennium BC, dromedary domestication in Arabia might have been inspired by contacts across the Gulf. Such contacts are well documented by finds from Muweilah (Magee 1999; 2002). It should, however, also be mentioned that the site also yielded evidence for contacts with South Arabia in the form of an inscribed name on a large jar using the Old South Arabic alphabet (Magee 1999). Thus, Muweilah is on the crossroads of the early history of dromedary domestication not only with regard to its animal remains. However, in spite of the early occurrence of domestic dromedaries at sites in the UAE this does not necessarily indicate that this area was the geographic place of origin for this important domestic animal. Only a major intensification of archaeozoological research throughout the Arabian Peninsula will provide a solid answer to the most interesting question where dromedaries were tamed and domesticated for the first time.

Notes

1 Indicated by the 'whiskers' of the graphs for each site.
2 Indicated by the open boxes in the graphs.
3 Indicated by the black bars in the graphs.

References

Becker, C. 2008, ‚Die Tierknochenfunde aus Tall Šēh Hamad/ Dūr-Katlimmu: Eine zoogeographisch-haustierkundliche Studie', in H. Kühne (ed.), *Umwelt und Subsistenz der Assyrischen Stadt Dūr-Katlimmu am unteren Habur. Berichte der Ausgrabung von Tall Šēh Hamad/Dūr-Katlimmu* 8, 61–132. Harrassowitz Verlag, Wiesbaden.

Beech, M., Mashkour, M., Huels, M. & Zazzo, A. 2009, 'Prehistoric camels in south-eastern Arabia: the discovery of a new site in Abu Dhabi's Western Region, United Arab Emirates'. *Proceedings of the Seminar for Arabian Studies* 39, 17–30.

Compagnoni, B. & Tosi, M. 1977, 'The camel: its distribution and state of domestication in the Middle East during the third millennium B.C. in light of finds from Shahr-I Sokhta', in R. H. Meadow & M. A. Zeder (eds), *Approaches to Faunal Analysis in the Middle East*, 92–105. Harvard University Press, Cambridge MA.

Fedele, F. G. 2009, '*Sabaean animal economy and household*

consumption at Yalā, eastern Khawlan al-Tiyal, Yemen'. Proceedings of the Seminar for Arabian Studies 39, 135–154.

Magee, P. 1996, 'Excavations at Muweilah. Preliminary report on the first two seasons'. *Arabian Archaeology and Epigraphy* 7, 195–213.

Magee, P. 1999, 'Writing in the Iron Age: the earliest South Arabian inscription from southeastern Arabia'. *Arabian Archaeology and Epigraphy* 10, 43–50.

Magee, P. 2002, 'The indigenous context of foreign exchange between south-eastern Arabia and Iran in the Iron Age'. *Journal of Oman Studies* 12, 161–168.

Magee, P., Thompson, E., Mackay, A., Kottaras, P. & Weeks, L. 2002, 'Further evidence of settlement complexity: report on the 2001 excavations at the Iron Age site of Muweilah, Emirate of Sharjah, United Arab Emirates'. *Arabian Archaeology and Epigraphy* 13, 133–160.

Ozgul, A., Tuljapurkar, S., Benton, T. G., Josephine, M., Pemberton, J. M., Clutton-Brock, T. H. & Coulson, T. 2009, 'The dynamics of phenotypic change and the shrinking sheep of St. Kilda'. *Science* 325, 464–467.

Uerpmann, H.-P. 1987, *The Ancient Distribution of Ungulate Mammals in the Middle East*. Beihefte zum Tübinger Atlas des Vorderen Orients, Dr. Ludwig Reichert, Wiesbaden.

Uerpmann, H.-P. 2008, 'Animal domestication', in D. M. Pearsall (ed.), *Encyclopedia of Archaeology* 1, 434–445. Academic Press, New York.

Uerpmann, H.-P. & Uerpmann, M. 2002, 'The appearance of the domestic camel in SE-Arabia'. *Journal of Oman Studies* 12, 235–260.

Uerpmann, M. 2001, 'Remarks on the animal economy of Tell Abraq (Emirates of Sharjah and Umm al-Quwain, UAE)'. *Proceedings of the Seminar for Arabian Studies* 31, 227–233.

Uerpmann, M. & Uerpmann, H.-P. 2008, 'Animal economy during the Early Bronze Age in South-East Arabia', in E. Vila, L. Gourichon, H. Buitenhuis & A. Choyke (eds), *Archaeozoology of the Near East* VIII, 465–485. TMO, Maison de l'Orient et de la Méditerranée, Lyon.

von den Driesch, A. & Obermaier H. 2007, 'The hunt for wild dromedaries during the 3rd and 2nd millennia BC on the United Emirates coast. Camel bone finds from the excavations at Al Sufouh 2, Dubai, UAE'. *Documenta Archaeobiologiae* 5, 133–167.

22. Fish exploitation at Qal'at al-Bahrain from the Early Dilmun period (3rd millennium BC) to the Middle Islamic period (13th–16th centuries AD): preliminary results

Justine Vorenger

Excavations have taken place at the site of Qal'at al-Bahrain in Bahrain since the 1950s. The site, located in the Persian Gulf, has been excavated by the French Archaeological Mission since 1977. Excavations have revealed significant architectural remains as well as a rich assortment of material, including ceramics, metal objects, botanical and faunal remains, which include a large number of fish bones. Exceptional stratigraphy, that extends from the Early Dilmun (3rd millennium BC) to the Middle Islamic period (13th–16th centuries AD), provides the possibility for studying successive occupations at the site.

The preliminary results discussed here are based on the analysis of fish bones from the French excavations. A total of 5763 fish bones have been studied, out of which 2552 can be quantified which have precise stratigraphic dating. Five families dominate the assemblage during the occupation at the site: groupers (Serranidae), jacks/trevallies (Carangidae), seabream (Sparidae), emperors (Lethrinidae) and tuna/mackerel (Scombridae). The taxonomic variety increases however to a total of 11 families during the Islamic period. The preliminary results discussed here provides evidence of the focus on marine exploitation, in comparison to the consumption of mammals (cf. the previous study carried out by Carine Tomé).

Keywords Bahrain, archaeo-ichthyology, fishing, fishes, Dilmun period, Tylos period, Islamic period, palaeo-economy

Introduction

Fishing in the Persian Gulf has played an important role in the economy of local populations since the earliest times. Since more than 20 years there have been a number of studies and syntheses of this material (Beech 2004). The site of Qal'at al-Bahrain, located in the island of Bahrain, occupies a key position (Fig. 22.1).

Excavations at the settlement of Qal'at al-Bahrain began in 1954, and have revealed a remarkable stratigraphy and chronology. Favourable environmental and geographic conditions led to the development of an Early Dilmun settlement (ca. 2200–1750 BC) which continued until the Islamic period (15th century AD). This includes a number of significant buildings, performing military (the fort), political and religious (the palace) and civilian (the houses) functions, epitomising the economic expansion of the ancient Dilmun capital (Figs 22.2 & 22.3).

The fauna from the site has already been subject to a number of studies. Material from the Danish excavations at Qal'at al-Bahrain has been studied by Uerpmann (1993) and Uerpmann and Uerpmann (1994; 1997). They determined that the majority of the remains consisted of not only groupers (Serranidae), jacks/trevallies (Carangidae), emperors (Lethrinidae) and seabream (Sparidae), but also

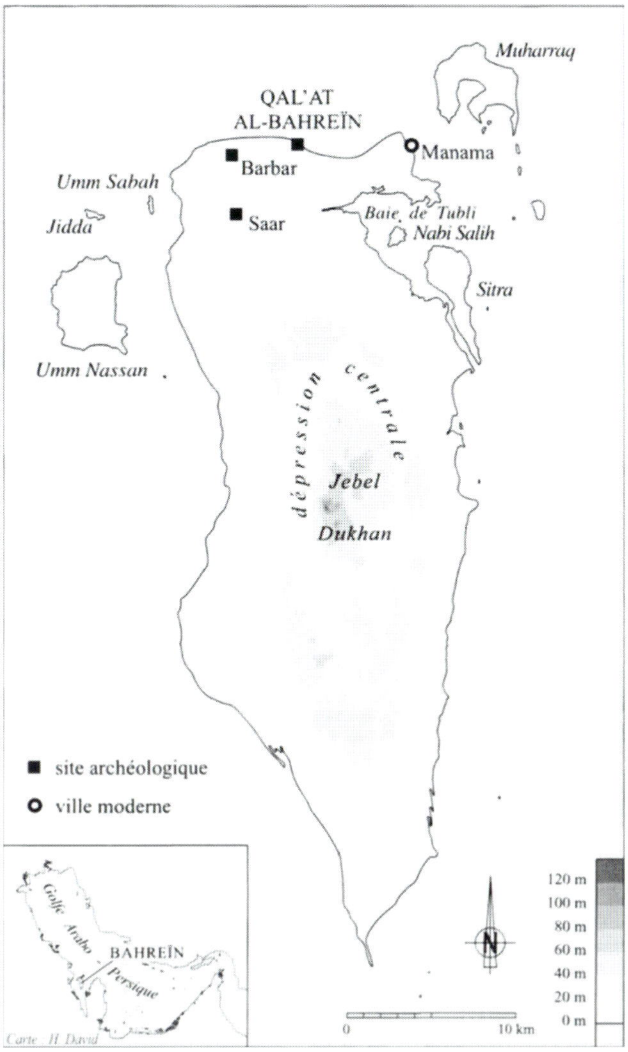

Fig. 22.1. *Location of Qal'at al-Bahrain and other key sites in Bahrain (source: French Archaeological Mission, drawn by H. David)*

Fig. 22.2. Aerial view of Qal'at al-Bahrain (source: French Archaeological Mission in Bahrain)

Fig. 22.3. View of the excavations at Qal'at al-Bahrain (source: J. Vorenger. French Archaeological Mission in Bahrain)

of cartilaginous fishes, i.e. sharks and rays. For the Middle and Late Dilmun periods, Van Neer and Uerpmann (1994) also remarked upon the presence of sharks and rays, and of bony fishes from at least a dozen families, principally groupers, jacks/trevallies, emperors and seabream. The material collected from the first excavation campaigns of the French Archaeological Mission were studied by Desse-Berset (1995). Her results complement those obtained previously, however they were qualitative in nature and did not provide precise quantifications. Nineteen families were determined belonging to both bony and cartilaginous fishes (Table 22.1). The same four families dominated the assemblage, to which could be added rabbitfish (Siganidae), needlefish (Belonidae), sardines/herrings (Clupeidae), grunts (Haemulidae), parrotfish (Scaridae), mullets (Mullidae), snappers (Lutjanidae) and tuna/mackerel (Scombridae).

With respect to other taxa present at the site, one should mention the previous study of the mammalian faunal remains from the site (Tomé 2003). This study remarked upon the predominance of sheep/goat (7667 elements out of 9100 determined bone fragments). From the Early Dilmun to Tylos periods the fauna was dominated by domestic species, representing 90.7% of all animals consumed (Fig. 22.4). Additional species noted there included cattle, dog, pig, equid, camel, chicken, cormorant, sea turtle, dugong, gazelle, hare, cat, mongoose, pink flamingo and grey heron.

The consumption of shellfish is also attested by the discovery of numerous fragments (Cataliotti-Valdina, 1994). An interesting variety of shells have been identified including *Turbo* sp., *Strombus* sp., *Conus* sp., *Marcia* sp., *Cypraeae* sp., and of course, pearl oyster, *Pinctada* sp. Shellfish were exploited for food, however after their consumption, shells were used as raw material for the manufacture of artefacts.

Some additional environmental studies have also been carried out at the site, such as the analysis of archaeobotanical remains from the Dilmun and Tylos period levels (Tengberg & Lombard 2002). Three types of remains were recognised, wood charcoal, carbonised

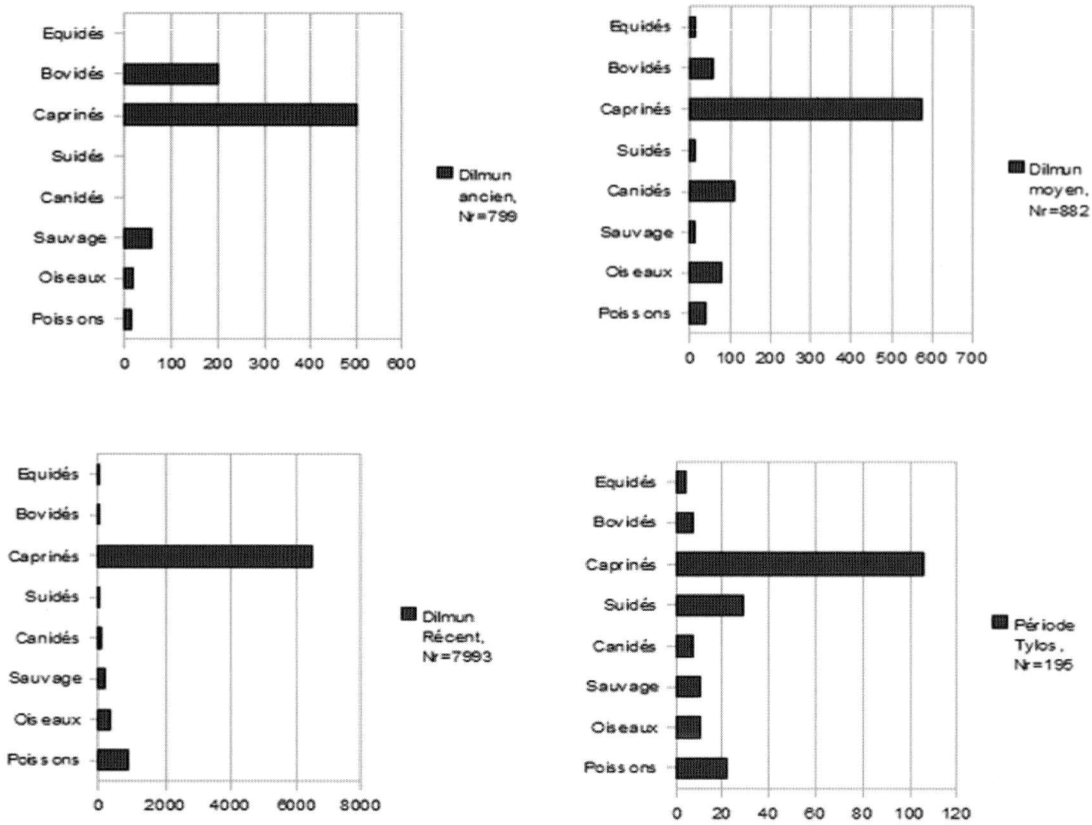

Fig. 22.4. Evaluation of different types of animal taxa by occupation period from the central excavation area at Qal'at al-Bahrain (after Tomé 2003)

seeds and plant impressions in mud-brick fragments. Some species noted are indigenous, whilst others result from imported items. It is important to note the presence of date palm (*Phoenix dactylifera*), jujube (*Zizyphus spina-christi*), barley (*Hordeum vulgare*) and wheat (*Triticum durum/aestivum*).

The approach taken here aims to give a diachronic view of the occupation of Bahrain from the Early Dilmun to Islamic periods. Here, we present preliminary results on the fish species identified and the habitats from which they originate, as compared with previously published faunal studies.

Qal'at al-Bahrain

Located in the Persian Gulf, on Bahrain Island, Qal'at al-Bahrain was a large harbour placed in a strategic geographic location, in the middle of the Gulf between Europe and Asia (Fig. 22.1). It has been a significant commercial and maritime location for several millennia. This tell or occupation mound, situated facing the sea, has provided archaeologists with an exceptionally deep stratigraphic sequence of deposits, unique within the Arabian peninsula. Beyond its clear regional status, Qal'at al-Bahrain represents a true testament to the historical and economic development of Dilmun, one of the most important cultures of the ancient Gulf. Moreover, Qal'at al-Bahrain was the capital of ancient Dilmun from the 3rd millennium BC to the 18th century AD.

The site is exceptional because of its coastal position and the presence of artesian springs that enabled the development of date palm gardens (Sanlaville, 1988), so that it was certified by the Sumerians as 'paradise on earth'. Its coastal location led to it playing a key role in trade relations with Mesopotamia, the Iranian and Arab coasts, with Oman, the Indian subcontinent or the Far East, as can be demonstrated by various import items found in archaeological contexts at the site (Lombard, 1999).

Excavations at the site began in 1954 under the direction of a Danish team led by Geoffrey Bibby and Peter Vilhelm Glob. Twenty-four years of excavation permitted the identification of successive occupations and an initial chronology. In 1978, Monik Kervran founded the French Archaeological Mission to Bahrain to continue the previous work. Her research focused on fortified buildings from the Islamic era and then investigated the coastal fortress and the 'Portuguese Fort'. In 1989, Pierre Lombard took over and turned more towards the earlier periods. He concentrated excavations on the southeast side of the Portuguese fort, and its interesting building complex, known as the 'Ouperi Palace', dating from the early Dilmun (ca. 2300–1800 BC.), Middle Dilmun (mid-2nd millennium BC.) to Late Dilmun period (first half of the 1st millennium BC). This work refined the chronological development of the site proposed

Table 22.1. Comparison of fish taxa from the different archaeological missions at Qal'at al-Bahrain.

FAMILY	*SPECIES*	*COMMON NAME*	*Danish excavations 1954–1978 (Van Neer & Uerpmann 1994)*	*French excavations 1989–1994 (Desse-Berset 1994; 1995)*	*French excavations 1989–2004 (Vorenger, this study)*
Carcharhinidae		*Shark*	■		
Sphyrnidae		*Hammerhead Shark*	■		
Pristidae		*Sawfish*	■		
Rhinobatidae		*Guitarfish*	■		
Clupeidae	*Sardinella sp.*	*Sardine*		■	
Ariidae	*Netuma spp.*	*Sea Catfish*	■	■	■
Belonidae	*Ablennes hians*	*Needlefish*		■	
Belonidae	*Tylosurus spp.*	*Needlefish*			■
Triglidae	*Lepidotrigla sp.*	*Gurnard*		■	
Platycephalidae	*Platycephalus indicus*	*Indian Flathead*		■	
Serranidae	*Epinephelus spp.*	*Grouper*	■	■	■
Echeneidae	*Echeneis naucrates*	*Remora*		■	
Carangidae	*Alectis indicus*	*Threadfish*	■		
Carangidae	*Alepes mate*	*Scad*		■	
Carangidae	*Atule mate*	*Yellowtail scad*			■
Carangidae	*Carangoides bajad*	*Orange Spotted Jack*	■		■
Carangidae	*Carangoides chrysophrys*	*Longnose Trevally*	■		
Carangidae	*Gnathanodon speciosus*	*Golden Trevally*	■	■	■
Carangidae	*Scomberoides commersonianus*	*Talang Queenfish*	■		■
Carangidae	*Seriola dumerili*	*Greater Amberjack*	■		■
Carangidae	*Seriola spp.*	*Amberjack*			■
Carangidae	*Trachinotus spp.*	*Pompano*	■		
Lutjanidae	*Lutjanus sp.*	*Snapper*		■	
Haemulidae	*Diagramma pictum*	*Painted Sweetlips*			■
Haemulidae	*Plectorhinchus sordidus*	*Sordid Thicklip*		■	
Haemulidae	*Plectorhinchus spp.*	*Sweetlips*			■
Haemulidae	*Pomadasys stridens*	*Striped Grunt*		■	
Lethrinidae	*Lethrinus sp.*	*Emperor*	■	■	■
Sparidae	*Acanthopagrus bifasciatus*	*Double-barred Seabream*	■	■	■
Sparidae	*Acanthopagrus berda*	*Twobar Seabream*		■	
Sparidae	*Argyrops spinifer*	*King Soldierbream*	■	■	■
Sparidae	*Diplodus sargus*	*Onespot Porgy*		■	
Sparidae	*Rhabdosargus spp.*	*Goldstriped/Haffara Seabream*	■	■	■
Sparidae	*Sparidentex hasta*	*Sobaity*		■	
Mugilidae	*Valamugil sp.*	*Mullet*			■
Mullidae	*Parupeneus sp.*	*Goatfish*		■	
Scaridae	*Scarus sp.*	*Parrotfish*		■	
Sphyraenidae	*Sphyraena spp.*	*Barracuda*	■	■	
Scombridae	*Scomberomorus commerson*	*Narrow-barred Spanish Mackerel*		■	
Scombridae	*Scomberomorus spp.*	*Mackerel*			■
Scombridae	*Euthynnus affinis*	*Kawakawa/ Little Eastern Tuna*	■	■	■
Scombridae	*Thunnus spp.*	*Tuna*	■		■
Siganidae	*Siganus spp.*	*Rabbitfish*		■	■

by the Danish team. After the abandonment of the palace, the construction of Tylos period (ca. 300 BC–AD 300) domestic structures, also provided rich material. Finally, the excavation campaigns led to the discovery of the first levels of the Islamic town (13th–16th centuries AD).

Excavation campaigns conducted since 1954 have thus enabled the uncovering of various architectural components, both administrative (the palace), military (fortresses), or domestic, but also a lot of materials (ceramics, metals, vegetation, wildlife) that inform us about the lifestyles of island populations successive Bahrain (Højlund & Andersen 1997).

The majority of the occupation levels at the site have been investigated and their study provides precise chronological information. Thus, the Early Dilmun period city is succeeded by a phase of public buildings during the Middle Dilmun and Late Dilmun periods (characterised by pre-Achaemenid and Achaemenid occupation). During the Tylos period, the city is influenced by the Seleucid-Parthians, which precedes the establishment of an Islamic city. The rulers of Hormuz construct at the site a vast fortress in the fourteenth and fifteenth century. In the 16th century, the Portuguese, by reinforcing the corner bastions, give it the name 'the Fort of Bahrain' – Qal'at al-Bahrain (Lombard & Kervran 1989).

Archaeological contexts and retrieval methods

THE SAMPLE

The bone material collected during the campaigns led by Pierre Lombard since 1989 is abundant. The bones are currently being studied from various excavations conducted from 1989–1996, and 2000–2004. All periods of occupation of the site are represented in the material. Thus, the large building complex, called the 'Ouperi Palace' provided a substantial amount of material. The Late Dilmun period comprised the majority of the material (4054 out of the 5763 bone remains studied). The domestic occupation material from the Tylos period and the Islamic era complete the remainder of the sample (Table 22.2). Only a total of 2552 can, however, be so far quantified which have precise stratigraphic dating. Smaller quantities of bones are noted for the Tylos period (NR = 160) and Early Dilmun period (NR = 237), since the full extent of the occupation horizons for these particular periods were not uncovered during the excavation.

RETRIEVAL METHODS

The bones were mostly collected in the field by manual retrieval. Their relative abundance and good preservation conditions have facilitated their collection. However, systematic screening of part of the sediment was carried out for all Dilmun period levels. In the case of later deposits, such as the Tylos and Islamic period periods, sieving was only carried out for some stratigraphic units. These samples provide smaller fish anatomical elements that are not necessarily found by hand retrieval, such as otoliths and scales. In addition, the sieved samples help reveal the presence of bones from smaller species like sardines and herrings (Clupeidae). In general, the material was collected from within particular structures and by stratigraphic unit, which allows its designation to particular occupation periods.

CONSERVATION

The excellent preservation of the material facilitates the identification of many of the bones and their abundance provides a broad spectrum of taxa for all periods. However, some material, mostly dating from the Middle Dilmun period, was extremely fragmented making its identification problematic. On the majority of the remains, cut marks, where available, could be clearly witnessed (Fig. 22.5). Finally, the generally good state of conservation of material provides a remarkable anatomical diversity (vertebrae, skull bones, otoliths, scales).

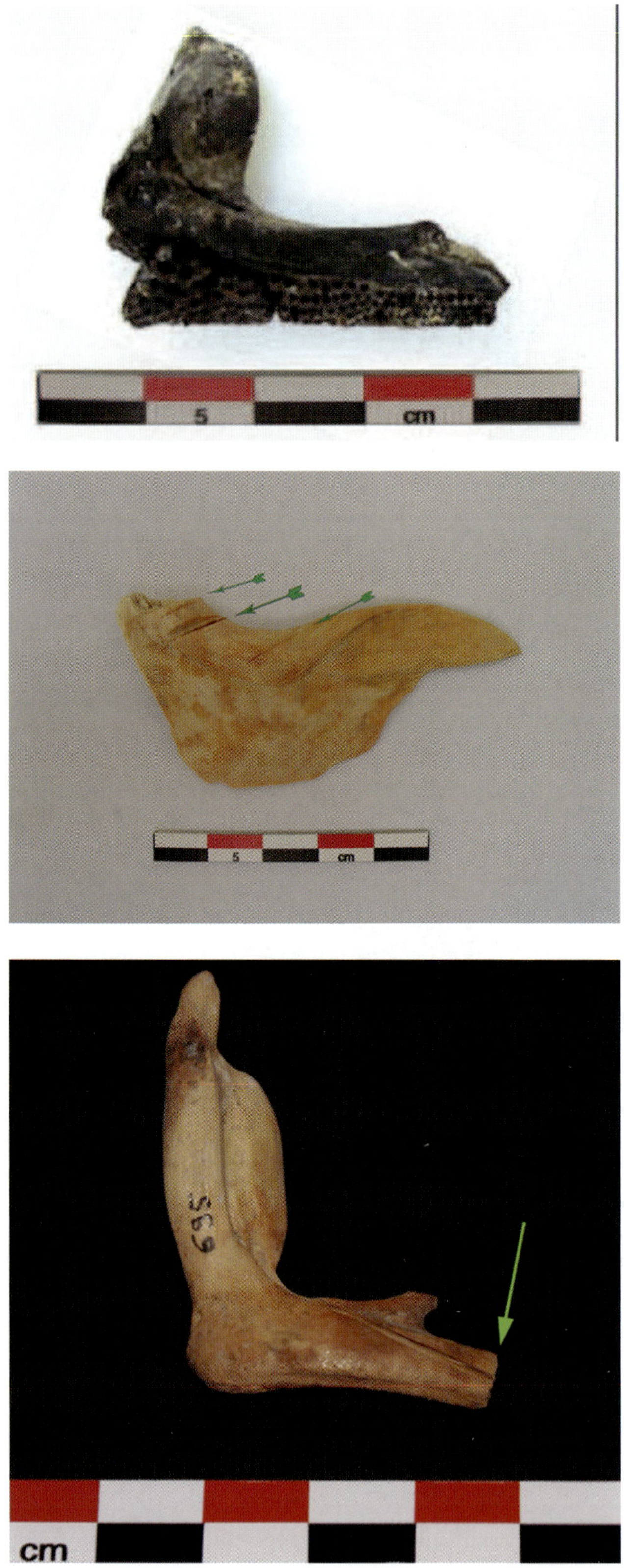

Fig. 22.5. Butchered and burnt fish bones from Qal'at al-Bahrain: a. Burnt Grouper (Epinephelus spp.) premaxillar, right side (Early Dilmun period, US 1898); b. Grouper (Epinephelus spp.) cleithrum, left side, with traces of cut marks, US 2113); c. Golden Trevally (Gnathanodon speciosus) premaxillar, left side, with chop mark (Late Dilmun period, US 695)

Fig. 22.6. Different fishing methods utilised in Bahrain: a. 'Haddrah', traditional tidal fish traps located near to Qal'at al-Bahrain (Source: National Museum of Bahrain); b. fishing boat loaded with 'gargoor' basket traps; c. a fisherman using a net in shallow coastal waters; d. examples of stone net sinkers in the National Museum of Bahrain

LIMITATIONS TO THE STUDY

The list of taxa identified must be viewed with caution because of the absence of systematic sieving for each occupation level (less sieving was carried out from the Tylos and Islamic period levels), thus giving an incorrect view of the fish consumed at the site. Indeed, the high biodiversity of fishes present in Gulf waters are poorly represented amongst the bones of Qal'at al-Bahrain. The bones recorded are, indeed, the strongest anatomical parts that are collected and therefore larger fish have been identified. Many smaller species may have been also caught which are not represented. This bias could possibly be partly rectified by analysis of the sieved samples. This may offer a wider variety of species qualitatively, but this cannot be quantitatively compared, as small fish are only present in a few samples, in comparison to hand retrieved bones from larger fish species which are much more abundant.

Major problems in undertaking this study lie both in the unequal amounts of material, but also in the irregular sieving carried out within each chronological period, which does not facilitate easy comparisons between levels. Finally, some fish species are not identified to species, either because of lack of comparative skeletons, or because of their taxonomic proximity. For example, the choice to limit the identification of grouper to the genus, *Epinephelus*, is due to the morphological similarity of different species within the same genus, as well as the limited differences in their associated habitats. We believe that specific species determination is not necessary here to understand the occupation of the site.

STUDY CONDITIONS

The bone material was studied in the archaeozoology laboratory of CEPAM,[1] using the fish skeletal comparative collections made by Jean Desse and Nathalie Desse-Berset from fish caught in Bahrain and throughout the Gulf. Diversity typical of the region is well represented through the quantity of reference skeletons. Different sized fish present in the collections allow the estimation of sizes and weights for several species.

DIACHRONIC ANALYSIS OF THE FISH BONES FROM QAL'AT AL-BAHRAIN

To date, 5763 bones have been studied, including 3692, which can be determined to family, genus or species, which is an identification rate of 64%. This percentage is quite high due to the good preservation of the material. The rich faunal spectrum includes 27 genera, 18 species from 12 different families. The results presented here are based only on the hand-collected samples (Table 22.2). The elements from the sieved samples are not yet studied in detail, although some comments can be provided on them. The best represented families during all periods are the emperors (Lethrinidae: Lethrinus sp.) with 1310 remains, jacks/trevallies (Carangidae: *Alecta indicus*, *Alepes mate*, *Carangoides bajad, Carangoides chrysophrys, Carangoides malabaricus, Gnathanodon speciosus, Scomberoides commersonnianus, Seriola dumerili*) with 816 remains, groupers (Serranidae: Epinephelus sp.) with 747 remains and seabream (Sparidae: *Acanthopagrus bifasciatus, Acanthopagrus latus, Argyrops spinifer, Cheimerius nufar, Crenidens crenidens, Rhabdosargus* sp., *and Sparidentex hasta*) with 690 remains. Eight other families are represented by smaller quantities of bones, such as mackerel, wahoo, tuna and bonito (Scombridae: *Rastrelliger kanagurta, Scomberomorus commerson, Euthynnus affinis* and *Thunnus* spp.) 53 bones, sea catfish (Ariidae: *Netuma* spp.) 27 bones, barracuda (Sphyraenidae: *Sphyraena* spp.) 20 bones, rabbitfish (Siganidae: *Siganus* spp.) 12 bones, grunts (Haemulidae: *Diagramma pictum* and *Plectorhinchus* spp.) 11 bones, needlefish (Belonidae: *Tylosurus* spp.) 4 bones, mullets (Mugilidae: *Valamugil* sp.) 1 bone, and parrotfish (Scaridae: *Scarus* sp.) 1 bone. Some cartilaginous fish are also present, as 15 vertebrae from sharks and rays were counted. Looking at the distribution of fish remains by period provides a very interesting diachronic view. The Early Dilmun period levels only contain a small amount of bones. Thus, 258 bones were analysed, including 230 which could be identified (89%). A total of seven families, 11 genera, including seven species were represented. In order of importance, are groupers (85 remains) with 37% of the total identified remains from the period, with an average size of 74 cm, emperors (62 bones or 27%), jacks/trevallies (43 bones or 19%) and seabream (36 bones or 15,5%). To these, can be added sporadic remains of tuna/mackerel and marine catfish.

In the Middle Dilmun period the number of remains are higher than that of the previous period, and taxonomic diversity is much more significant, despite the marked fragmentation. Thus, out of 582 bones studied, 237 could be identified (41%), with nine families, 14 genera and ten species being identified. The four dominant families are the same, but in different proportions. Thus, emperors dominate (104 bones or 44%), followed by jacks/trevallies (61 bones or 25,5%), groupers (30 bones or 12,5%) and seabream (33 bones ie 14%). Other families complete the spectrum, but in smaller proportions: grunts, mullets, barracuda, rabbitfish and tuna/mackerel. During the Late Dilmun period the vast amount of material (4054 items, including 2659 identified specimens (65%)) has allowed the identification of nine families with 20 genera and 13 species. Emperors dominate (955 bones identified, or 36%), then jacks/trevallies (eight species providing 657 bones, 25%) and seabream (7 species giving 572 bones or 21,5%) and groupers (415 bones or 15,5%), whose average size is 68 cm. Five other families are represented in lower proportions: the sea catfish (21 bones), tuna/mackerel (17 bones from three species), barracuda (16 bones), needlefish and grunts (1 and 2 bones). The Tylos period only had a relatively small quantity of bones (160 bones, out of which 104 bones or 65% could be determined). This small quantity can be explained by the lack of excavation of deposits dating to these particular occupation horizons. A total of six families, 11 genera and seven species were, however, noted. The same four families dominate the spectrum, emperors (38 bones, 36,5%), groupers (27 bones, 26%), jacks/trevallies (three species, 21 pieces, 20%) and seabream (three species, ten bones, 9,5%). Tuna/mackerel and sea catfish were also poorly represented. Finally, during the Islamic period, there were 709 remains studies, of which 462 could be identified (65%), and the range of fauna represented was as varied as that during the Middle and Late Dilmun periods, with 11 fish families represented. Thus, a total of 17 genera were identified and 8 species.

In order of importance the assemblage is as a whole dominated by groupers (190 bones, 41%), whose average size is 62cm, emperors (151 bones, 33%), jacks/trevallies

Table 22.2. Occurrence of fish species by occupation period at Qal'at al-Bahrain. Note: this only includes hand collected material from the 1989–2004 French excavations.

			Early Dilmun		*Middle Dilmun*		*Late Dilmun*		*Tylos*		*Islamic*		*Total*
Family	*Species*	*Common name*	*NR*	*%*	*NR*	*%*	*NR*	*%*	*NR*	*%*	*NR*	*%*	
Ariidae	*Netuma* spp.	Marine Catfish	1	0.4			21	0.8	1	1	4	0.9	27
Belonidae	*Tylosurus* spp.	Needlefish					1	0.04			3	0.6	4
Carangidae	*Alectis indica*	Indian Threadfish					1	0.04					1
Carangidae	*Atule mate*	Yellowtail Scad			1	0.4	1	0.04					2
Carangidae	*Carangoides bajad*	Orange spotted Jack	13	5.7	12	5.1	223	8.4	4	4	6	1.3	258
Carangidae	*Carangoides chrysophrys*	Longnose Trevally	1	0.4	1	0.4	18	0.7					20
Carangidae	*Carangoides malabaricus*	Malabar Trevally			2	0.8							2
Carangidae	*Carangoides* spp.	Trevally			3	1.3	11	0.4					14
Carangidae	*Gnathanodon speciosus*	Golden Trevally	17	7.4	34	14.3	259	9.7	12	11.5	26	5.6	348
Carangidae	*Scomberoides commersonianus*	Talang Queenfish	6	2.6			71	2.7	3	2.9			80
Carangidae	*Scomberoides* spp.	Queenfish					20	0.8					20
Carangidae	*Seriola dumerili*	Greater Amberjack					6	0.2					6
Carangidae	*Seriola* spp.	Amberjack	1	0.4			5	0.2					6
Carangidae	Indeterminate		5	2.2	8	3.4	42	1.6	2	2	2	0.4	59
Total Carangidae													*816*
Epinephelinae	*Epinephelus* spp.	Grouper	85	37	30	12.7	415	15.6	27	26	190	41.1	747
Haemulidae	*Diagramma pictum*	Painted Sweetlips			1	0.4							1
Haemulidae	*Plectorhinchus* spp.	Sweetlips					1	0.04			8	1.7	9
Haemulidae	*Pomadasys* spp.	Grunt					1	0.04					1
Lethrinidae	*Lethrinus* spp.	Emperor	62	27	104	43.9	955	35.9	38	36.5	151	32.7	1310
Mugilidae	*Valamugil* spp.	Mullet			1	0.4							1
Scaridae	*Scarus* spp.	Parrotfish									1	0.2	1
Scombridae	*Scomberomorus commerson*	Narrow-barred Spanish Mackerel			1	0.4			4	4	1	0.2	6
Scombridae	*Scomberomorus* spp.	Mackerel			1	0.4					1	0.2	2
Scombridae	*Euthynnus affinis*	Kawakawa/ Little Eastern Tuna	1	0.4	1	0.4	4	0.2	1	1	14	3.0	21
Scombridae	*Thunnus* spp.	Tuna					13	0.5			3	0.6	16
Scombridae	*Rastrelliger kanagurta*	Indian Mackerel									1	0.2	1
Scombridae	Indeterminate	Tuna/ Mackerel	1	0.4			3	0.1	2	2	1	0.2	7
Total Scrombidae													53
Shyraenidae	*Sphyraena* spp.	Barracuda	1	0.4	2	0.8	16	0.6			1	0.2	20
Siganidae	*Siganus* spp.	Spinefoot/ Rabbitfish			2	0.8					10	2.2	12
Sparidae	*Acanthopagrus bifasciatus*	Double-barred Seabream	10	4.3	20	8.4	103	3.9	2	2	15	3.2	150
Sparidae	*Acanthopagrus latus*	Yellowfin Seabream					3	0.1					3
Sparidae	*Argyrops spinifer*	King Soldierbream	12	5.2	1	0.4	337	12.7	5	4.8	14	3	369
Sparidae	*Cheimerius nufar*	Santer Seabream					1	0.04					1

Family	*Species*	*Common name*	*Early Dilmun* NR	%	*Middle Dilmun* NR	%	*Late Dilmun* NR	%	*Tylos* NR	%	*Islamic* NR	%	*Total*
Sparidae	*Crenidens crenidens*	Karanteen Seabream					2	0.1					2
Sparidae	*Sparidentex hasta*	Sobaity Seabream									1	0.2	1
Sparidae	*Rhabdosargus* spp.	Goldstriped/ Haffara Seabream	12	5.2	7	3	84	3.2	2	2			105
Sparidae	Indeterminate	Seabream	2	0.9	5	2.1	42	1.6	1	1	9	2	59
Total Sparidae													690
Identified			230	89	237	40.7	2659	65.6	104	65	462	65.2	3692
Unidentified			28	11	345	59.3	1395	34.4	56	35	247	34.8	2071
Total			258	100	582	100	4054	100	160	100	709	100	5763

(34 bones, 7%), seabream (39 bones, 8,5%), tuna/mackerel (21 bones, 4,5%), rabbitfish (ten bones, 2,2%), and grunts (eight bones, 1,7%). Sea catfish, needlefish, barracuda and rabbitfish represent the remainder.

Discussion

Comparison of the results with previous studies shows that throughout the occupation, four families dominate the fisheries of the inhabitants of Qal'at al-Bahrain. Their order of importance varies according to the chronological periods, these are emperors (Lethrinidae), groupers (Serranidae), jacks/trevallies (Carangidae) and seabream (Sparidae).

This dominance has already been demonstrated in previous studies (Table 22.1). However, a comparison can be made with the results obtained with material from the Danish excavations. Indeed, they are the only ones to have led to the publication of quantitative chronologically calibrated results.

In the case of the Early Dilmun period, the results obtained by Van Neer and Uerpmann (1994), with a total of 2899 identified remains, highlight the presence of groupers (36%), emperors (30%), seabream (15 %) and jacks/trevallies (15%). Three other families represent the remainder of the fauna: marine catfish (0.6%), tuna/mackerel (0.6%) and barracuda (0.1%). Our study confirms the preference for these particular fish populations. Despite the lack of remains identified from the French excavations, the percentages are broadly equivalent to earlier results. Furthermore, taxonomic diversity is better explained in our sample. Some species, such as *Carangoides bajad*, *Scomberoides commersonnianus*, and *Acanthopagrus bifasciatus* are specifically determined, thus providing further clarification. It is likely that this diversity may be further expanded once the study of the skeletal remains retrieved from sieving is carried out.

For the following periods of occupation at both Qal'at al-Bahrain and Saar (cf. Uerpmann & Uerpmann 1997; 1999) the Middle Dilmun and Late Dilmun periods, the preference for grouper is less obvious, and the local inhabitants seem to have preferred emperors (33%), then groupers (29%), followed by jacks/trevallies (17%), seabream (14%), rabbitfish (1%), barracuda (1%), tuna/mackerel (0.6%), sea catfish (0.2%), and finally, grunts and drums (0.1% each). Our results confirm the preponderance of emperors, while groupers find themselves in fourth position, after jacks/trevallies and seabream. Two new families complement the previous studies, needlefish and mullets. In contrast, the drums present in previous studies (Uerpmann & Uerpmann 1997; 1999) do not appear in our material. The meeting of these two studies has enriched the faunal spectrum by an additional family, and leads to a total of 11 families. In total for this period there are 15 species, seven genera and an additional family highlighted by this new study.

The preference for emperors (Lethrinidae) is confirmed during the following Tylos period. Their remains are more frequent than those of groupers, jacks/trevallies and seabream. The remaining fish represented are tuna/mackerel and marine catfish. During the Islamic period, groupers come first in the species consumed, followed by emperors, jacks/trevallies, seabream, tuna/mackerel, rabbitfish, grunts, sea catfish, needlefish, parrotfish and barracuda.

In general, we find throughout the occupation the same type of fish caught and eaten by the people of Bahrain island. Analysis of the fish bones from Dilmun period levels at the Saar Temple (Moon & Irving 1997) and from Islamic period levels at Bilad al Qadim (Beech 2005) identified a similar range of fishes.

Taking into account the habitat of each species, and the reconstruction of the size of the fishes, it allows us to locate possible fishing areas and to propose techniques which may have been used to capture them, as well as to quantify the relative importance of fish in the diet of the inhabitants.

Biotopes of identified fishes

Eight species of jacks/trevallies (Carangidae) were determined: *Alectis indica*, *Atule mate*, *Gnathanodon speciosus*, *Carangoides bajad*, *Carangoides chrysophrys*, *Carangoides malabaricus*, *Scomberoides commersonianus*, and *Seriola dumerili*. These all evolve in the same habitat, that is to say, in coral reefs, near to shore, generally in shallow waters.

The main species of emperors (Lethrinidae) attested at the present time is *Lethrinus nebulosus*. This lives in seagrass beds and mangroves, particularly in shallow water, at least during its subadult stage. Four other species of the same genera may be determined at the present time: *L. lentjan*, *L. mahsena*, *L. microdon* and *L. borbonicus*.

Several species of groupers, which all inhabit very similar habitats (coral reefs), are present in the coastal waters of Bahrain. The choice was made for the time being, to treat these all under the generic name, *Epinephelus* spp. Their living conditions are quite similar, and accurate data on their ecology too fragmentary (Desse & Desse-Berset 2000a).

Eight species of seabream (Sparidae) are present at the site: *Acanthopagrus bifasciatus*, *Acanthopagrus latus*, *Argyrops spinifer*, *Cheimerius nufar*, *Crenidens crenidens*, *Rhabdosargus haffara*, *Rhabdosargus sarba* and *Sparidentex hasta*). The first and last live in shallow (<60m) and on sandy bottoms. *Argyrops spinifer* and *Cheimerius nufar* also live near the bottom, but respectively up to depths of 150–300m.

In the Scombridae, four major species are currently determined. *Scomberomorus commerson* and *Rastrelliger kanagurta* live in very deep water (<70m). *Euthynnus affinis* and *Thunnus* sp. are predators of the open sea, preferring greater depths up to 250m. Thus, for exploitation of these species, it may be possible that the coastal fisheries seasonally fished in offshore waters. Among the six families found sporadically during the successive occupations, the marine catfish (Ariidae: *Netuma* spp.) are quite often demersal catch and can be found in brackish water lagoons along the coast, whilst needlefish (Belonidae: *Tylosurus* spp.) and barracuda (Sphyraenidae: *Sphyraena* spp.) are associated with reefs and shallow waters. Rabbitfish (Siganidae: *Siganus* spp.), mullets (Mugilidae: *Valamugil* spp.) and grunts (Haemulidae: *Diagramma pictum* and *Plectorhinchus* spp.) are all coastal fish, also living in shallow waters.

Thus, the majority of species exploited came from coral reefs in shallow inshore coastal waters (Belonidae, Carangidae, Serranidae, Sphyraenidae, Mugilidae and Haemulidae), mangroves and lagoons (Ariidae, Lethrinidae and Siganidae), and sandy bottoms (Sparidae).

Fishing techniques in Bahrain

The presence of both coastal and pelagic species such as tuna, suggests that different fishing techniques may have been utilised (Fig. 22.6). Archaeozoological analyses and ethno-archaeological work carried out by Nathalie Desse-Berset during her fieldwork provides useful information on the techniques used by coastal populations in Bahrain (Desse-Berset 1995). Fish traps, known locally as '*haddrah*', made from date palm fronds, have a very specific form which can vary but which are usually arrow-shaped. These catch fish at low tide, and are still very much in use at the present day (Desse-Berset 1995).

Dhows, more specifically boats such as Jaliboot, Sambuk, etc., allow the local fishermen to move along shallow coastal waters (from 2–10m in depth) for distances up to around 50km. The fishermen use large basket traps of different sizes, known locally as '*gargoor*', to catch fish. All of these techniques allow the capture of all species present in the levels of Qal-at al-Bahrain. Moreover, fishing with lines near to Qal-at al-Bahrain is attested by the discovery of fish hooks and stone net sinkers, particularly during the Tylos period.

It is likely that some of the techniques mentioned above were used during the successive occupations of the site. All that is certain is that the diversity of techniques utilised attests to a highly developed fishing industry which is justified by the strategic position of Bahrain island in the Persian Gulf.

Conclusion

The diachronic study of fish bones from the Early Dilmun to Islamic periods at the site of Qal'at al-Bahrain provides valuable information concerning the economy of the capital of ancient Dilmun. One notes the pervasiveness of four main families throughout the occupation, emperors, groupers, jacks/trevallies and seabream. These fish provide the greatest amount of bones at the site, and denote various food preferences during successive periods of occupation. Other taxa complement these, but in smaller quantities.

Ongoing studies complement the earlier Danish and French studies, and confirm the results. Taking into account that some of the excavated sediment was subjected to sieving, the future analysis of this material will undoubtedly enhance the diversity of the identified assemblage, since most of the previous studies were based on hand retrieved material.

The current study provides additional qualitative and quantitative data, and allows a diachronic view of exploitation of the sea by the island populations of Bahrain. In particular, a transition from sedentary species such as groupers, towards species living in groups like emperors, along with occasional exploitation of pelagic fishes from deeper waters. The study highlights that fishing took place in mangroves/lagoons, on the open seas, on coral reefs, as well as along the coast, i.e. all aquatic environments were exploited in the vicinity of the site.

Acknowledgments

I wish to sincerely thank Mark Beech and Marjan Mashkour who allowed me to participate in this conference. Thank also go to all the sponsors: the Service of Cooperation and

Cultural Action (SCAC) and the Office of the Cooperation for France (BCF) of the French Embassy in Abu Dhabi, in particular to Prof. Didier Gazagnadou; His Excellency Mohammed Khalaf Al Mazrouei, Director General of the Abu Dhabi Authority for Culture and Heritage (ADACH) and Mohammed Amer Al Neyadi, Head of the Historic Environment Department at ADACH in Al Ain; and to all the other sponsors.

I would like to express my appreciation to Pierre Lombard for his help and support in preparing this presentation for the conference. A big thank you goes to the late Jean Desse and Nathalie Desse-Berset who supervised this work, and to Philippe Bearez for his advice. Finally, special thanks go to Mark Beech for his work on translating and substantial editing of my paper.

Note

1 CEPAM, UMR 6130, 250 rue Albert Einstein, Sophia Antipolis, 06560 Valbonne, France.

References

Beech, M. 2004, *In the Land of the Ichthyophagi: Modelling fish exploitation in the Gulf of Oman from the 5th millennium BC to the Late Islamic period.* Abu Dhabi Islands Archaeological Survey Monograph 1, British Archaeological Reports International Series S1217, Archaeopress, Oxford.

Beech, M. 2005. 'The faunal and botanical remains. Part 3 – the fish bones & Appendix 5.3 – fish bone data tables', in T. Insoll, *The Land of Enki in the Islamic Era – Pearls, Palms and Religious Identity in Bahrain*, 240–251, 463–479. Kegan Paul, London, New York, Bahrain.

Cataliotti-Valdina J. 1994, 'Les coquillages, Qala'at al-Bahrain', in H. Hellmuth Andersen, F. Højlund & O. Callot (eds), *The Northern City Wall and the Islamic Fortress, The Carlsberg Foundation's Gulf Project* 1, 455–458. Jutland Archaeological Society Publications 25, Moesgaard, Aarhus.

Desse, J. & Desse-Berset, N. 2000a, Préhistoire du mérou, in *Symposium international sur les mérous de Méditerranée. Proc. of a Symposium, 5–7 nov. 1998, at île des Embiez, France*, Mémoires de l'Institut océanographique Paul Ricard 45–57.

Desse-Berset, N. 1995, 'La pêche est au bout du jardin... Deux îles, hier et aujourd'hui', IVe Colloque International de l'Homme et l'Animal, Société de recherche interdisciplinaire, November 1994'. *Anthropozoologica* 21, 7–20.

Højlund, F. & Andersen, H. 1997, *Qal'at al-Bahrain, volume 2: The Central Monumental Buildings,* 10. The Carlsberg Foundation's Gulf Project, Jutland Archaeological Society Publications 30(2), Moesgaard, Aarhus.

Lombard, P. 1999, 'Qal'at al-Bahrein, mémoire d'une île et d'une civilisation'. *Dossiers d'Archéologie* 7, 4–13.

Lombard, P. & Kervran, M. (eds). 1989, *Bahrain National Museum Archaeological Collections, Volume I: A selection of pre-Islamic Antiquities from excavations 1954–1975.* Directorate of Museums and Heritage, Ministry of Information, State of Bahrain.

Moon, J. & Irving, B. 1997, 'Faunal remains', in H. Crawford, R. G. Killick & J. A. Moon (eds), *The Dilmun Temple at Saar. Bahrain and its Archaeological Inheritance*, 81–83. Kegan Paul International, London.

Sanlaville, P. 1988, 'Des mers au milieu du désert: mer Rouge et Golfe Arabo-Persique', in J.-F. Salles, *L'Arabie et ses mers bordières. I. Itinéraires et voisinages*, 9–26. Travaux de la Maison de l'Orient 16, Lyon.

Tengberg, M. & Lombard, P. 2002, 'Paléoenvironnement et économie végétale à Qal'at al-Bhareïn aux périodes Dilmoun et Tylos. Premiers éléments d'archéobotanique'. *Paléorient* 27(1), 167–181.

Tomé, C. 2003, 'Les mammifères et les oiseaux dans l'économie des sociétés protohistoriques et historiques du Golfe Arabo-Persique (IIIe millénaire avant J.-C. – VIIe siècle après J.-C.)'. Doctoral thesis, University of Aix-Marseille.

Uerpmann, M. 1993, 'Animal remains from Qala'at al-Bahrain. A preliminary report', in H. Buitenhuis & A. T. Clason, *Archaeozoology of the Near East, volume I*, 60–66. Universal Book Services, Leiden.

Uerpmann, M. & Uerpmann, H.-P. 1994, 'Animal bone find from Excavation 520 at Qala'at al-Bahrain', in *Qala'at al-Bahrain, volume I: The Northern City Wall and the Islamic Fortress*, The Carlsberg Foundation's Gulf Project, Højlund, Flemming & Andersen, Hellmuth, Jutland Archaeological Society Publications XXV: 1, Moesgaard, Aarhus, pp. 417-444.

Uerpmann, M & Uerpmann, H-P 1997, 'Animal bone find from Excavation 519 at Qala'at al-Bahrain', in *Qala'at al-Bahrain, volume II : The Central Monumental Buildings*, The Carlsberg Foundation's Gulf Project, Højlund, Flemming & Andersen, Hellmuth, Jutland Archaeological Society Publications XXX: 2, Moesgaard, Aarhus, pp. 235-264.

Uerpmann, M & Uerpmann, H-P 1999, "The animal economy of Ancient Dilmun in the light of faunal remains from excavations at Saar and Qala'at al-Bahrain. *ISIMU. Revista sobre Oriente proximo y Egipto an la antigüedad* 2 : 635-646.

Van Neer, W & Uerpmann, M 1994, 'Fish remains from Excavation 520 at Qala'at al-Bahrain', in H. Hellmuth Andersen, F. Højlund & O. Callot (eds), *The Northern City Wall and the Islamic Fortress, The Carlsberg Foundation's Gulf Project* 1, 445–454. Jutland Archaeological Society Publications 25, Moesgaard, Aarhus.

23. Evidence for deep-sea fishing and cultural identity during the Neolithic period at Akab Island, Umm al-Qaiwain, United Arab Emirates

Mark J. Beech, Vincent Charpentier and Sophie Méry

Analysis of the fish bones from the 5th millennium BC settlement on Akab Island in Umm al-Qaiwain emirate in the United Arab Emirates provides evidence of open ocean fishing. The discovery of numerous bones of tuna, as well as the presence of shell fish hooks at the site, suggests that fishing was at least part of the time carried out from boats in the open sea, beyond the shallow waters of the local lagoon. Some fishing was also carried on in these sheltered waters, and analysis of the molluscan and crab remains indicates that mangrove areas were also exploited. Some comments are made concerning the presence of shell fish hooks at the site, as well as certain types of beads and jewellery, which reinforce the idea of a coherent regional cultural entity during the 5th–4th millennia within this region.

Keywords deep-sea fishing, tuna, shell fish hook, 5th millennium BC, United Arab Emirates

Introduction

Between the Gulf and the Arabian Sea all the Neolithic settlements which have been excavated within the Oman peninsula are coastal sites. This is probably explained by the fact that this type of site is more easily detected and suffers less from deflation due to aeolian processes and the mobility of sand dunes on those sites located within the interior. These coastal habitats sometimes have significant stratigraphy, such as more than 2m thick, for example, in the case of the 5th–4th millennium BC site of Suwayh 1 in the Sultanate of Oman (Charpentier 2008). Due to the presence of large quantities of marine shells at these sites, the relatively high calcium carbonate content ensures that animal bones are much better preserved within these 'shell-middens'. This is particularly the case for stratigraphic levels associated with the 5th millennium BC settlement on Akab Island (UAE), which is characterised by a high degree of mineralisation.

Today, all of these Neolithic coastal sites reveal a material culture which is often a reflection of human activities related to the sea (Beech 2002; 2004). At the settlement of Akab, the presence of ceramic remains provide evidence of long distance maritime trade with the northern Gulf and the Ubaid Culture of Lower Mesopotamia, in particular. This article presents another aspect of the relationship between populations in the Gulf during the 5th millennium BC: not only fish production, but the conquest of the sea.

Akab Island, Umm al-Qaiwain emirate, United Arab Emirates

Akab Island is located 50km north of Dubai in the large lagoon of Umm al-Qaiwain in the United Arab Emirates (Fig. 23.1). The archaeological site, first investigated by a palaeontologist at the beginning of the 1990s, was initially interpreted as a butchering area for dugongs, and thus became known as the oldest known site of dugong hunting within the Gulf (Prieur & Guérin 1991; Jousse *et al.* 2002).

Subsequent excavations carried out by the French Archaeological Mission to the United Arab Emirates

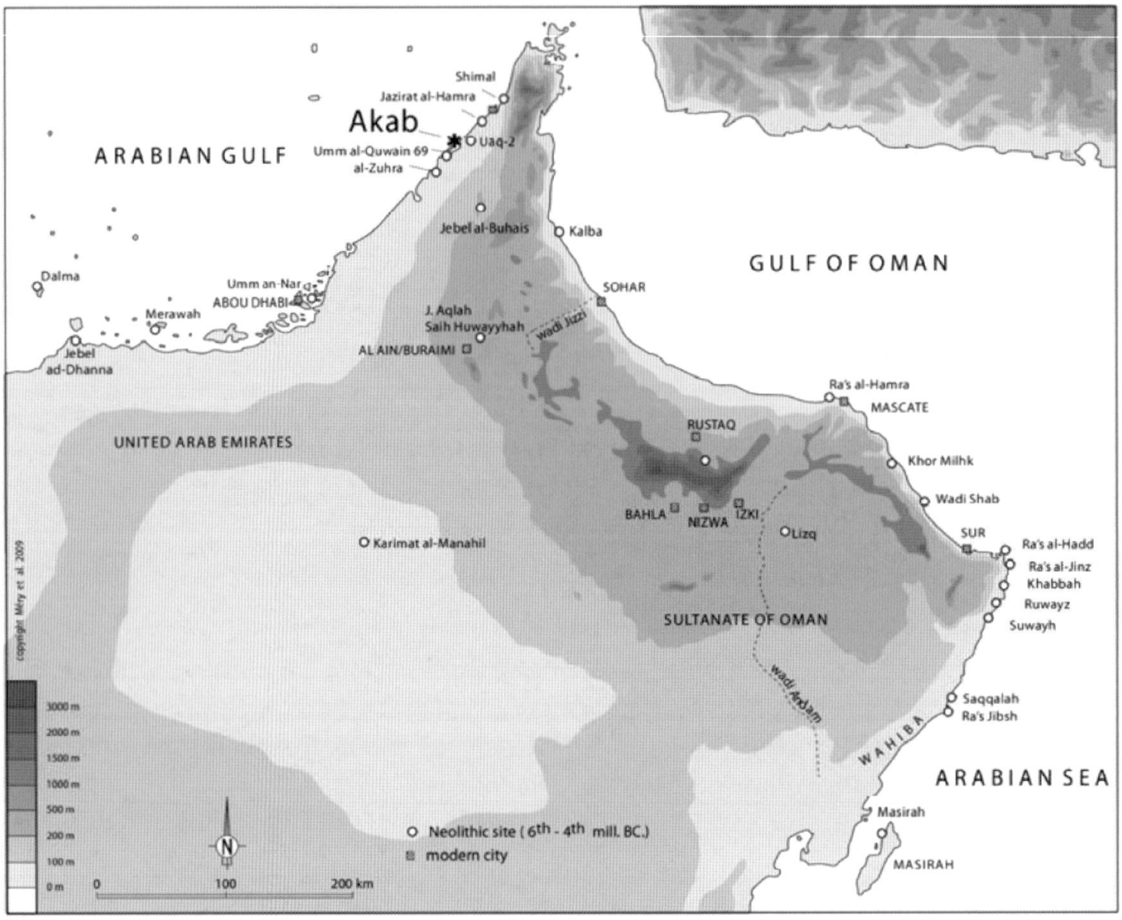

Fig. 23.1. Location of the site of Akab (source: French Archaeological Mission in the UAE).

(FAMUAE) revealed that the dugong mound of Akab was only a small part of a much larger Neolithic site, periodically occupied in the 5th millennium BC, which included the remains of circular habitations (Charpentier & Méry 2008). Radiocarbon dating indicates that this settlement was occupied between around 4700–4100 BC. In the 4th millennium BC, except for the dugong mound, traces of human occupation were more ephemeral, probably because they had been partially destroyed by deflation.

In 2006–8 the excavation of the dugong mound was resumed by a multi-disciplinary team of prehistorians and faunal experts (Charpentier & Méry 2012; Méry *et al*. 2009, Méry & Charpentier 2012). The hypothesis of a butchering site, which had been advanced previously, was rapidly put into question when the mound was discovered to be a *structured* accumulation of bones, a complex arrangement whose layout had been accomplished in stages. This structure was built on a hillock of wind-blown sand and consisted of at least two stratified levels, about 40cm in height. Radiocarbon dating of a dugong bone sample attributed it to the 2nd half of the 4th millennium BC (5140±55 BP, Pa-2433, ca. 3568–3116 cal BC). The dugong bone platform contained the remains of about 40 dugongs. Its base consisted of mandibles laid flat, wedged by ribs. This provided a foundation upon which the skulls were placed upright in anatomical position. The skulls were carefully wedged by ribs (Fig. 23.2), and were arranged in a deliberate manner, with a row of eight aligned at the front. Adults, as well as juvenile, including very young, dugongs were well represented. No animal appears to have been deposited whole in the structure. Certain elements such as ribs, vertebrae and limbs were under-represented, which means that there is evidence of intentional selection (Méry *et al* 2009; cf. also Beech 2010, regarding the selection of particular elements of dugongs).

Several hundred objects were deposited in or inserted into this mound of dugong bones. These mainly consisted of ornamental jewellery in the form of beads (*Spondylus* sp, *Pinctada sp*., *Ancilla* sp., etc). The Neolithic dugong bone mound at Akab is now interpreted as a monument with pre-conceived organisation which represents the oldest known ritual site in Arabia (Charpentier & Méry 2012; Méry *et al*. 2009; Méry & Charpentier 2012).

The purpose of this paper, however, is to discuss the preliminary analysis of the environmental remains associated with the earlier 5th millennium BC settlement site.

Fig. 23.2. Part of the ritual dugong bone platform – the dugong skulls at Akab are placed in anatomical position, aligned and wedged by ribs (source: French Archaeological Mission in the UAE).

The 5th millennium settlement site

Five excavation campaigns, carried out since 2002 by the French Archaeological Mission to the UAE, showed that the site of Akab had multiple occupations, dated by radiocarbon, to between 4750 and 3120 BC. Most of the excavated levels date back to the 5th millennium BC (Charpentier & Méry 2008). The site, which exceeds more than an acre (0.45ha), includes anthropogenic deposits without a trace of any major discontinuity for more than half a millennium. The remains of architectural traces in the form of post-holes indicate the repeated construction of structures during the 5th millennium BC.

No significant abandonment phase has been identified at the site, which was not occupied until after 3100 BC. The site was abandoned during a period corresponding to the late Neolithic in the Oman Peninsula, a period consistent with a phase of aridification. No significant later remains have been discovered on Akab Island, but some remains dating to the Islamic period are attested, including fire places.

Akab 2002 excavations – some preliminary results of the environmental remains

The first author was asked to undertake an analysis of the archaeozoological remains from the 2002 excavations at the Akab settlement site. All material was identified using the first author's comparative osteological collection of the Gulf mollusca, crabs, fish and marine shells from the region, which is curated in Abu Dhabi in the United Arab Emirates. Quantification of the material was made using a rapid semi-quantitative system as indicates in Table 23.1.

Molluscs

At least six families of mollusc are present at the site (Fig. 23.3 & Table 23.1). These include an as yet undetermined species of Cerithiidae, *Terebralia palustris*, *Cypraea* sp., an undetermined species of Dentaliidae, *Pinctada* sp. and *Spondylus* sp.

The presence of *Terebralia palustris* in three levels at the site (S1 L1; S2 L6; and S5 G9 L6) is of some interest here. This large gastropod, known as a mudcreeper, prefers intertidal habitats on the surface of mud in mangroves and soft water logged soil. Large quantities of these have been found within the Gulf at a number of prehistoric sites in the northern Emirates. They used to form a major food resource in ancient times. This species is, however, now extinct within the Gulf. It can only be found in a small number of locations on the east coast of the UAE at the present time. Its extinction within the Gulf may be due to the loss of suitable mangrove habitats along the coast due to the over-exploitation of mangroves for timber and fodder.

Some of the pearl oyster shell fragments (*Pinctada* sp.) are very large (S5 L5), and appear to belong to the species, *Pinctada margaritifera*. A total of 18 pearls was found during the excavation of the settlement, when sieving the sediments (Charpentier *et al.* 2012). Pearl oysters would clearly have been gathered as a food item, as well as for their pearls.

Crabs

Two types of crabs were identified amongst the Akab material (Fig. 23.3 and Table 23.1). The most common type represented was swimming crabs, Portunidae, from the genus *Portunus* sp. The remains of these predominantly consisted of their characteristic elongated *chelae* (pincers). The second less common type was the mud or mangrove crab, *Scylla serrata*. This had much more massive *chelae* with large molariform teeth.

Large quantities of *Portunus* crab remains were noted in some levels of the site (S1 L1 and S2 L6), and moderate levels in others (S2 L5; S5 G8 L6; S5 G9 L6; and S5 L6-7). The remains of *Scylla serrata* were only discovered

Table 23.1. Marine Mollusca and crabs identified from the 5th millennium BC settlement at Akab

COMMON NAME	FAMILY	TAXON	S1 L1	S1 L1 C1	S1 L2 C3	S1 L1 D1	S1 L1-2 D2	S1 L1 D3	S1 L1 D4	S1 L1 E1	S1 L2 C4	S2 L5	S2 L6	S3 L1A	S5 L1	S5 2e passe 22/01/2002	S5 3e passe	S5 4	S5 L2 C4	S5 L5	S5 G8 L6	S5 G9 P5-6 93cm	S5 G9 L6 75-90cm	S5 G9 L6	S5 L6-7 93cm	S5 H9 L6 93cm
Ceriths	Cerithiidae	indeterminate	*										+										+			
Mudcreeper	Potamididae	*Terebralia palustris* (Linnaeus, 1767)	+										*										+			
Cowrie	Cypraeidae	*Cypraea* sp.																					+			
Elephant tusk shell	Dentaliidae	indeterminate																							+	
Pearl oyster	Pteriidae	*Pinctada* sp.																					*	*		
Spiny oyster	Spondylidae	*Spondylus* sp.																		*						
Mollusc indeterminate		Mollusc, indeterminate																		*	*		*			
Swimming Crab	Portunidae	*Portunus* sp.	***									**	***		*	*					**	*	**	*	**	
Mud/Mangrove Crab	Portunidae	*Scylla serrata* (Forskål, 1775)	*										+													
Shark		Chondrichthyes, indeterminate	+																							
Marine Catfish	Ariidae	*Arius* sp.	*									+	+			+				+			*			
Needlefish	Belonidae	indeterminate							+								*			*	+	+	+	*	+	
Grouper	Serranidae	*Epinephelus* sp.														+				+		*	+			
Golden Trevally	Carangidae	*Gnathanodon speciosus* (Forskål, 1775)																					+		*	
Jack/Trevally	Carangidae	indeterminate	*										*			*				*	*	*	*	+	**	
Seabream	Sparidae	*Rhabdosargus* sp.	*									+				+					+	*				
Emperor	Lethrinidae	*Lethrinus* sp.	+																							
Barracuda	Sphyraenidae	*Sphyraena* sp.																							+	
Tuna/Mackerel	Scombridae	Thunninae, indeterminate	*												*		*			*	*		**	*	*	
Fish indeterminate		Pisces, indeterminate										*			*		*			*			*		*	*
Sheep/Goat/Gazelle	Bovidae	*Ovis* or *Capra* or *Gazella* indeterminate																			+				*	
Gazelle	Bovidae	Gazella sp.											+	+												
?Dog	?Canidae	*?Canis lupus familiaris* (Linnaeus, 1758)															+									
Dugong	Dugongidae	*Dugong dugon* (Müller, 1776)	*	+	+		+	+	+	+	+			*		*	*	*		+			*		*	+
Mammal indeterminate		Mammalia, indeterminate																	+	+						
?Human		? *Homo sapiens*				+																				

+ = single example, * = 2–9 examples, ** = 10–49 examples, and *** = 50+ examples in two levels at the site (S1 L1 and S2 L6). These same levels also contained examples of the gastropod species, *Terebralia palustris* (see above), confirming that mangrove environments were being exploited during these particular occupation horizons.

Fishes

The majority of the environmental remains retrieved from the excavations consisted of fish bones (Table 23.1 & Fig. 23.4). This demonstrates the importance of fishing to the peoples of the lower Gulf during the Neolithic period. At least nine families of fishes are present at the Akab site including sharks (Chondrichthyes, indet.), marine catfish (Ariidae), needlefish (Belonidae), groupers (Serranidae), jacks/trevallies (Carangidae), seabream (Sparidae), emperors (Lethrinidae), barracudas (Sphyraenidae) and tuna (Scombridae: Thunninae).

A single large shark vertebra was identified. This came from a requiem shark (Carcharhinidae). Judging from the relative sise of the vertebra the shark must have been at least 2m in length. Marine catfish (Ariidae) were represented by neurocranial fragments as well as by otoliths. Needlefish (Belonidae) were identified in six levels by dentaries, premaxillae and vertebrae fragments. Groupers (Serranidae) occurred in three levels. These all belonged to the genus *Epinephelus*. Jacks/trevallies (Carangidae) were quite common in the material, occurring in nine levels. Some of these remains belonged to the golden trevally (*Gnathanodon speciosus*). Seabreams (Sparidae) were represented in five levels by the genus *Rhabdosargus*, identified on the basis of its characteristic oval rear molar, present in both its dentary and premaxilla. Emperors (Lethrinidae) were represented by a single otolith from the genus *Lethrinus*. A single dentary from barracuda (Sphyraenidae) was noted. Bones from tuna (Scombridae: Thunninae) occurred in no less than eight levels at the site, being common in level S5 G9 L6. These all consisted of vertebrae, which judging from their relative sise probably belonged to fishes around a metre or less in size. The species *Thunnus tonggol*, known as longtail tuna, is the type of tuna commonly sold in fish markets in the UAE at the present day. This generally attains a maximum length of only 140cm (Randall 1995). It may well be this relatively small tuna which was also caught in the past.

Mammals

At least three types of mammals were recorded (Table 23.1). Some small bone fragments could only be classified as belonging to either domestic sheep/goat or gazelle. At least two specimens, a calcaneum and 2nd phalanx, could be definitively assigned to gazelle. An almost complete 3rd metatarsal was identified as belonging to dog (from level S5 3^{e} passe). The commonest mammal occurring in the deposits at Akab was dugong. Traces of this animal in the form of skull, rib and vertebrae fragments occurred in 16 out of the 24 units or levels excavated at the site. This confirms the importance of Akab for the hunting and exploitation of dugong.

Finally, a small fragment of what appears to be a human vertebra was noted in level S1 L1 D1. Special care should be given during future excavations at the site to check whether further human skeletal material is present in this part of the site. It should be noted that the nearby broadly contemporary site excavated by Carl Phillips consisted of midden material in association with a cemetery (Phillips 2002).

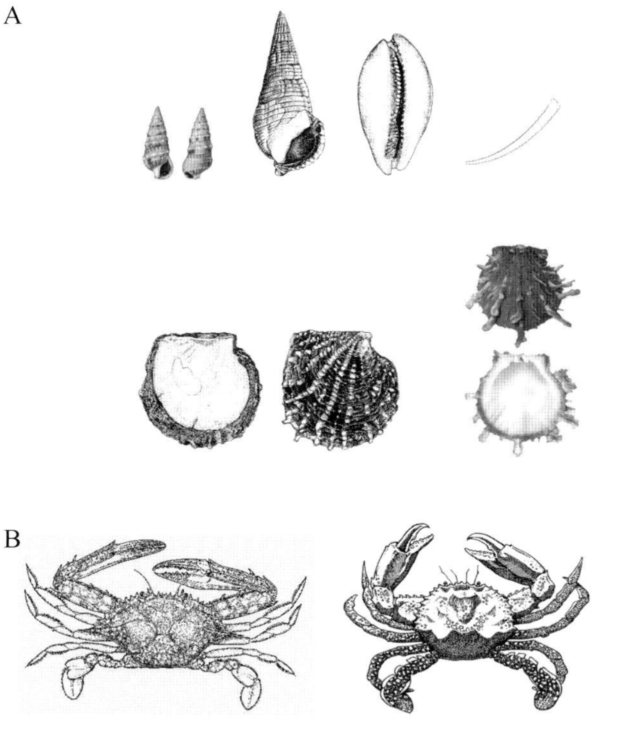

Fig. 23.3. Marine molluscs and crabs identified from the 5th millennium BC settlement at Akab. From left to right: A. Top: Cerithium, Terebralia palustris, Cypraea sp. and Dentaliidae; Centre: Pinctada margaritifera and Spondylus sp. B. Bottom: Portunus pelagicus and Scylla serrata.

Discussion

During the 5th millennium BC on Akab Island, it is clear that resource exploitation primarily concentrated on mangroves and on the local lagoon environment, but the inhabitants also fished for tuna on the high seas. This meant that they had to use boats (although we do not have any direct remains of these), and fishing gear suitable for deep water fishing. Although the net weights are quite small and few in number, it was noted that the inhabitants of the site were making shell fish hooks (Mery *et al.* 2008). This is contrary to the hypothesis raised by Margarethe and Hans-Peter Uerpmann (1996), who suggested that these

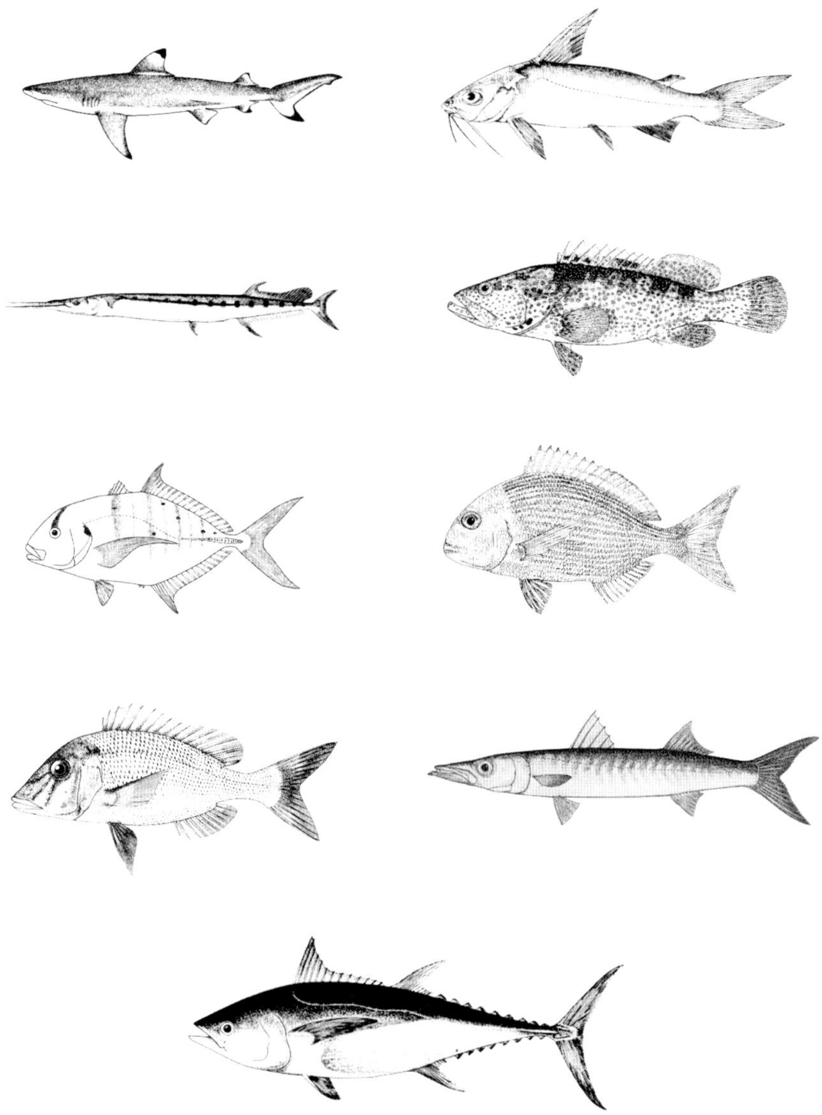

Fig. 23.4. Marine fishes identified from the 5th millennium BC settlement at Akab. From left to right: Top: Carcharhinidae: Carcharhinus melanopterus and Ariidae: Arius thalassinus; 2nd row: Belonidae: Ablennes hians and Serranidae: Epinephelus coioides. 3rd row: Carangidae: Gnathanodon speciosus and Sparidae: Rhabdosargus sarba; 4th row: Lethrinidae: Lethrinus nebulosus and Sphyraenidae: Sphyraena putnamiae; Bottom: Scombridae: Thunnus tonggol.

technologies were absent during the time the lagoon of Umm al-Qaiwain was first occupied.

Assuming that much of the environmental material from Akab was retrieved locally, the presence of the gastropod species, *Terebralia palustris*, as well as the mud crab, *Scylla serrata*, suggests that the site was located in the vicinity of mangroves. Neither of these species are now present in the Umm al-Qaiwain area. Although some mangrove cover occurs in this area at the present time, this is largely concentrated on the eastern side of Akab island and towards the central area in the Umm al-Qaiwain lagoon (Al-Ghais & Cooper 1996, 411, fig. 1). It may be the case that mangrove habitats were more widespread in the past (Beech & Hogarth 2002).

A modern study of crabs in the Umm al-Qaiwain lagoon noted three species of swimming crabs (Portunidae), *Portunus pelagicus*, *Thalamita crenata* and *Thalamita poissoni* (Al-Ghais & Cooper 1996, 423–425). It was reported that *Portunus pelagicus* was the largest and most commonly encountered species in the area. Much of the archaeological crab remains from Akab probably belong to this species judging from their relative sise. *P. pelagicus* is commonly found in shallow sublittoral waters and can even be caught offshore in locally-made gargoor fish traps by traditional fishermen.

It is clear from an examination of the types of fish caught at Akab that fishing largely took place in shallow water habitats. A study of the modern fishes present within the Umm al-Qaiwain lagoon reported moderate numbers of requiem shark, *Carcharhinus dussumieri* and marine catfish, *Arius thalassinus* (Department of Fisheries 1984). At least three species of grouper were noted, *Epinephelus areolatus*, *E. coioides* and *E. polylepis*, although only *E. coioides* was abundant. A number of jacks/trevallies were common within the lagoon, *Gnathanodon speciosus* being particularly abundant. The seabream species *Rhabdosargus sarba*, emperors (*Lethrinus* spp.) and barracudas (*Sphyraena* sp.), were all reported as being abundant within the lagoon. Thus, the majority of the species represented amongst the archaeological material are still present in the lagoon today. Such fish could have been caught using various techniques including tidal traps (known locally as '*hadrah*'), nets, basket traps, as well as the occasional use of hook and line.

Similar evidence that much fishing was carried out in local shallow waters was obtained from other 5th millennium BC excavations in the vicinity of Umm al-Qaiwain (Beech 2003; 2004; Uerpmann & Uerpmann 1996), as well as at the nearby later site of Ed-Dur (Van Neer & Gautier 1993).

Whilst the majority of the fish could have been caught in the neighbouring Umm al-Qaiwain lagoon, tuna (Scombridae – Thunninae) were probably caught outside the lagoon in open waters. Their presence was not noted in the study of modern fishes caught within the lagoon (Department of Fisheries 1984). An important point to bear in mind is that whilst small quantities of tuna can be caught all year round they are far more abundant at certain times of year in the Gulf waters of the northern emirates. For example, in Ras Al-Khaimah waters most of the annual catch of tuna was caught during the months of April and May during 1982 (Ali & Cherian 1983). The catching of tuna may have therefore been seasonal.

Bead making and jewellery

The presence of numerous traces of *Spondylus* shell bead production in all settlement levels was another important finding. It indicates that some sites within the Gulf were specialised in craft production (Charpentier & Méry 2008; and cf. the production of *Spondylus* beads at As-Sabiyah in Kuwait; Carter & Crawford 2010).

The bead-making remains are indeed extremely abundant and occur throughout the stratigraphy of the site. In addition, even if we have no indication of manufacture of the famous 'Akab type' of tubular beads (made from chlorite and murex), their discovery in the 5th millennium BC settlement at Akab suggests that they are not a chronological marker at the very end of the 4th millennium, but rather a cultural marker. Indeed, they appear from around 4600 BC and persist until about 3100 BC, such beads having been discovered on sites ranging from Qatar to Oman. It should be noted that this is the first time, in southeast Arabia, that we can observe such a wide distribution of Neolithic ornamental elements. Their distribution (with the exception of Qatar) corresponds with two other types of ornamental elements very characteristic of the Neolithic of this region; namely, laurel leaf-shaped pendants made from *Pinctada margaritifera* and composite bracelets carved from large Conidae (Charpentier & Méry 2008).

Regional cultural identity

The archaeological excavations at Akab are not the only excavations to be carried out on coastal Neolithic sites in the UAE, but they do represent the first investigation of this type of site that has been conducted over such a large area, with $70m^2$ total surface being excavated down to virgin soil. Data on the Neolithic period are scarce in the Gulf, where only a few coastal sites have been excavated or surveyed, for instance Dalma and Marawah in the United Arab Emirates, Al-Markh in Bahrain, Khor and Shagra in Qatar, and As-Sabiyah in Kuwait (Roaf 1976; Desse 1988; Flavin & Shepherd 1994; Uerpmann & Uerpmann 1996; Beech & Elders 1999; Beech *et al.* 2000; 2005; Beech & Glover 2005; Carter & Crawford 2010). Important issues concerning the chronology of the Neolithic across southeast Arabia remain to be resolved, as well as the lifestyles of local people and their trading networks. The role of indigenous cultures in the appearance of these early societies remains to be determined, as well as the influence of the PPNB of the Levant in the emergence of the Neolithic within the Oman Peninsula.

If one considers the bead making and other jewellery items discovered on coastal Neolithic sites, these observations support the hypothesis of a common cultural entity within the Oman Peninsula during the 5th–4th millennia BC. Recent studies of flint arrowheads have also helped to define a similar geographic area, two distinct areas being recognised with distinct regional techno-cultural characteristics for the period 6500–3800 BC (Charpentier 2008). The fact that some technology, including the use of pearl shell fish hooks, was shared by the coastal populations of the UAE and the Sultanate of Oman, reinforces the image of a coherent regional cultural entity during the 5th–4th millennia within this region.

This new evidence that the people living in the Akab settlement were able to venture out on the high seas, fishing beyond the safety of their own lagoon for tuna, shows that Neolithic peoples had indeed developed ocean-going boats. The fact that shell fish hooks are only known within the Gulf from two locations, Akab in Umm al-Qaiwain and Shimal in Ras Al-Khaimah, both located in the northern Emirates, may be a reflection that both these areas have deeper water in close proximity, compared with other areas with much shallower waters in the southern and western sides of the Gulf. This may also have helped to reinforce connections between these people and those living on the coast of the Sultanate of Oman, who also fished for tuna using similar technology.

Acknowledgements

Thanks go to H.H. Sheikh Saud bin Rashid al Mualla, UAE Supreme Council Member and Ruler of Umm Al Quwain, H.E. Sheikh Khaled bin Humed al Mualla, Director, Manager of the Department of Archaeology and Heritage – Umm al Quwain, and to Ms Alyaa Mohammad al Ghufly, Director of the Museum of Umm Al Quwain, for their kind support of this archaeological work.

References

Ali, R. M. & Cherian, T. 1983, *Environmental Conditions of the Coastal Waters of the United Arab Emirates During 1982*. Annual Report – Technical Report 9, Department of Fisheries, Ministry of Agriculture and Fisheries, United Arab Emirates.

Beech, M. 2002, 'Fishing in the 'Ubaid: a review of fish-bone assemblages from early prehistoric coastal settlements in the Arabian Gulf'. *Journal of Oman Studies* 12, 25–40.

Beech, M. 2003. 'The development of fishing in the United Arab Emirates: a zooarchaeological perspective', in D. T. Potts, H. Naboodah & P. Hellyer (eds), *Archaeology of the United Arab Emirates: Proceedings of the First International Conference on the Archaeology of the UAE*, 289–308. Trident Press, London.

Beech, M. J. 2004, *In the Land of the Ichthyophagi: modelling fish exploitation in the Arabian Gulf and Gulf of Oman from the 5th millennium BC to the Late Islamic period*. Abu Dhabi Islands Archaeological Survey Monograph 1, British Archaeological Report S1217, Archaeopress, Oxford.

Beech, M. J. 2010, 'Mermaids of the Arabian Gulf: archaeological evidence for the exploitation of dugongs from prehistory to the present. *Liwa (Journal of the National Center for Documentation and Research)* 2(3), 3–18.

Beech, M. & Elders, J. 1999, 'An 'Ubaid-related settlement on Dalma Island, United Arab Emirates'. *Bulletin of the Society for Arabian Studies* 4, 17–21.

Beech, M. & Glover, E. 2005, 'The environment and economy of an Ubaid-related settlement on Dalma island, United Arab Emirates'. *Paléorient* 31(1), 97–107.

Beech, M. & P. Hogarth. 2002, 'An archaeological perspective on the development and exploitation of mangroves in the United Arab Emirates', in S. Javid & A. G. de Soyza (eds), *Research and Management Options for Mangrove and Salt Marsh Ecosystems – Proceedings of the 2nd International Symposium and Workshop on Arid Zone Environments (22–24 December 2001, Abu Dhabi, UAE)*, 196–198. Environmental Research and Wildlife Development Agency (ERWDA), Abu Dhabi, UAE.

Beech, M., Elders, J. & Shepherd, E. 2000, 'Reconsidering the 'Ubaid of the Southern Gulf: new results from excavations on Dalma Island, U.A.E'. *Proceedings of the Seminar for Arabian Studies* 30, 41–47.

Charpentier, V. 2008, 'Hunter-gatherers of the "empty quarter of the early Holocene" to the last Neolithic societies: chronology of the late prehistory of south-eastern Arabia (8000–3100 BC)'. *Proceedings of Seminar for Arabian Studies* 38, 59–82.

Charpentier, V. & Méry, S. 2008, 'A Neolithic settlement close by the Strait of Hormuz: Akab Island – United Arab Emirates'. *Proceedings of the Seminar for Arabian Studies* 38, 83–102.

Charpentier, V. & Méry, S. 2012. Un sanctuaire marin de l'Arabie néolithique', in N. Schlanger & L. Taylor (eds), *La préhistoire des Autres*, 337–349. Musée du quai Branly, La Découverte. Paris.

Charpentier, V., Phillips, C. S. & Méry, S. 2012, 'Pearl fishing in the ancient world: 7500 BP'. *Arabian Archaeology and Epigraphy* 23, 1–6.

Department of Fisheries. 1984, *Study on Mariculture Environment of Umm al-Quwain Lagoon and the Experimental Rearing of Shrimp, Rabbitfish and Mullet*. Annual Report - Technical Report 8. Department of Fisheries, Ministry of Agriculture and Fisheries, United Arab Emirates.

Desse, J. 1988, 'Khor «P», Khor «F.B.» et «Shagra». Les faunes. Le rôle de la pêche (Fish remains and micrommalian fauna from Khor and Shagra. Methodology and preliminary results)', in M.-L. Inizan (ed.), *Préhistoire à Qatar*, 157–165, 225–226. Mission Archéologique française à Qatar, Paris.

Flavin, K. & Shepherd, E. 1994, 'Fishing in the Gulf: Preliminary investigations at an Ubaid site, Dalma (U.A.E.)'. *Proceedings of the Seminar for Arabian Studies* 24, 115–134.

Al-Ghais, S. & Cooper, R.T. 1997, 'Brachyura (Grapsidae, Ocypodidae, Portunidae, Xanthidae and Leucosiidae) of Umm al Quwain mangal, United Arab Emirates'. *Tropical Zoology* 9(2), 409–430.

Jousse, M., Faure, M., Guerin, C. Prieur, A. & Desse, J. 2002, 'Exploitation des ressources marines au cours des Ve–IVe millénaires: le site à dugongs de l'île d'Akab (Umm al-Qaiwain, Emirats Arabes Unis)'. *Paléorient* 28(1), 43–60.

Méry, S. & Charpentier, V. 2012, 'Akab island, a Neolithic sanctuary in the Gulf', in D. T. Potts & P. Hellyer, *Fifty Years of Emirates Archaeology. Proceedings of the Second International Conference on the Archaeology of the United Arab Emirates*, 69–77. Ministry of Culture, Youth and Community Development, Abu Dhabi, United Arab Emirates.

Méry, S., Charpentier, V. & Beech, M. 2008, 'First evidence of shell fish hooks in the Gulf'. *Arabian Archaeology and Epigraphy* 19, 15–21.

Méry, S., Charpentier, V., Auxiette, G. & Pellé, E.. 2009, 'A dugong bone mound: the Neolithic ritual site on Akab in Umm al-Quwain, United Arab Emirates'. *Antiquity* 83, 696–708.

Neer, W. van & Gautier, A. 1993, 'Preliminary report on the faunal remains from the coastal site of ed-Dur, 1st–4th Century AD, Umm al-Quwain, UAE', in H. Buitenhuis & A.T. Clason (eds), *Archaeozoology of the Near East: Proceedings of the First International Symposium on the Archaeozoology of Southwestern Asia and Adjacent Areas*, 110–118. Universal Book Services / Dr. W. Backhuys, Leiden.

Phillips, C. 2002, 'Prehistoric middens and a cemetery from the southern Arabian Gulf'. in S. Cleuziou, M. Tosi & J. Zarins (eds), *Essays on the Late Prehistory of the Arabian Peninsula*, 169–186. Istituto Italiano per l'Africa e l'Oriente, Rome.

Prieur, A. & Guerin, C. 1991, 'Découverte d'un site préhistorique d'abattage de dugongs à Umm al-Qaiwain (Emirats Arabes Unis)'. *Arabian Archaeology and Epigraphy* 2(2), 72–83.

Randall, J. E. 1995, *Coastal Fishes of Oman*. Crawford House Publishing, Bathurst.

Roaf, M. 1976, 'Excavations at Al Markh'. *Proceedings of the Seminar for Arabian Studies* 6, 144–160.

Uerpmann, M. & Uerpmann, H.-P. 1996, 'Ubaid pottery in the eastern Gulf – new evidence from Umm al-Qaiwain (U.A.E.)'. *Arabian Archaeology and Epigraphy* 7, 125–139.

Part 6

Rituals and animal deposits

24. Elite equids 2: seeing the dead

Jill A. Weber

Cuneiform documents dating to the mid- to late- 3rd millennium refenced a special type of equid, a kunga, *that inhabited the elite world of gods, kings, and diplomats. Textual references to the animals' mixed parentage elicited expectations that these were hybrid equids – likely a cross of the domestic ass x wild onager, both of which were mentioned along with the* kunga. *Despite these references, clear evidence for such animals had never been found in archaeological contexts. Analysis of more than 30 complete and nearly-complete skeletons from an elite burial complex dating to the 2nd half of the 3rd millennium at the northern Syrian site of Umm el-Marra has changed that. These animals, which form a single population, conform neither to populations of ass, hemione, nor horse. Hybridity is indicated by the unique combination of morphological features, the absolute size and proportions of the bones, and the large degree of dental malocclusion. Their burial context and demographic patterning further link these hybrids to the ancient* kunga.

Keywords Hybrids, equids, Syria, 3rd millennium, Umm el-Marra, kunga

An elite burial complex dating to ca. 2600–2200 BC has been uncovered at the site of Umm el-Marra in northern Syria (Schwartz *et. al.* 2003; 2006) (Fig. 24.1a). Nine stone-built tombs have been identified to date in this complex, which occupies the site's central acropolis (Fig. 24.1b). The monumental tombs contained multiple human inhumations and grave goods, including precious metals, ceramic vessels, and animal bones. Animal bones inside the human-occupied tombs included sheep, goat, cattle, geese, ducks and dog. A unique feature of Umm el-Marra's burial complex is the separate inhumation of at least 40 male, draft equids (Schwartz *et. al.* 2006; Weber 2008). Initial analysis indicated that these skeletons – of which 25 are complete – were distinct from known populations of *Equus asinus, E. hemionus* (and sp.), and *E. przewalski.* Instead, they displayed mixed morpho-metric characteristics of both donkey and hemione. The context in which the animals were found – an elite burial complex dating to the mid- to late 3rd millennium, BC – suggested correlation to the elite *kunga* equid documented in contemporary texts from Syro-Mesopotamia. Large-sised males of that breed were specifically used to pull the vehicles of the ruling elite and the earthly representatives of the gods (Heimpel 1994, 11; Maekawa 1979, 111–13; Zarins 1976, 450–561). The sterile *kunga* is largely considered to have been a hybrid produced from donkey and hemione parentage. The mixed morpho-metric traits of Umm el-Marra's skeletons and their culturally-appropriate identification with the *kunga* support their identification as hemione x donkey hybrids.

An initial report on the earliest of these animals excavated was given at ASWA 8 in Lyon, and subsequently published in its proceedings (Weber 2008). In the three years since the ASWA 8 conference, more skeletons have

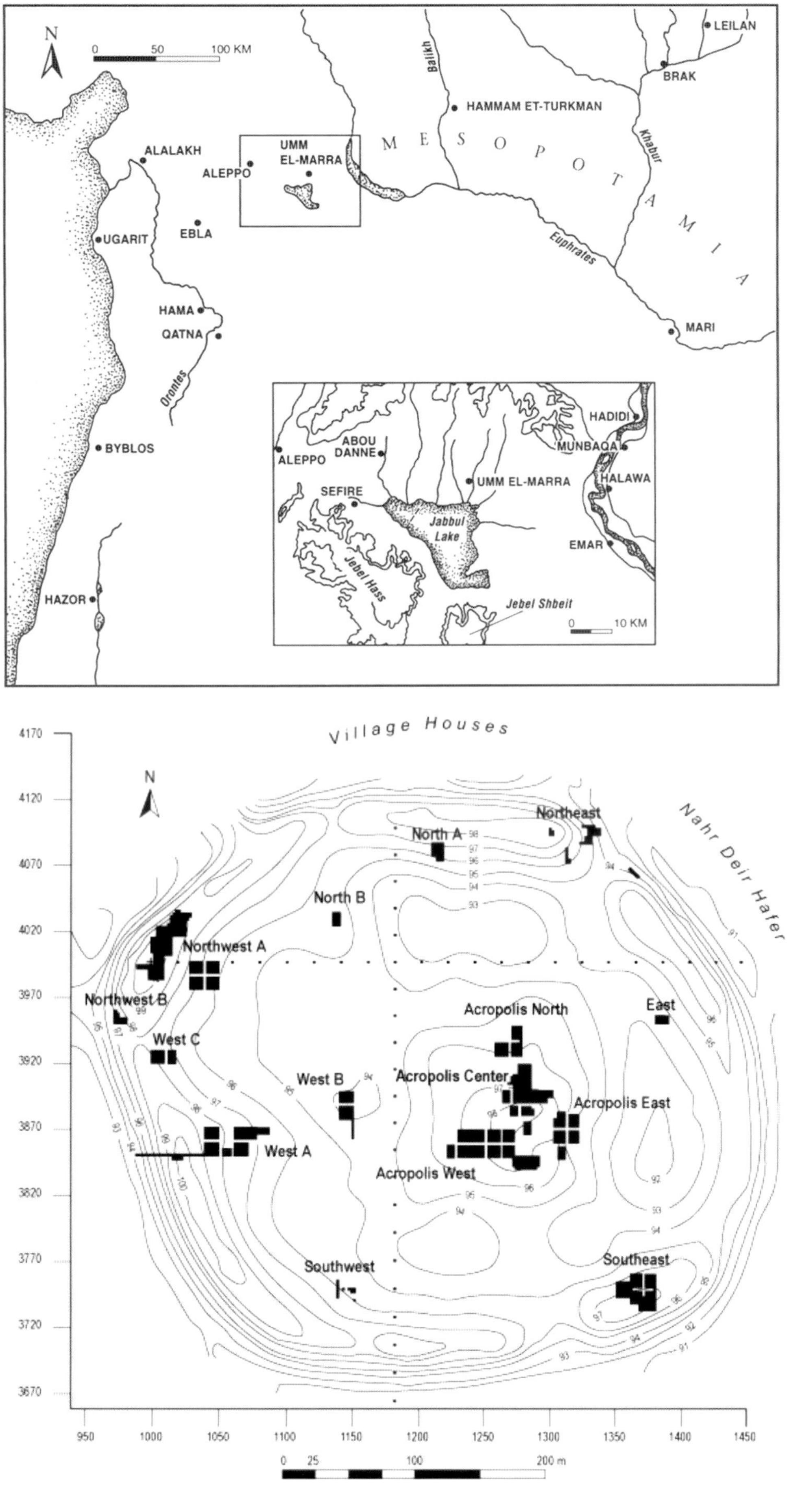

Fig. 24.1. a. Location of Umm el Marra; b. Location of different excavated sectors on the site.

been excavated and greater effort has been devoted to providing taxonomic clarity to these finds. This work has involved the documentation and morpho-metric analysis of a larger number of bones. The results of these latest efforts are a better understanding of their intra-population variation, and a better ground for comparison between the Umm el-Marra archaeological population and other known equine populations. As this dataset grows, so too does the likelihood that it is representative of the ancient population of which the animals were part (e.g. the Law of Large Numbers). The fact that there are 30 or more independent examples of most elements makes normative statistics possible (e.g. Central Limit Theorem) and thus adds significance to statistical analyses and comparisons with other populations. Such a large collection – among the largest collections of complete equid skeletons from a single taxon of either modern or archaeological origin – enables unprecedented examination of intra-population variability of equids from the ancient world.

The last few years have also heralded the initiation of a 3-D modelling and archiving programme involving 3-D laser scanning of all of the equid bones from the burial complex. This will culminate in the complete, virtual preservation of the skeletons with the resulting ability to broadly disseminate the 'raw' bones as well as their derived documentation. The resulting models retain the superficial appearance of the originals, and impart extraordinary topographical information that may aid in the identification of new areas for equine morpho-metric discriminations. The 3-D data is manipulable and infinitely reproducible, allowing broad access to primary materials for present and future examination. This is of extreme importance as the bones remain in Syria, and will do so for the foreseeable future.

Results obtained from the more recent excavations and analyses will be emphasised here. Description of the population's traits and proper classification continue, and we await preliminary results of genetic testing.

New excavations

Excavations in 2008 brought clarity to some installations that had been incompletely excavated, and also increased the number of complete skeletons to 25, with partial skeletons of a minimum of 15 more individuals (Table 24.1). An example is the so-called 'Grey Lined Pits' that were partially exposed in 2006. It now appears that these

Table 24.1. Skeletons per installation.

	Primary, complete individuals	*Secondary and partial individuals*
Installation A	4	1
Installation B	2	1
Installation C	2	0
Installation D	2	4
Installation E	4	2
Installation F	4 (?)	Unclear
Installation G	6	≥5
Single skeletons	1	2*

*possibly disturbed individuals from Installation F

Table 24.2. Summary of select measurements of equine taxa

	Umm el-Marra burial equids				*E.* hemionus onager				Equus przewalski			
Element	*n*	*mean*	*sd*	*V*	*n*	*mean*	*sd*	*V*	*n*	*mean*	*sd*	*V*
Humerus	17	245.3	11.93	4.86	31	241.9	8.14	3.365	42	263.2	9.88	3.75
Radius	25	301.4	10.43	3.46	30	295.1	10.73	3.64	40	312.2	10.795	3.45
MC	32	209.2	5.65	2.70	33	210.2	7.35	3.50	50	216.6	7.29	3.37
Femur	11	328.8	9.99	3.04	29	328.9	10.03	3.05	43	357.6	13.64	3.81
Tibia	18	320.6	9.10	2.84	30	312.6	9.80	3.13	44	326.1	12.18	3.73
MT	30	245.7	5.82	2.37	32	246.3	8.87	3.60	50	257.9	8.035	3.115
PH1 A	31	79.1	2.70	3.41	31	76.3	2.485	3.26	43	79.1	3.46	4.38
PH1 P	32	72.7	3.14	4.32	31	70.6	2.47	3.50	41	74.9	3.81	5.09
Mc/Rad	21	69.4	2.76	3.98	30	71.3	2.86	4.01	40	69.5	2.23	3.20
MC slend	28	14.0	0.69	4.97	33	12.7	0.72	5.67	50	14.45	0.80	5.56
MT slend	26	11.5	0.63	5.46	32	10.55	0.625	5.92	50	11.55	0.58	5.06

	Equus asinus				Equus africanus				*Mules*			
Element	n	mean	sd	V	n	mean	sd	V	n	mean	sd	V
Humerus	31	234.6	36.28	15.47	12	248.3	11.91	4.80	16	284.7	35.17	12.35
Radius	33	284.8	43.15	15.15	12	296.0	11.94	4.03	16	331.5	40.18	12.12
MC	50	189.4	23.17	12.23	14	201.6	7.27	3.60	19	227.25	27.51	12.105
Femur	32	319.9	46.20	14.44	13	337.9	18.97	5.61	17	387.6	48.55	12.53
Tibia	33	294.0	65.48	22.27	13	312.6	16.41	5.25	16	355.25	39.53	11.13
MT	44	225.5	25.11	11.14	15	237.1	8.95	3.78	19	268.9	28.18	10.48
PH1 A	38	73.9	11.22	15.19	14	76.5	4.13	5.40	16	88.3	13.475	15.26
PH1 P	36	68.65	10.33	15.0	15	72.4	4.40	6.07	17	83.65	12.59	15.05
Mc/Rad	31	66.7	2.67	4.00	12	68.0	1.22	1.79	15	68.1	2.06	3.02
MC slend	50	13.7	1.08	7.88	14	13.4	0.63	4.69	19	14.3	1.35	9.43
MT slend	44	10.8	0.95	8.74	15	11.1	0.63	5.72	19	11.5	1.25	10.87

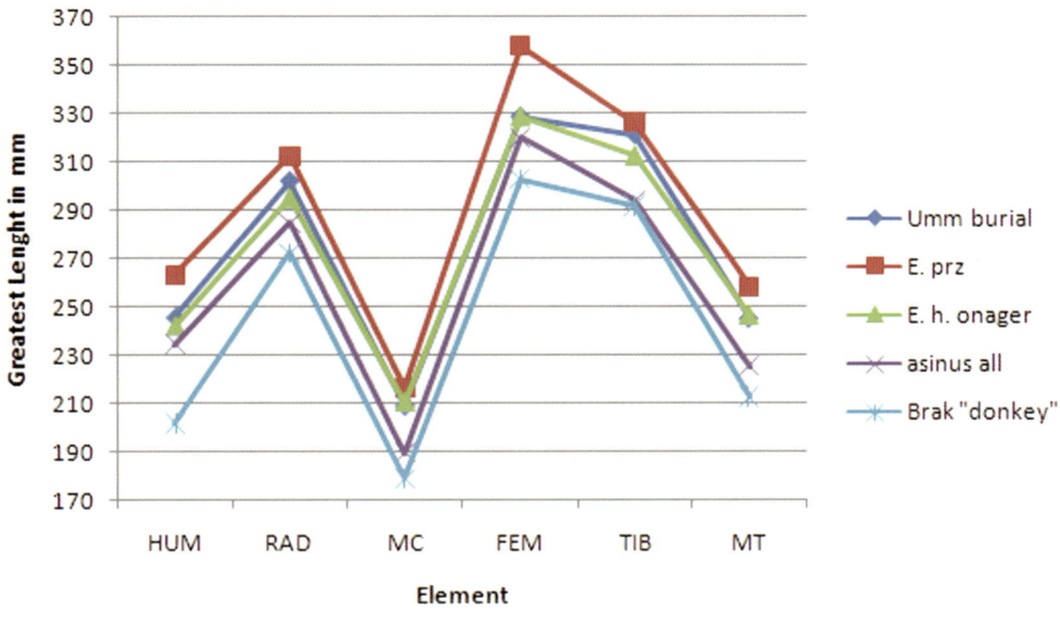

Fig. 24.2. Mean greatest length of limb-bones, by taxa.

are two separate installations superimposed one above the other. The earliest is the single equid skeleton discovered in 2006 (Weber 2008). That individual was laying on its right side with its head in the baulk to the west; its skull has not yet been recovered. Directly above this individual was a larger pit – called Installation G[1] – that contained two levels of at least four individuals each. The first, lower level contained fully articulated bodies lying on their left sides. All skeletal elements were recovered except for the skull of the northernmost individual. This was disturbed or removed when the second layer of equids was deposited immediately above the lower layer. The later, upper layer contains two fully articulated skeletons, as well as two groupings of disarticulated bones. Each grouping contains the bones of at least two individuals, but these may represent the same two individuals. The newly excavated animals do not differ from those already excavated: all are male, draft equids.

Absolute size and proportional differences of limb bones between taxa

In order to work out some of the more gross metrical characteristics of the Umm el-Marra burial skeletons, collections of measurements from other equine populations were compiled. This job was made much easier by the existence of a website established by Vera Eisenmann (vera-eisenmann.com) containing thousands of measurements from samples of multiple modern and extinct equine populations. Long-bone measurements (and select others) from a selection of those populations plus persian onager measurements from Eisenmann and Mashkour (2000), as well as the author's Umm el-Marra measurements, are summarised in Table 24.2. Unless otherwise noted, all figures contain measurements collected from those sources.

Relative lengths

Initial examination of the Umm el-Marra burial material suggested that these were relatively large animals – larger than the contemporary donkeys found at Umm el-Marra and roughly the size of onager found at the same site. Measurements from additional skeletons have further validated this supposition. The mean values of the greatest lengths of the long bones of the burial equids are intermediate between Przewalski's horse and domestic ass, and are roughly equivalent to, though slightly larger than, the Persian onager (Fig. 24.2). They have greater means than do modern donkeys, which includes animals bred for large size, and are much larger than the six mid-3rd millennium 'donkeys' excavated at Tell Brak (see Clutton-Brock 1989; Clutton-Brock & Davies 1993).

As is evident from Figure 24.2 limb proportions are roughly similar across equid taxa, though there are measureable differences. According to Groves and Willoughby (1981, 328), on average the metatarsus is always shorter than the humerus in horses and asses, but longer in hemiones; mules also appear to have relatively shorter metatarsus bones (Willoughby 1974, 436, table 32; Fig. 24.3, here). The mean greatest lengths of these two elements have nearly the same value for Umm el-Marra's burial population; the average metatarsus GL is only 0.4mm longer than the average humerus GL. Only 13 individuals have GL preserved for both the humerus and matched metatarsus. Of these, seven (54%) have metatarsus bones averaging 6.4mm longer than their matched humerus, five (38%) have metatarsus bones averaging 7.9mm shorter, and one has both bones of equal length. In this feature, Umm el-Marra's burial equids show greater variation than do other populations, which may be a result of the small sample size. It falls closest to the variation expressed by

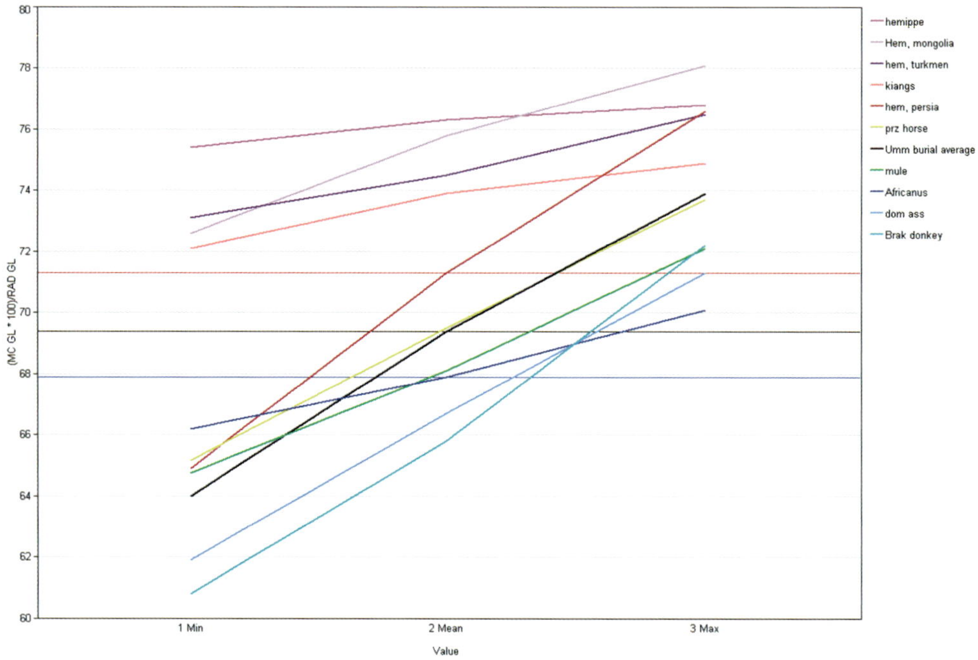

Fig. 24.3. Comparison of values between taxa of the proportional length of the Comparison of values between Taxa of the Proportional Length of the Metacarpus Relative to the Radius

Table 24.3. Comparison of Group Means via Student's t-test

Abs(Dif)-LSD	E.h. onager	E. prz	*Umm burial*	*Mules*	E. africanus	E. asinus
E.h. onager	–1.2840	0.5698	0.4605	1.5606	1.6186	3.5638
E. prz	0.5698	–1.1120	–1.2356	–0.1434	–0.0905	1.8809
Umm burial	0.4605	–1.2356	–1.5347	–0.4234	–0.3577	1.5508
Mules	1.5606	–0.1434	–0.4234	–1.8158	–1.7419	0.1288
E. africanus	1.6186	–0.0905	–0.3577	–1.7419	–2.0302	–0.1856
E. asinus	3.5638	1.8809	1.5508	0.1288	–0.1856	–1.1409

Positive values show pairs of means that are significantly different at $\alpha = 0.05$, t=1.97591.

the Persian onager, of which 21 of 30 (70%) individuals had, as expected, longer metatarsus than humerus bones. Measurements of *E. przewalski* show 75% (30 of 40) conformity to the expected larger greatest length of the humerus. The domestic ass expressed still less variation in this regard – even despite the inclusion of animals bred to a wide range of sizes – with 25 of 30 (83%) individuals having the expected longer humerus/shorter metatarsus pairing. A wild ass population showed no individuals (n=12) deviated from the expected result: 100% had a longer humerus than metatarsal bone.

One of the more well-known proportional distinctions among the equids is between the GL of the radius and metacarpus bones; the metacarpus bones of hemiones are longer relative to their radius bones than are those of asses (Boessneck 1976; Clutton-Brock 1986; von den Driesch & Amberger 1981; Eisenmann 1995, 8), horses (Willoughby 1974, 436, table 32). The largest differences are between asses and hemiones. Boessneck (1976) had asserted that the greatest length of the metacarpus bones would always be less than 70% the greatest length of the radius for asses, and always more than 70% for hemiones. According to Figure 24.3, such precise discrimination is unfounded. The union of the ranges for individual donkeys and hemiones is quite large: 64.1–71.3%. This indicates that a simple formula cannot be used to discriminate *individuals*. However, the axiom does hold true for *population means* of asses (<70%) and hemiones (>70%). The largest mean for this ratio among published measurements for modern domestic ass is 67.2% for 27 animals in Willoughby (1974, 436, table 32), which compares well to 66.7% for 31 animals in Eisenmann (2007a,b). Six archaeological 'donkeys' from Tell Brak averaged 65.8% (after Clutton-Brock 1989; Clutton-Brock & Davies 1993), well under the 70% population maximum. The mean for 12 *E. africanus* individuals in Eisenmann (2007c,d) was slightly greater at 68.0%. Of all of the hemione taxa whose measurements are recorded in Eisenmann and Mashkour (2000), the Persian onager have the lowest population mean at 71.3%. This mean, as well as the highest ass-population mean (68.0%), are plotted as

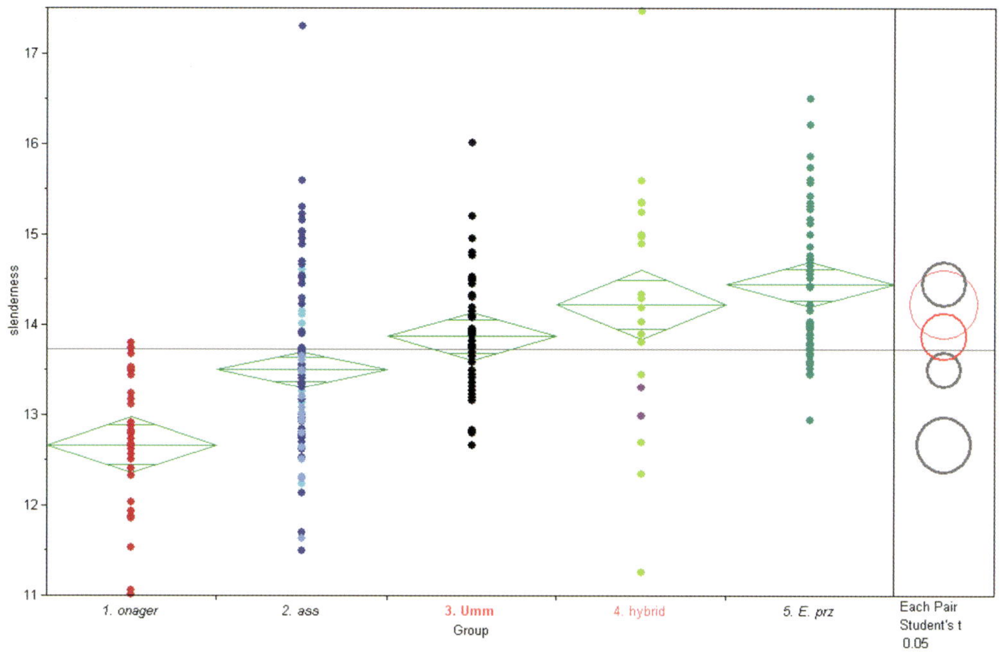

Fig. 24.4. Slenderness in the metacarpus.

horizontal lines on Figure 24.3 and demonstrate the clear separation *at the population level* of those taxa.

Populations of Przewalski's horse have a mean value for this ratio that falls midway between the asses and hemiones; a mean of 69.5% (65.2–73.7%) was derived from measurements of 40 individuals published by Eisenmann (2013a,b), while Willoughby (1974, 436, table 32) reported a mean of 69.6%[2] for 13 individuals. The 68.1% mean for a sample (n=15) of *E. caballus* × *E. asinus* mules (Eisenmann 2007e,f) falls midway between populations of horse and ass, suggesting that hybrids variously draw from both parents in this trait. The mean for 21 individuals from Umm el-Marra's burial complex is 69.5% with a range of 64.0–73.9%. This is a near match to the Przewalski's horse in both mean and range. Moreover, the Umm el-Marra animals (and Przewalski's horse) fall almost exactly midway between asses and half-asses in expression of this ratio, as is indicated by the black horizontal line representing the Umm el-Marra mean, and which falls between the ass and hemione blocks. Comparing group means at a 95% level of confidence (alpha = 0.05) (Table 24.3) shows that the domestic ass population is significantly different from all non-ass taxa and the onagers are significantly different from all other non-hemione taxa. However, przewalski's horse, the mules, the wild asses, and the Umm el-Marra population are *not* significantly different from each other. Overall, donkeys, onagers, Przewalski's horses, mules, and individuals from Umm el-Marra's burial population all overlap within the range 65.2–71.3% making most individual identifications based upon this proportion untenable. These ratios fall into three groups with relatively clear hemione and domestic ass values, but a third group with horses, mules and the Umm el-Marra burial population that are internally less distinctive.

For other limb-bone greatest lengths, Umm el-Marra's burial population has the shortest radius relative to humerus, but the longest tibia relative to other long bones. For example, the burial individuals have, on average, tibia GL nearly 98% of their femur GL.

Individual element proportions

Shape is another characteristic aspect of bones, and indices meant to express shape are used to distinguish between equid taxa. One of the best known of these is the 'slenderness index', which expresses the ratio of the mid-shaft breadth to the greatest length of metapodial bones. Typically, half asses have the slenderest metapodials relative to horses and asses, while horses have the thickest (Eisenmann 1986, ; Hilzheimer 1941; Stehlin & Graziosi 1935). This is demonstrated for the metacarpus and metatarsus bones in Figures 24.4 and 24.5 respectively. The slenderness is plotted by taxon, and the green 'means diamond' marks the mean (the largest breadth of the diamond) and 95% confidence interval (the upper and lower horizontal lines) for each population examined. The solid, black horizontal line marks the overall mean for all groups. The grey and red circles on the right side replicate the mean's diamond and show group similarity via 95% confidence: circles that do not overlap show means of significant difference.

For each mean tested, the red circles indicate means that are *not* significantly different. In each case, the means diamond for the Persian onager lies below the overall

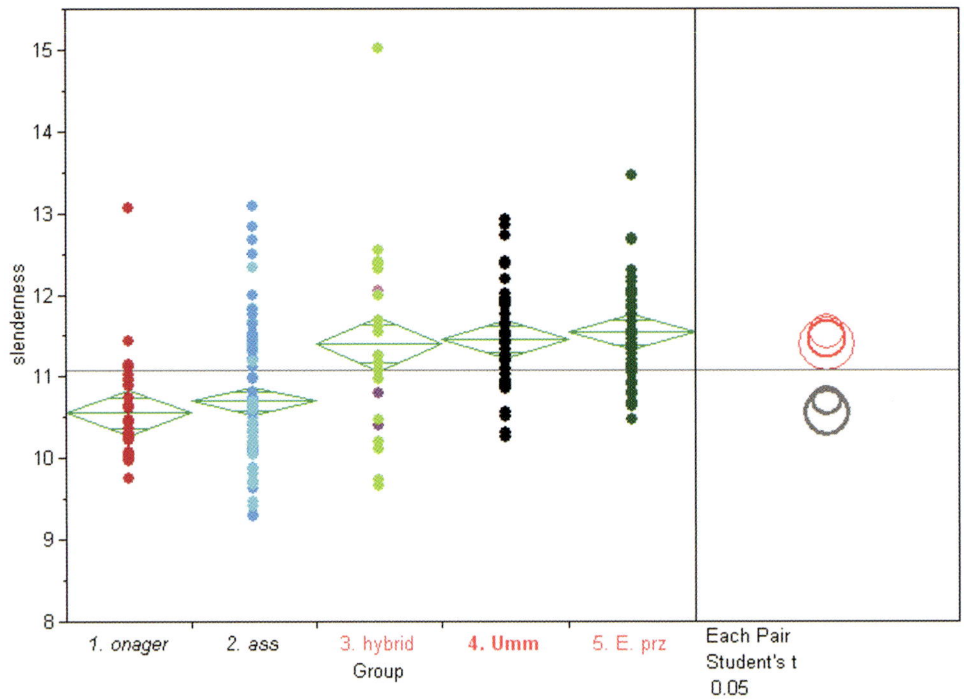

Fig. 24.5. Slenderness in the metatarsus.

average indicating its greater slenderness, while that of Przewalski's horse lies well above indicating its greater robustness. Ass individuals are quite variable (see also Davis 1981, 308, fig. 13) but as a group they are among the more slender of the populations examined. The group means for each of these taxa indicate that they are significantly different from each other at a 95% level of confidence. The hybrids[3] are similarly variable individually, though their group mean places them among the more robust; they are *not* significantly different from the horses. The results for the Umm el-Marra burial population differ by element. For the metacarpus bone, its mean lies quite close to the overall group mean, and the confidence interval overlaps it – placing the burial population in the middle of all those examined. Comparison of means shows that it is significantly different from the hemione, ass, and horse populations, but *not* from the hybrids. For the metatarsus bone, the Umm el-Marra burial population lies further from the group average and is relatively more robust. The population is significantly different from the hemiones and asses, but *not* from the horses or hybrids.

Intra-population variation

One of the major problems in examining ancient animal populations is the difficulty with measuring their intra-group variability due to their typically small numbers of complete elements or skeletons. If the extent of individual variation is unknown, classification is made more difficult and inter-group comparisons lack statistical weight. Umm el-Marra's burial population is both numerous and relatively complete, enabling assessment of its coherence through the coefficient of variation (standard deviation x mean, 'V' in Table 24.2). For the measurements of greatest length and proportion summarised above, the coefficient of variation ranges from 2.37 to 5.46. This is low – even relative to the other equine taxa. Values between 4 and 10 are considered standard for variability contained within a single population; numbers below four suggest that the entire range of variability is not present within the population being tested while numbers above ten indicate the likelihood of individuals from more than one population (Simpson *et al.* 1960, 91). The low V might indicate that the complete range of variation is not present for all of the measured traits of the burial population. The animals are all males, which may provide a source of constraint in the data, but the general lack of sexual dimorphism in size in equids (MacFadden 1992, 295) makes this less likely. Instead, it may be that equines have lower intra-group variation in limb lengths and skeletal proportion. Table 24.2 shows similarly low values for the 'wild' onagers, asses, and horses – despite the fact that many of these are zoo specimens and/or stem from different breeding populations (cf Dive & Eisenmann 1991, 284). It is only the domestic asses and the mules that show variations indicative of population non-homogeneity. This should be expected given the inclusion of donkeys bred to specific size, of mule offspring having donkey and horse mares, and of wide distribution in their geographic origins.

Regardless, the low coefficients of variation for the Umm el-Marra burial population – especially in comparison to the 'wild' taxa – indicate the appropriateness of treating it as a homogeneous population for purposes of taxonomic comparison and classification.

Discriminant analysis

The above discussion indicates that while all equids have very similar limb proportions, there are statistically-significant characteristic differences that might help discriminate between taxa. When individual elements or traits of slenderness and proportion are compared, the Umm el-Marra population does not fit neatly into any of the other taxonomic groups, but rather shares a few traits with each population. Discriminant analysis is a way to quantify unique or variable features in order to characterise different classes or taxa (cf Albrecht 1980, 681). Using the resulting classes, attributes of individuals of unknown identity can be assessed and their group affinity predicted. Analysis of the Umm el-Marra material has been criticised to date for *not* incorporating multivariate methods for comparing with, and discriminating from, known equid taxa. This is a valid critique that has resulted in the commencement of such research. I report here on its very earliest results.

To begin, I have used the simple measures of greatest length and slenderness from known groups of *E. przewalski, E. h. onager, E. asinus*, and mules. Correlation between these different variables was assessed using step-wise variable selection to create maximum distance between groups and thus the greatest predictive potential. The greatest lengths of the femur, humerus, tibia, radius, metacarpus, and the anterior proximal phalange, along with the slenderness of the metacarpus and the ratio of its GL to that of the radius proved to yield maximum distance, and the grouped variables successfully predicted class in 92 of 106 (87%) of the known cases examined (Table 24.4). However, because only seven of the Umm el-Marra skeletons have complete information for all of those variables, I eliminated the femur, humerus and tibia from the tests increasing the number of usable skeletons from the burial population to 19. Excluding these variables resulted in only a slight decrease (83%) in predictive success of the known taxa, chiefly among the mules (Table 24.5). The 19 individuals from the burial complex were not consistently predicted to be of any of the known groups. Despite their low intra-population variation, individuals were predicted into each of the groups to varying degrees (Table 24.5).

The largest proportion (42%) was predicted to be mules, followed by onagers, przewalski's horses, and domestic asses. In essence, there is no common class into which the Umm el-Marra equids are predicted to reside. By contrast, all six of the 'donkey' skeletons from Tell Brak were predicted to be *E. asinus*. If the burial equids are assessed as their *own* class using the same variables, a greater percentage of misclassification results (24%); the burial population had the lowest proportion of 'correct' predictions at 68%, but all groups experienced a small percentage decrease as some individuals were "reassigned" to the burial group (Table 24.6).

Preliminary multivariate testing suggests that discriminant analysis is promising for classificatory exploration of the Umm el-Marra material (Fig. 24.6). That population did not fit well into any single group other than its own. However, interpretations of the results have not been fully explored and further testing involving variables of dimensions other than length, and the inclusion of other bones or teeth may add to canonical distances in order to

Table 24.4. Actual and predicted classes using discriminant analysis with eight variables

Actual class	*Predicted class*				*%*
	E. asinus	E. prz	E. h. onager	*Mules*	*correct*
E. asinus	24	1	1	2	86
E. prz	0	32	0	3	91
E. h. onager	2	0	24	1	89
Mules	1	1	2	10	71

Table 24.5. Actual and predicted classes using discriminant analysis with five variables

Actual class	*Predicted class*				*%*
	E. asinus	E. prz	E. h. onager	*Mules*	*correct*
E. asinus	28	1	1	1	90
E. prz	0	31	1	3	89
E. h. onager	2	0	24	2	86
Mules	2	3	2	7	50
Umm burial	2	4	5	8	

Table 24.6. Actual and predicted classes using discriminant analysis with 5 variables, including the burial population as a known class

Actual class	*Predicted class*					*%*
	E. asinus	E. prz	E. h. onager	*Mules*	*Umm burial*	*correct*
E. asinus	27	1	1	1	1	87
E. prz	0	30	0	3	2	86
E. h. onager	2	0	20	1	5	71
Mules	1	3	2	6	2	43
Umm burial	1	1	1	3	13	68

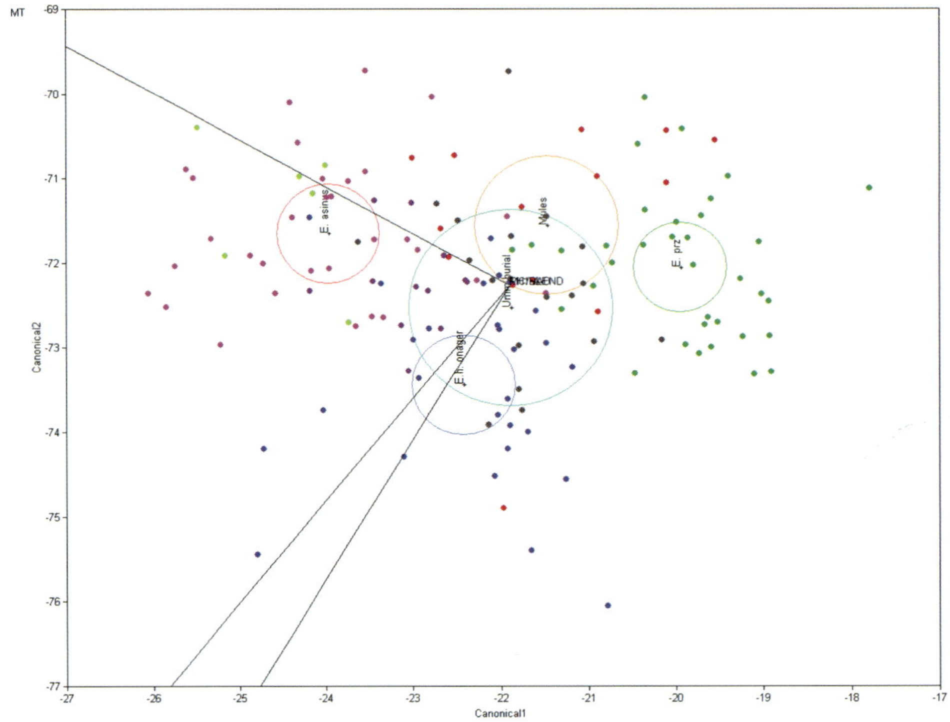

Fig. 24.6. Canocical analyis of mateatrasal.

create stronger, unique classes. In particular, there is a whole range of distinctive morphological characteristics whose quantification could prove useful in creating class attributes for discriminant analysis.

Morphology

Morphological analysis of the Umm el-Marra burial population continues; a summary of a few pertinent traits can be found in the ASWA 8 proceedings (Weber 2008).

Dentition

Teeth of all taxa are quite varied at an intra-population level. Tooth traits are not definitive, but are suggestive particularly as a means to characterise a population (cf Eisenmann 1986, 75–76; Zeder 1986). The teeth from Umm el-Marra's burial population are still being studied, but they are characterised by their diversity, lack of conformation with any known taxa, and large proportion of malocclusion. Despite great intra-population variability of most taxa, there are some generally agreed-upon traits. Caballines are characterised by maxillary teeth with well-developed *pli caballin* and long, posteriorly projecting protocones, and by mandibular teeth with u-shaped *sinus externalis* and an asymmetrical 'double knot'. Asinine maxillary teeth are characterised by weak or absent *pli caballin* and short, symmetrical protocones, while their mandibular teeth show v-shaped *sinus externalis* and symmetrical 'double knots'. There is less agreement regarding hemiones, though they typically display features somewhere between caballine and asinine traits; they have weak or developed *pli,* asymmetrical protocones of medium-length, and rounded-v *sinus externalis*.

In these descriptive traits, the Umm el-Marra burial population does not consistently conform to any single group (Figs 24.8 & 24.9). The maxillary teeth have protocones that are variably long and asymmetrical or short and symmetrical, and *pli* are absent or poorly developed in ca. 60% of the individuals, while they are very well-developed in the other 40%. Those *pli,* however, intrude into a distal orientation from the mesial edge of the peninsula making them more like *pli* found in hemione teeth than in caballine teeth, which intrude lingually from buccal (Meadow & Patel 1997, 312–314). Mandibular ectoflexids (Fig. 24.7) are mainly relatively short (reaching but not penetrating the stem of the 'double knot') – but longer than is typical for asses – and 'double knots' are chiefly asymmetrical. Eisenmann (1986, 94, fig. 19; 97, fig. 27) quantified variation between taxa in the presence of isolated hypoglyphs and open post-fossettes on upper 3rd molars, and the proportion of short ectoflexids (one that does not reach the preflexid) on mandibular molars. Her findings are presented in Table 24.7, which also provides comparison to the Umm el-Marra material for the same traits.

The Umm el-Marra burial population has un-asinine proportions of upper 3rd molar open post-fossettes and 'short' mandibular ectoflexids, but differs from hemiones in its higher frequency of the same. While they do not

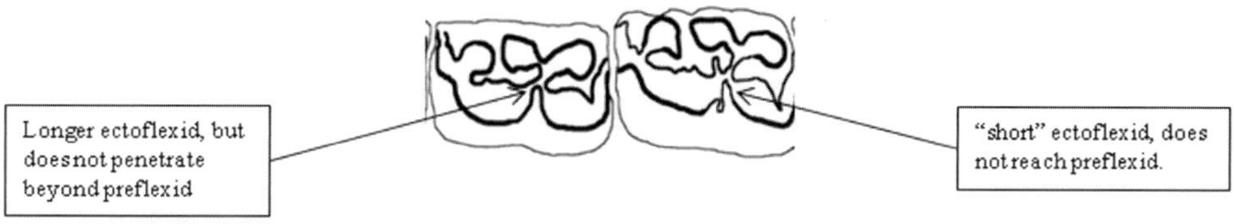

Fig. 24.7. Ectoflexid length.

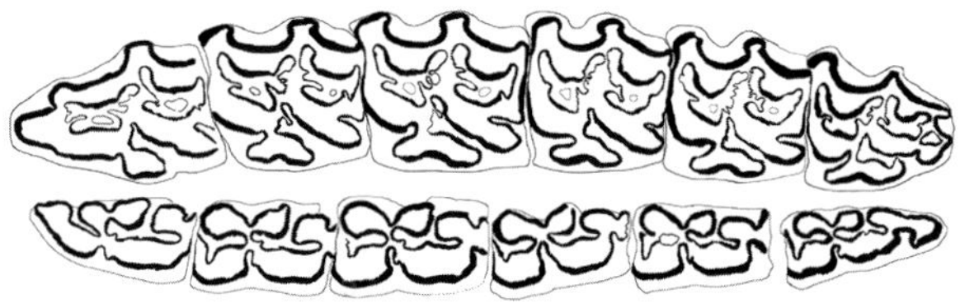

Fig. 24.8. Dentition from Individual 2, Installation G.

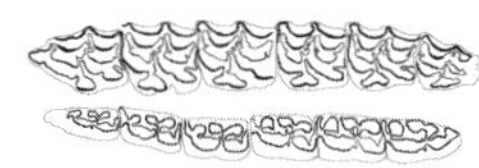

Fig. 24.9. Dentition from Individual A, Installation E.

substantially differ from horses in these quantitative measures, the teeth simply do not share most caballine traits.

One very common feature among the Umm el-Marra animals' dentititon is the presence of malocclusion – which is found in nearly 100% of the individuals. This takes several forms, including rostral hooks on either the upper or lower P2, and caudal hooks on either the upper or lower M3, These usually are opposing in the jaw – an upper rostral hook will occur with a lower caudal hook, as in Figure 24.10. But, there are also cases where both caudal and rostral hooks have developed on the upper teeth. These are primary malocclusions that are generally hereditary (Nordby, *et al.* 1945; Stoddard 1947). 'Step' and 'wave' type formations on the occlusal surface – such as are shown in Figure 24.9– are secondary malocclusions that may result from primary problems.

The degree of malocclusion did not seem to impact life-expectancy – as many animals lived quite long (e.g. the ca. 20-year-old animal in Figure 24.10) – but is notable for its frequency.

Morphological characterisation and 3-D modelling

Morphological characterisation of the Umm el-Marra burial population shows a great deal of variation between individuals, and lack of conformity to any single, known taxon (for early results, see Weber 2008). Long-term morphological examination has been made possible through the creation of virtual 3-D models of most complete elements from the 30+ individuals recovered so far. Scans have been created through capture from a NextEngine 3-D Laser Scanner; an earlier 'standard' version has now been replaced with a high-definition scanner. The virtual 3-D models aid in identification of often subtle, distinctive, morphological characteristics through the capture of bone textures and topography, which can then be digitally

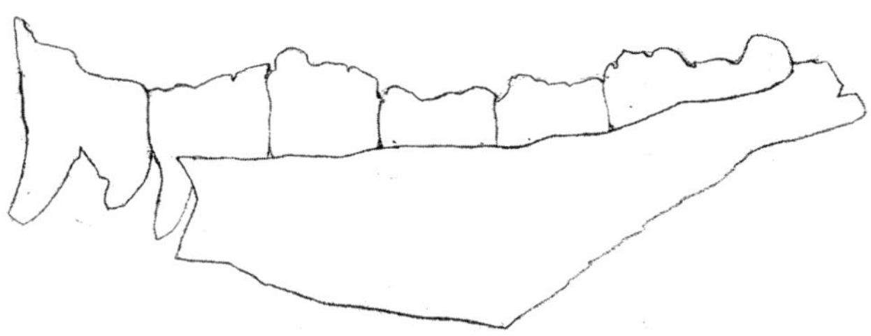

Fig. 24.10. Dentition from A1, Installation D.

enhanced and analysed. Importantly, they preserve all such data for future examination, improving the chance that new discoveries of morphological distinction and detail can be applied at any present or future time. These models can be cheaply and easily transferred between researchers, preserving and sharing this important database.

For instance, I was unaware of many of the traits useful for distinguishing between donkeys, horses and mules (see Johnstone 2004; Peters 1998) until *after* much initial work had been completed with Umm el-Marra's burial equids. But it is possible to re-examine the scanned works and assess the elements for those criteria. Figure 24.11 is a model of a radius that shows torsion of the shaft, the presence and depth of the *christa transversa*, and the presence of a 'line' across the distal, lateral condyle (cf Uerpmann 1991).

Some degree of subjectivity can be eliminated and standardisation between researchers improved by easily documenting morphological variability without the aid of a skilled illustrator or photographer. For example, the three astragali in Figure 24.12 document varying levels of protrusion by the medial tubercle (after Uerpmann 1986). Figure 24.13 shows the shape and position of the muscle scars on the volar surfaces of two separate proximal, posterior phalanges (Fig. 24.14).

Fig. 24.11. Dentition from Individual C, Installation G.

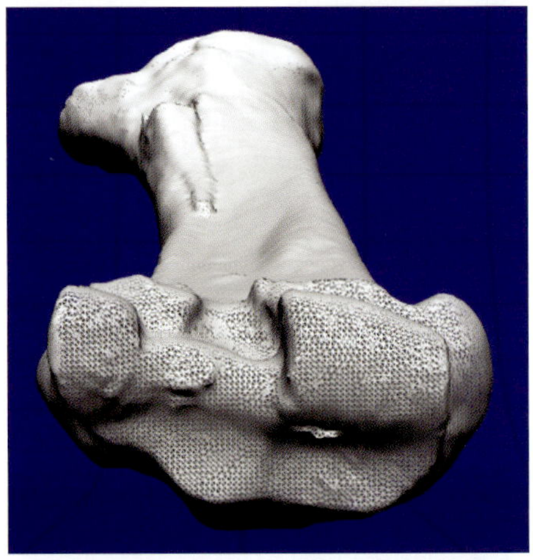

Fig. 24.12. Image of 3-D model of radius, Individual 405.

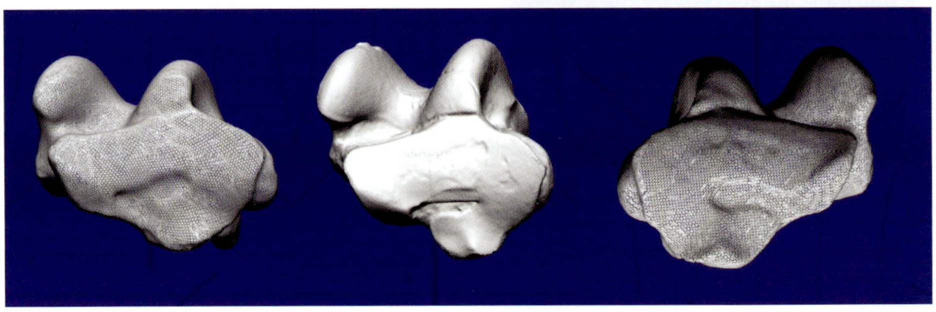

Fig. 24.13. Images of 3-D models of astragali from a. Installation A; b. Individual 405; c.

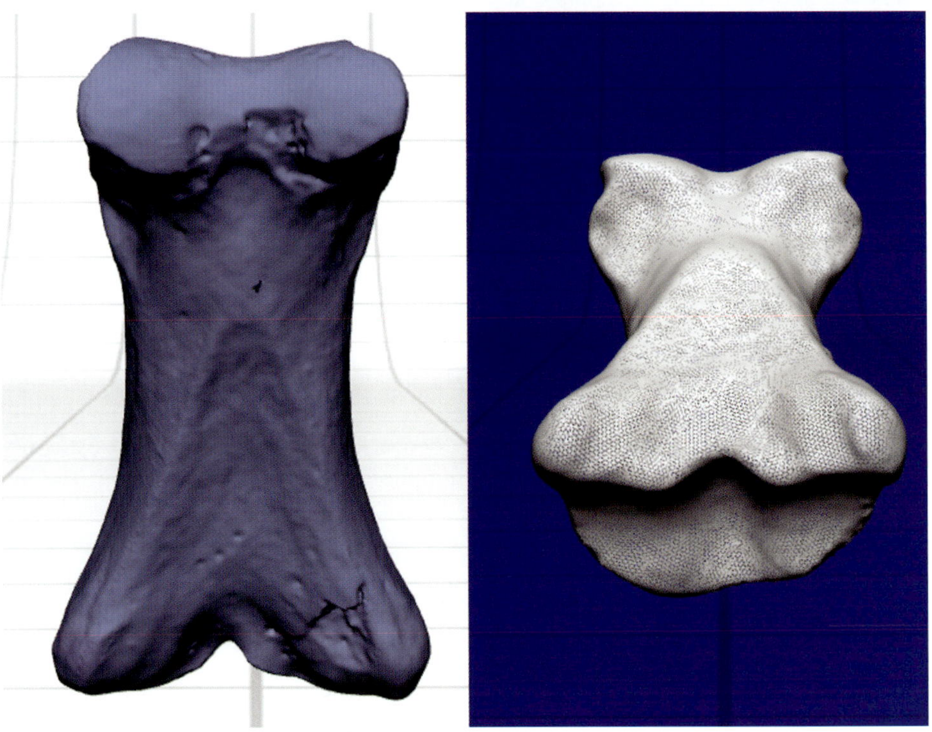

Fig. 24.14. Image of 3-D model of posterior, proximal phalanges from a. Individual 405; b. Installation A (not to same scale) Installation F (not to same scale).

Discussion

The metrical analyses shown above, while still preliminary, indicate significant differences between the Umm el-Marra burial population and known populations of ass, onager, and horse; these suggest that the burial population is not to be classified as one of those taxa. As what, then, are they to be classified?

Overall we have an animal whose height and limb-bone lengths are mainly similar to those of onager, whose robusticity in the metapodiae, and relative metacarpus to radius length is reminiscent of przewalski's horse, and whose dental characteristics bear no strict relationship to any single taxon. Initial discriminant analysis also indicates that the burial population is more internally consistent than it is similar to other, known taxa. Furthermore, analysis of variation indicates that the burial individuals constitute a single, highly homogeneous population.

We know that part of the reason for that homogeneity is the extreme conformity of their demographic: they are all male, were used for draft from a young age, were controlled/confined, shared the same environment in life and death, and were bred within a few hundred years of each other. The draft status of the animals also suggests that the animals were under human control, and thus were domestic – or at least tamed.

A domestic status for the Umm el-Marra population seems to stand in contrast to the low variation (V) expressed in measures of long-bone length and robusticity. As a result of domestication, itself, 'total variation increases in almost every respect including animal size and proportions …' (Boessneck & von den Driesch 1978, 26). Artificially maintained, captive, domestic populations typically have greater genetic variability due to the removal of natural barriers and more 'relaxed' selection (Price 2002, 74). But, traits that have explicit selection have lower total variation, while those with low selection have much higher variation (Simpson 1944, 83). The majority of the metric traits examined are phenotypic expressions of height and proportion that could be targeted in selection and which are largely independent of activity. On the other hand, there is much greater (apparent) variation in the expression of non-metric morphological traits of the individual bones and dentition. Such traits may not have experienced explicit selection – whether because they had no outwardly-visible phenotypic expression or were not considered important, resulting in much greater intra-population variation. Dental malocclusion, however, is a visible phenotypic trait that is hardly considered beneficial in animal breeding, but which is very frequent in the Umm el-Marra burial population.

The totality of these features – a single population that does not conform to populations of ass, hemione, or horse, with very low variation in size, high variation in non-metric traits, and a large degree of dental malocclusion –supports their identification as hybrids, and as ancient *kunga*. In the wild, 'natural' hybrids are characterised by increased trait variation that is typically expressed as 'non-metric anomalies' (Ackerman *et al.* 2006, 642). Hybrid trait variability often exceeds that of its parent populations, does not always reflect equal influence by both, and can result in innovative traits not found in either parent (Ackerman *et al.* 2006, 642). Darwin (1900, 16–17) had already observed such trends in equid hybrids and, more recently, Eisenmann and Bekouche (1986, 124, 126–127) noted variability in the degree to which hybrid offspring resembled one or both parents in their morphology. Significantly, dental abnormalities and anomalies are quite common features of interspecific hybrids (Ackerman *et al.* 2006, 641; Goodwin 1998; Mengel 1971).

The cultural context of the skeletons – buried in the midst of an elite cemetery – already suggests that these were elite animals. Their high-status and their specific demographic as large-sised, all male, draft animals specifically links them to the ancient *kunga* (Weber 2008).

Sise was an important feature of the ancient *kunga*, which were grouped into 'female', 'male' and 'superior male' categories (Zarins 1976, 450–461). It was the latter group of large-sised males that were specifically selected for use by royalty in pulling their wheeled vehicles.

The extreme conformity in height (greatest lengths) displayed by the burial population suggests that this was a phenotype that was specifically selected, adding further support to its identification as the *kunga*.

The most parsimonious explanation for the data currently available is that the Umm el-Marra burial population is composed of hybrid animals. Their cultural context and demography equate these animals to the ancient *kunga*, often considered to be a hybrid of ass and onager parentage (Archi 1998, 9, n. 48; Heimpel 1994; 1995, 89–91; Postgate 1986; Zarins 1986, *contra* Maekawa 1979; 2006). I suggest that classification is correct, and that the burial population considered here is, in fact, a population of ass and onager hybrids.

Notes

1 Installation F is the entity formerly referred to as the 'Stone Platform' (Weber 2008).
2 A range was not published by Willoughby.
3 The 'hybrids' group includes two, non-traditional hybrids: a hemione × horse cross and a hemione × ass cross, measurements from Eisenmann (2007g,h).

References

Ackermann, R. R., Rogers, J. & Cheverud, J. M. 2006, 'Identifying the morphological signatures of hybridisation in primate and human evolution'. *Journal of Human Evolution* 51, 632–645.

Albrecht, G. H. 1980,'Multivariate analysis and the study of form, with special reference to Canonical Variate Analysis'. *American Zoologist* 20, 679–693.

Boessneck, J. 1976, *Tell el-Dab'a III: Die Tierknochenfunde, 1966–6.* Österreichische Akademie der Wissenschaften, Wien.

Boessneck, J. & von den Driesch, A. 1978, 'The significance of measuring animal bones from archaeological sites', in

R. H. Meadow & M. A. Zeder (eds), *Approaches to Faunal Analysis in the Middle East*, 25–39. Harvard University, Cambridge, MA.

Clutton-Brock, J. 1989, 'A dog and a donkey excavated at Tell Brak'. *Iraq* 51, 217–224.

Clutton-Brock, J. & Davies, S. 1993, 'More donkeys from Tell Brak'. *Iraq* 55, 209–221.

Darwin, C. 1900, *The Variation of Animals and Plants Under Domestication* 2. Appleton, New York.

Davis, S. 1981, 'Late Pleistocene and Holocene Equid Remains from Israel'. *Zoological Journal of the Linnean Society* 70, 289–312.

Dive, J. & Eisenmann, V. 1991, 'Identification and discrimination of first phalanges from pleistocene and modern equus, wild and domestic', in R. H. Meadow, R. H. Uerpmann & H.–P. Uerpmann (eds), *Equids in the Ancient World* 2, 278–333. Ludwig Reichert Verlag, Wiesbaden.

Eisenmann, V. 1986, 'Comparative osteology of modern and fossil horses, half-asses, and asses', in R. H. Meadow, R. H. & H.–P. Uerpmann (eds), *Equids in the Ancient World* 1, 67–116. Ludwig Reichert Verlag, Wiesbaden.

Eisenmann, V. 1995, 'L'origine des Anes: questions et reponses paleontologiques'. *Ethnozootechnie*. 56, 5–26.

Eisenmann, V. www.vera_eisenmann.com.

Eisenmann, V. & Bekouche, S. 1986, 'Identification and discrimination of *Equus* metapodials', in R. H. Meadow, R. H. Uepermann & H.–P. Uerpmann (eds), *Equids in the Ancient World* 1, 117–163. Ludwig Reichert Verlag, Wiesbaden.

Eisenmann, V. & Mashkour, M. 2000, *Data Base for Teeth and Limb Bones of Modern Hemiones*, CNRS, Antibes.

Goodwin, H. T. 1998, 'Supernumerary teeth in Pleistocene, recent, and hybrid individuals of the Spermophilus richardsonii complex (Sciuridae)'. *Journal of Mammalogy* 79, 1161–1169.

Groves, C. P. & Willoughby, D. P. 1981, 'Studies on the taxonomy and phylogeny of the genus Equus, 1: subgeneric classification of the recent species'. *Mammalia* 45, 321–354.

Heimpel, W. 1994, 'Towards an understanding of the siKKum Institution'. *Revue d'Assyriologie* 88, 5–31.

Hilzheimer, M. 1941, *Animal Remains from Tell Asmar*, University of Chicago Press, Chicago.

Johnstone, C. J. 2004, 'A biometric study of equids in the Roman World'. PhD thesis, Department of Archaeology, University of York.

Jolly, C. J., Woolley-Barker, T., Beyene, S., Disotell, T. R. & Phillips-Conroy, J. E. 1997, 'Intergeneric hybrid baboons'. *International Journal of Primatology* 18, 597–627.

Macfadden, B. J. 1992, *Fossil Horses: systematics, paleobiology, and evolution of the family Equidae*. Cambridge University Press, Cambridge.

Maekawa, K. 1979, 'The ass and the onager in the late third millennium B.C'. *Acta Sumerological*, 35–62.

Meadow, R. & Patel, A. 1997, 'A comment on "Horse remains from Surkotada" by Sándor Bökönyi'. *Journal of South Asian Studies* 13, 308–315.

Mengel, R. M. 1971, 'A study of dog-coyote hybrids and implications concerning hybridisation in *Canis*'. *Journal of Mammalogy* 52, 316–336.

Nordby, J. E., Terrill, C., Hazel, L. N. & Stoehr, J. K. 1945, 'The etiology and inheritance of inequalities in the jaws of sheep'. *Anatomical Record* 92, 235–254.

Peters, J. 1998, *Römische Tierhaltung und Tierzucht: eine Synthese aus archäozoologischer Untersuchung und schriftlich-bildlicher Überlieferung*.Verlag Marie Leidorf, Rahden/Westfalen.

Price, E. O. 2002, *Animal Domestication and Behavior*. CABI Publishers, New York.

Schwartz, G., Curvers, H., Dunham, S. & Stuart, B. 2003, 'A third-millennium B.C. elite tomb and other new evidence from Tell Umm el-Marra, Syria'. *American Journal of Archaeology* 107, 325–361.

Schwartz, G. M., Curvers, H. H., Dunham, S. S., Stuart, B. & Weber, J. A. 2006, 'A third-millennium B.C. Elite mortuary complex at Umm el-Marra, Syria: 2002 and 2004 excavations'. *American Journal of Archaeology* 110, 603–641.

Simpson, G. G. 1941, 'Large Pleistocene felines of North America'. *American Museum Novitates* 1136, 1–27.

Simpson, G. G. 1944, *Tempo and Mode in Evolution*. Columbia University Press, New York.

Simpson, G. G., Roe, A. & Lewontin, R. C. 1960, *Quantitative Zoology*. Harcourt, Brace & Company, New York.

Stehlin, H. G. & Graziosi, P. 1935, *Ricerche Sugli Asinidi Fossili d'Europa*. E. Birkhaeuser,, Bale.

Stoddard, S. E. 1947, 'Inheritance of malocclusion'. *Journal of Heredity* 38, 117–120.

von den Driesch, A. & Amerberger, G. 1981, 'Ein Altbabylonisches Eselskelett vom Tell Ababra – Iraq'. *Bonner Zoologische Beiträge* 32, 67–74.

Weber, J. 2008, 'Elite equids: redefining equid burials of the mid- to late-3rd millennium BC from Umm el-Marra, Syria', in E. Vila, L. Gourichon, A. Choyke, & H. Buitenhuis (eds), *Archaeozoology of the Southwest Asia and Adjacent Areas VIII, Actes du 8e colloque international de l'A.S.W.A. tenu à Lyon du 28 juin au 1er juillet 200*, 00–00. Travaux de la Maison de l'Orient, Lyon.

Willoughby, D. P. 1974, *The Empire of Equus.*, A. S. Barnes and Co, Cranbury NY.

Zarins, J. 1976, *The Domestication of Equidae in Third Millennium B.C. Mesopotamia*. University of Chicago Press, Chicago.

25. An unusual cattle burial at Dayr al-Barshā (Late Period, Middle Egypt)

Veerle Linseele, Wim Van Neer, Harco Willems & Bart Vanthuyne

During excavations in the desert plain near Dayr al-Barshā in Middle Egypt, a circular pit filled with cattle bones was found. The pit, with a diameter of about 1.5m, contained the almost complete skeletons of 15 Egyptian Longhorns, dated to 410–360 cal BC. The material was in a remarkably good state of preservation, but few elements were found in articulation. Apart from the more-or-less complete skeletons, some fragmentary remains of at least three additional individuals have been collected. The age at death, sex and size of the animals has been established and the pathologies and traces on the bones have been recorded. No parallels are known for the cattle burial at Dayr al-Barshā. The animals were probably kept in the context of a local temple economy. In line with common practices at that time, their bodies were macerated and the obtained skeletons ceremonially buried later on.

Keywords cattle, burial, ritual, maceration, Herodotus

Introduction

Dayr al-Barshā is an archaeological site in the Nile Valley of Middle Egypt at about 300km south of Cairo (Fig. 25.1). Its main features are cemeteries and limestone quarries. Since 2002, the Katholieke Universiteit Leuven (KU Leuven) has been conducting archaeological fieldwork at the site, concentrating on its cemeteries (Willems *et al.* 2004). One of the principal research aims is to understand the spatial organisation of the cemeteries, which mainly date to the Old Kingdom, the First Intermediate Period, and the Middle Kingdom. In this way, it is hoped to obtain a clearer understanding of the social stratification of the community buried there, at least insofar as this reflects itself in burial practices. In 2006, excavations were started in the cemeteries in zone 9B at Dayr al-Barshā, which is part of the desert plain. That year, cattle bones were noted in one of the trench walls (Sector 120). When the trench was enlarged to investigate the nature of these cattle bones, it appeared that they were part of a large concentration. As the excavation season was coming to an end, and since no faunal specialist was present at the site at that time, it was decided to postpone further excavation of this feature to the following season. In 2007, the first and last authors of this contribution carried out the excavation of this locus.

The cattle bone concentration turned out to be deposited in a circular pit with a diameter of about 1.5m. At the time this pit was dug, the top of the shaft of an older grave was cut through. A few potsherds and some human remains were found mixed with the cattle bones. The cattle pit was divided in four quadrants and was emptied by removing the bones by artificial layers and by quadrant. Before a new layer was taken out, a picture was taken, on which the bones that were subsequently removed were indicated. Isolated long-bones or articulating long-bones were bagged separately while all miscellaneous bones, ribs, vertebrae, phalanges etc., were placed together in one find bag. The skulls were numbered, and all bones clearly associated with them were given the same number. Apart from standard sieving of all sediment on 4mm meshes, several soil samples were taken for archaeobotanical analyses during the excavations. Besides cattle bones, the pit contained very few other finds. There were some

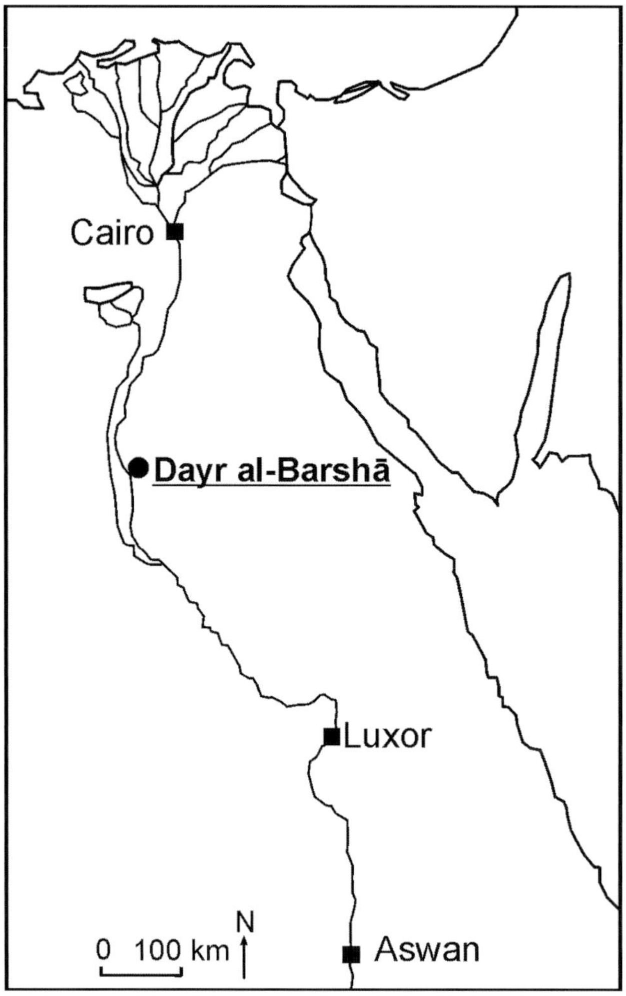

Fig. 25.1. Map of Egypt with location of Dayr al-Barshā.

pieces of textile, and some textile imprints, which have not been analysed yet. A few potsherds were also associated with the pit. These were tentatively dated to the end of the New Kingdom, and the beginning of the Third Intermediate period, which corresponds to around 1000 BC. However, the fragments were small and much eroded, and ceramics of this period are hard to discern from later Third Intermediate Period ceramics (Aston 1996). Moreover the sherds may already have been lying about when the cattle pit was dug. Therefore, 1000 BC is merely to be considered a *terminus ante quem non*. In an attempt to date the deposit more securely, a cattle bone was submitted for radiocarbon dating. The obtained date of 2140±40 BP or 410–360 cal BC (two sigma calibrated) is significantly younger than expected. No archaeological traces of similar age have been found up to now in the cemeteries at Dayr al-Barshā, but it was a period of large scale exploitation of the stone quarries, under the reign of pharaohs Nectanebo I and II (Willems *et al.* 2004; 2006; 2009; Depauw 2008).

Fig. 25.2. The concentration of cattle bones at Dayr al-Barshā halfway through its excavation.

Description of the bone remains

Minimum number of individuals and state of preservation

The faunal analyses have shown that the pit at Dayr al-Barshā contained the remains of 15 more-or-less complete cattle individuals. Although some articulating joints occur, most of the material was found disarticulated. Based on their horn shape and size (see below), the animals were identified as Egyptian Longhorns, the most common cattle type of ancient Egypt (Boessneck 1988, 69). The material is in an excellent state of preservation, often with hooves, horns, hair and sometimes even skin and tendons still adhering to the bones. Because of this exceptional state of preservation, the hair colour of the animals could also be observed. On the skulls dark-brown hair, sometimes with paler spots indicating piebald animals, as well as reddish hair is present. These colours correspond with the ones that can be seen in the iconography, for example in the three wooden models from the Middle Kingdom tomb of Meketre at Thebes that represent the management and slaughter of cattle (Gilbert 1988). The skulls of the animals, sometimes in association with the corresponding mandibles, hyoid bones or first cervical vertebrae, seemed to be approximately arranged along the western, southern and eastern edge of the pit (Fig. 25.2). With the exception of skull 2, all skulls were roughly facing east. The other bones were placed in the pit without any visible logic. The skeletons are apparently not entirely complete, since for most skeletal elements a few specimens are missing (Table 25.1). In addition to the 15 well-preserved animals, there are fragmentary postcranial remains of at least three other cattle individuals. These remains differ from the other material by their black colour and more fragile nature. The blackish bones occurred in concentrations in the pit, and were especially numerous at the bottom. This is also where other faunal remains and the aforementioned pot sherds were found. The faunal elements have the same

Table 25.1. Fusion data for the long-bones of the complete cattle found at Dayr al-Barshā, with in the right column the elements that are missing to arrive at 15 complete skeletons

Element	*Age of fusion*	*left*			*right*			*left or right*			*Missing bones*
	(Silver 1963)	*NF*	*fus*	*F*	*NF*	*fus*	*F*	*NF*	*fus*	*F*	
Distal humerus	12–18 months	2	1	12	2	2	11				
Proximal radius	12–18 months	1	–	14	1	–	13				1 right radius
Phalanx 1	18 months							8	–	95	37 phalanges 1
Phalanx 2	18 months							5	3	99	33 phalanges 2
Distal metacarpal	24–30 months	3	–	12	2	–	12				1 right metacarpal
Distal tibia	24–30 months	2+1?	2	10	2	2	11				1 left tibia
Distal metatarsal	37–36 months	3	–	12	3	–	12				
Calcaneus	36–42 months	4	4	5	4	4	7				2 left calcanei
Proximal femur	42 months	4	9	1	4	9	1				1 left, 1 right femur
Proximal humerus	42–48 months	5	9	1	5	9	1				
Distal radius	42–48 months	2	8	5	3	5	6				1 right radius
Proximal ulna	42–48 months	11	–	4	9	1	4				1 right ulna
Distal femur	42–48 months	4	8	2	4	7	3				1 left, 1 right femur
Proximal tibia	42–48 months	7	6	1	7	6	2				1 left tibia

NF=Not Fused, fus=fusing, F=Fused

appearance as the additional cattle bones and include a fragment of a large freshwater bivalve, an equid distal metapodial (attributed to donkey because of its small size), a dog or jackal first phalanx, an ovicaprine lower third molar and distal humerus were also recorded at the bottom of the pit. The colour of the blackish remains is due to the sediment adhering to them. Analysis of its composition (30% clay, 45% silt and 25% sand) has shown that it is loam-clay loam (Bert Dusar pers. comm.). This type of sediment is characteristic for the alluvial plain and is not present naturally in zone 9B near Dayr al-Barshā. The plant remains found inside this sediment are in agreement with such an origin, since they are all from sedges (Cyperaceae), associated with wet places like marshes, swamps, muddy river banks or wet sand dunes (Elena Marinova pers. comm.).

Age at death

Several criteria were used to estimate the age at death of the animals. First of all the eruption of the mandibular teeth was recorded. One animal has the 1st molar erupting, another one the 2nd molar, still another one the 3rd molar, whereas all the others have all their permanent teeth in place. Great variation in tooth eruption dates have been observed depending on breed, management and nutrition (Silver 1963). The better the housing and feeding of the animals, the earlier the teeth erupt. Since ancient Egyptian cattle were usually well taken care off, it is supposed that the ages for early maturing cattle may apply better to the Dayr al-Barshā specimens. For such animals an age of 0.5 years is accepted for the 1st molar to erupt, of 1.5 years for the 2nd molar, and of 2–2.5 years for the 3rd molar (Silver 1963). Once all teeth have erupted, tooth wear is used to determine (relative) age. A summary of the Grant (1982) wear stages of the mandibular teeth shows that all cheek teeth rows have moderate to rather heavy wear (Table 25.2).

Like the tooth eruption data, the fusion stages for the long bones point to the presence of three animals under the age of 2 or 2.5 years (Silver 1963) (Table 25.1). It appears that many of the sutures that close after 2 or 2.5 years of age are in the process of fusing. A striking observation is that fusion stages are different in several cases between the left and right side of the same skeletal element. It seems that only two or three individuals have epiphyseal sutures fused that usually close around the age of 4 years. Also the skull suture data point to only two individuals over the age of 3 years or older. Skull 9 is of an animal of at most 7 years of age, skull 13 is of an individual of maximum 10 years of age (Table 25.3). According to Habermehl (1975, 103) the ossification of the pubic symphysis takes place at an age of 3–5 years, although he adds that in some individuals ossification has not happened yet at the age of 7 years. None of the cattle from Dayr al-Barshā has its two pelvic halves fused. Fusion data of vertebrae have not been recorded, because of the inconsistencies that are frequently noted in these elements, often with different fusion states between the cranial and caudal side of the same vertebra. Another ageing method for cattle uses the number of growth rings visible on the horn sheath (Habermehl 1975, 85–86). Such rings are thought to be formed as a consequence of a seasonal change or a slowing down in the growth rate of domestic cattle. The actual age of the animals would then be the number of rings, plus two in late mature breeds, and plus one, in early mature breeds (Habermehl 1975, 89). The method has been applied by Chaix (2001) to determine the

Table 25.2. Tooth wear stages and eruption stage of the cement-enamel-junction (CEJ) for all mandibles from Dayr al-Barshā. Only the left mandible was recorded when both sides were found associated.

Number	*Side*	*Remark*	*Wear stages*[1]				
			dP4	*P4*	*M1*	*M2*	*M3*
AE4-AG8-AG18	L and R	with skull 15	e		E		
R5	L	with skull 1	h		c	V	
I2	R	with skull 1 and mand R5	h		c	E	
T7	L	with skull 14 and mand X3	k		l	j	1/2
X3	R	with skull 14 and mand T7	k		k	g	1/2
L3	L	with R4?		c	k	j	g
AG3	L and R			d	l	k	g
R4	R	with L3?		d	k	j	g
Skull 9	L and R			g	l	k	k
AC10	L			g	j	h	j
AG5	L			g	k	k	j
H1	L	with M3?		g	l	k	j
M12	L			g	o	l	k
AG17	L and R			g	l	k	k
M3	R	with H1?		g	k	k	h
P10	R			g	k	k	g
Skull 2	L and R			h	m	k	k
AC6	L	with Q5?		h	abn	m	k
Skull 13	R			h	p	k	l
K5	L and R			j	abn	e	k
AC9	L and R			j	abn	l	k
Skull 12	R			j	n	l	j
Q5	R	with AC6?			abn	l	j

[1]Grant (1982); abn=abnormal

Table 25.3. Fusion data for the skulls of the complete cattle found at Dayr al-Barshā.

Skull suture	*Age of fusion (Grigson 1982)*	*Skull number*												
		2	*3*	*4*	*5*	*6*	*7*	*8*	*9*	*10*	*11*	*12*	*13*	*14*
Basisphenoid-basiocciptal	2–3 years	NF	NF	F	fus	F	NF	fus	F	F	NF	NF	F	NF
Temporale-parietale	5–7 years	NF	NF	NF	NF	NF	NF	NF	fus	NF	NF	NF	F	NF
Parietale-frontale	5–7 years	NF	NF	NF	NF	NF	NF	NF	fus	NF	NF	NF	F	NF
Lacrimale-frontale (in orbit)	7–10 years	NF	NF	NF	NF	NF	NF	NF	NF	NF	NF	NF	F	NF
Frontal halves	7–9 years	NF	NF	NF	NF	NF	NF	NF	NF	NF	NF	NF	–	NF
Zygomatic processes of squamosal and jugal	10–15 years	NF	NF	NF	NF	NF	NF	NF	NF	NF	NF	NF	NF	NF
Maxilla-jugal	10–15 years	NF	NF	NF	NF	NF	NF	NF	NF	NF	NF	NF	NF	NF
Maxilla halves	15–40 years	NF	NF	NF	NF	NF	NF	NF	NF	NF	NF	NF	NF	NF

NF=Not Fused, fus=fusing, F=Fused

age of bucrania from Kerma in Soudan (Middle Kerma, ca. 2050–1750 BC). Growth rings are very difficult to distinguish on the horn sheaths of the cattle from Dayr al-Barshā. Most of them have about three or four such rings, while in skull 10 and skull 13, there may be up to ten. Only for skull 13, suture data had already indicated a relatively old age.

In summary, three out of the 15 complete cattle were younger than 2.5 years, namely 0.5, 1.5 and 2–2.5 years respectively. For the cattle older than that, the different types of data are not always consistent. Depending on the method used, a total is obtained of two or up to four individuals that are 'old', i.e. of an age over about 4 years. It is clear, however, that the majority of the individuals are prime adults. For the three additional, incomplete cattle only long-bone, fusion data can be used for age determinations. Since all epiphyses are fused the animals must have been at least 3.5–4 years.

Sex and height at the withers

For the sex identification both metrical data and morphological criteria can be used. The pelvic bones of the complete, adult animals from Dayr al-Barshā are male in shape, i.e. with a sturdy overall build and strong pubic branches. Some of the specimens seemed more slender and gracile, which would be indicative of castrated animals, but it was generally not easy to decide on the pelves whether they are from oxen or bulls. The method of Greenfield (2006), designed mainly to sex fragmentary ungulate acetabulae, was also tried on the pelvic bones of Dayr al-Barshā, but the results were inconclusive. The method relies much on the height of the medial wall of the acetabulum (H1) (see Table 25.7 below). Greenfield (2006) himself has noted a problem with this measurement. A variable morphological feature appears in the female of some taxa, including cattle, with the medial wall of the acetabulum characterized by a dip or depression in the centre. This renders it more difficult to measure the height of the acetabular wall and the measurement thus includes the dip. All of the pelvic bones from Dayr al-Barshā show this dip in the acetabular wall, even though their overall shape is male.

Other criteria for the sexing of cattle confirm the conclusions made on the basis of the shape of the pelvic bones from the complete skeletons from Dayr al-Barshā. Breadth measurements on the skulls, which allow separating skulls of bulls from those of cows of the same breed (Grigson 1982), do not indicate more than one group. They seem to fit best with the few published male and castrated Egyptian Longhorn cattle (Lortet & Gaillard 1903; Boessneck & von den Driesch 1987a) from the literature, rather than with the females (Jackson 1934). Grigson (1976) described a few non-metrical sexual differences in skulls of cattle. The intercornual ridge projects more strongly upwards in bulls than in cows and the supraorbital groove is deeper and longer, which tends to roof over earlier in bulls. The ring of nodules at the horn core base, which characterises old animals, develops earlier in bulls and so does the occipital flange. No such differences have been clearly noted between the individuals of Dayr al-Barshā. They all rather show the aforementioned characteristics for bulls, although not very pronounced (Table 25.4).

According to Sykes and Symmons (2007), two measurements at the horn core base, the Basal Circumference (BC) and the Basal Breadth (BB) are the best measurements for sexing horncores, although they do not allow differentiation between males and castrated males. The measurements from Dayr al-Barshā seem to form one homogeneous group. When they are compared with the few sexed ancient Egyptian Longhorns from the literature, they plot with bulls and oxen (Fig. 25.3). Based on iconographic evidence, the horn shape of female ancient Egyptian Longhorns has been described as lyre shaped, horns of males are shorter and thicker and crescent shaped and those of oxen are particularly long, and high, but less curved than in cows (Boessneck 1988, 69; Laudien 2000, figs. 16–18). The horns from Dayr al-Barshā do not seem to fit either of these descriptions perfectly. They are approximately crescent shaped, but they are too long and too slender to be bulls. The best match is with castrates.

Among the long-bones, the metatarsals and metacarpals are best suited for sex determinations. Male and female metapodia can easily be distinguished on distal breadth, and castrates can be separated from bulls on length

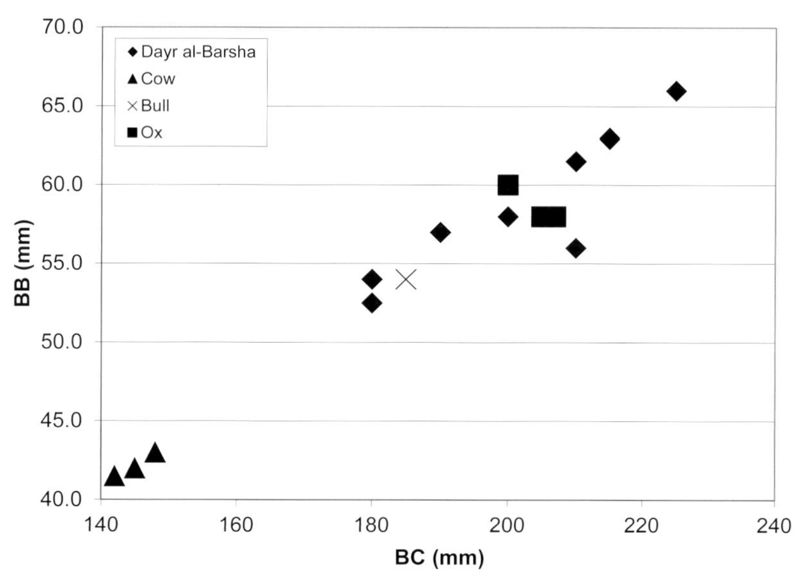

Fig. 25.3 Basal circumference (BC) and basal breadth (BB) of horn cores from cattle found at Dayr al-Barshā compared with published data. Cows and bull are from the temple at Elephantine (Boessneck & von den Driesch 1982). Oxen include the Münchner Ochsenmummie (Boessneck & von den Driesch 1987a) and two individuals from Tūna al-Jabal (Boessneck & von den Driesch 1987b).

Table 25.4. Description of cattle horns and skulls from Dayr al-Barshā

	Skull 2	*Skull 3*	*Skull 4*	*Skull 5*	*Skull 6*	*Skull 7*	*Skull 8*	*Skull 9*	*Skull 10*	*Skull 11*	*Skull 12*	*Skull 13*	*Skull 14*
Horn curve (classification Sykes & Symmons 2007)	4	3	3–4	4	3	3	2–3	3	3	3–4	3–4	3	–
Horn torsion (classification Sykes & Symmons 2007)	1	1	2	1–2	1	1	2	1–2	2	2	1	1–2	–
Left tip chopped off: yes (y), no (n)	–	n	y	n	y	n	–	n	y	y	–	y	–
Right tip chopped off: yes (y), no (n)	n	n	–	n	y	n	n	y	n	y	n	y	–
Non-measurable characters (Grigson 1976)*:													
Bone surface: slight roughening surface orbital rim (a), smooth (b)	a	a	a	a	a	b	b	a	a	a	b	a	a
'Perlen': present (y), absent or nearly absent (n)	n	n	n	n	n	n	n	n	n	n	n	n	n
Supraorbital groove (never roofed over): distinct (a), undistinct (b)	a	a	a	a	a	a	–	a	a	a	a	–	a
Occipital flange: present (y), absent (n)	y	y	y	y	y	y	y	y	y	y	y	y	y
Premaxilla touching nasal: yes (y), no (n)	y	y	y	y	y	–	y	y	y	–	n	y	y
Orbital rim: sharp (a), flat (b)	b	b	b	b	b	b	b	b	b	b	b	b	b
Suture frontale-lacrimale on face: bowed ventrally (c)	–	c	a	c	a	–	c	c	c	c	a	a	a
Suture lacrimale-jugale on face: straight (b), bowed dorsally (d), bowed ventrally (e)	–	e	e	e	e	–	d–e	e	e	e	e	e	e
Internasal suture open: yes (y), no (n)	y	y	y	y	–	–	y	y	y	y	y	y	y
Posterior end palate: straight & broad (a), convex & narrow (b), straight & narrow (c)	c	c	c	a	c	c	c	c	b	b	a	b	c
Frontale profile from above: convex (a), flat (b), slight boss (c)	c	b	b	b	b	b	b	b	b	a	b	b	c
Intercornual ridge: low single arch (a), low double arch (b), high single arch (c)	d	b	b	c	b	d	b	b	b	a	b	b	b
Horncore longitudinal grooves: strong (a), moderate/weak (b), absent (c)	a	b	–	a	b	b	c	–	a	a	b	–	b
Horizontal ridges or rings on horns: estimate of number	min 3?	min 3	min 4?	0	–	2–4	–	–	max 10?	3?	–	8–10?	0

* Sagittal profiles are not described since this involves making casts of the skulls. Without reference specimens at hand, it appeared difficult deciding whether the shape of the external auditory meatus was taurine or indicine-like. The shape of the nasal-frontal suture as well as the presence of vascular channels on the horncore surface were also hard to describe precisely. These three characteristics used by Grigson (1976) have therefore not been included.

measurements (Grigson 1982). The greatest length (GL) and the smallest diameter of the shaft (SD), taken on metacarpals of the complete cattle skeletons from Dayr al-Barshā, fall in the range of oxen and bulls when they are plotted against sexed Egyptian cattle remains from literature (Fig. 25.4). Measurements taken on a bone of one of the three additional animals, fall in the female range. Measurements on the metatarsals from the 15 cattle from Dayr al-Barshā fit with castrates and the upper range of females from literature, but not with the only bull (Fig. 25.5). The two metatarsals from the additional animals are smaller than those of the complete skeletons and fall, like the metacarpal, within the female range. None of the other sexing methods cited can be applied on the additional cattle because their remains are not preserved well enough.

In summary, all complete, adult cattle from Dayr al-Barshā are males. Although the pelvic bones did not clearly indicate so, the animals are considered as castrates on the basis of the horn shape and the metapodial measurements. A possible additional argument for an identification as

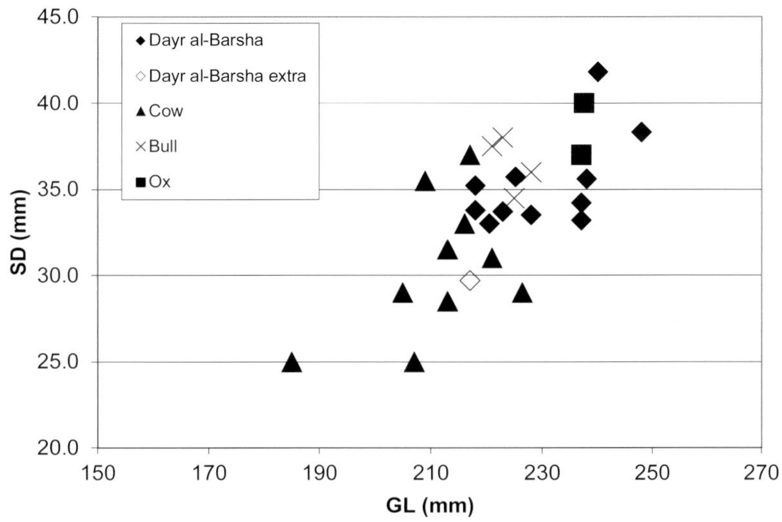

Fig. 25.4. Greatest length (GL) and smallest diameter of the shaft (SD) of right metacarpals from cattle found at Dayr al-Barshā compared with published data. Cows are from Tell el Dab'a (Boessneck & von den Driesch 1992), the temple at Elephantine (Boessneck & von den Driesch 1982), Tūna al-Jabal (Boessneck & von den Driesch 1987b) and Baqaria (Jackson 1934). Bulls are from Tell el Dab'a (Boessneck & von den Driesch 1992). Oxen include the Münchner Ochsenmummie (Boessneck & von den Driesch 1987a) and one individual from Tūna al-Jabal (Boessneck & von den Driesch 1987b).

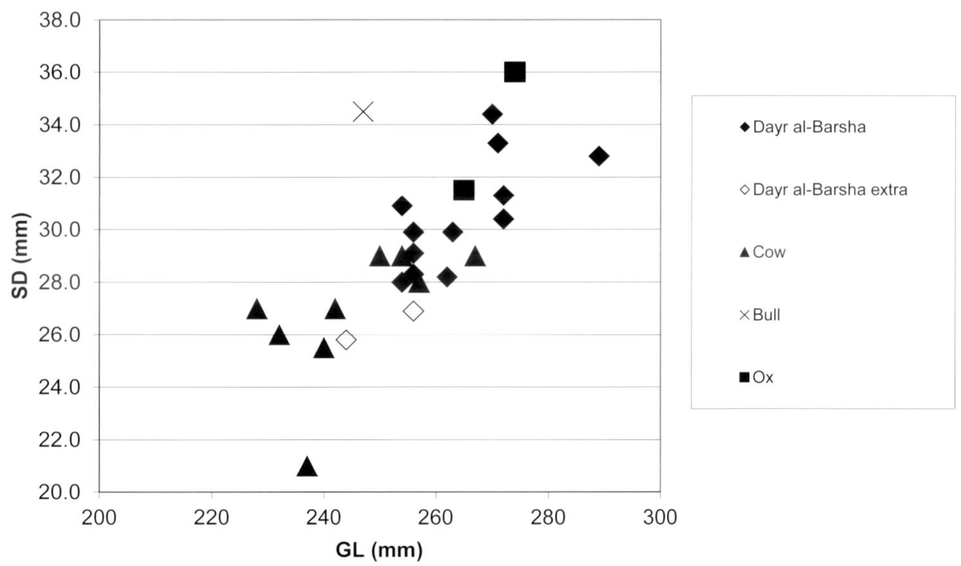

Fig. 25.5. Greatest length (GL) and smallest diameter of the shaft (SD) of right metatarsals from cattle found at Dayr al-Barshā compared with published data. Cows and bull are from Baqaria (Jackson 1934). Oxen include the Münchner Ochsenmummie (Boessneck & von den Driesch 1987a) and one individual from Tūna al-Jabal (Boessneck & von den Driesch 1987b).

oxen are the long-bone fusion data that are very variable in the sense that the fusion stages of left and right bones of the same element differ in several cases (Table 25.1). Many bones are, moreover, in the process of fusing. Both conditions could be a result of the effect of castration on the age of closure of (long-) bone sutures (cf. Davis 2000 for sheep). The metrical data of the additional bones indicate that these are from cows.

Withers height estimates were calculated from the long-bones of the complete adult cattle from Dayr al-Barshā, using the factors in von den Driesch and Boessneck (1974). The results are variable according to the bone, but they all give an average of about 140cm based on the factors for castrates (Fig. 25.6). This is rather tall, but cattle of ancient Egypt apparently grew up to 150cm at the withers and taller (Boessneck 1988, 69). For the additional animals the factors for females indicate shoulder heights of 117–130cm (Fig. 25.6).

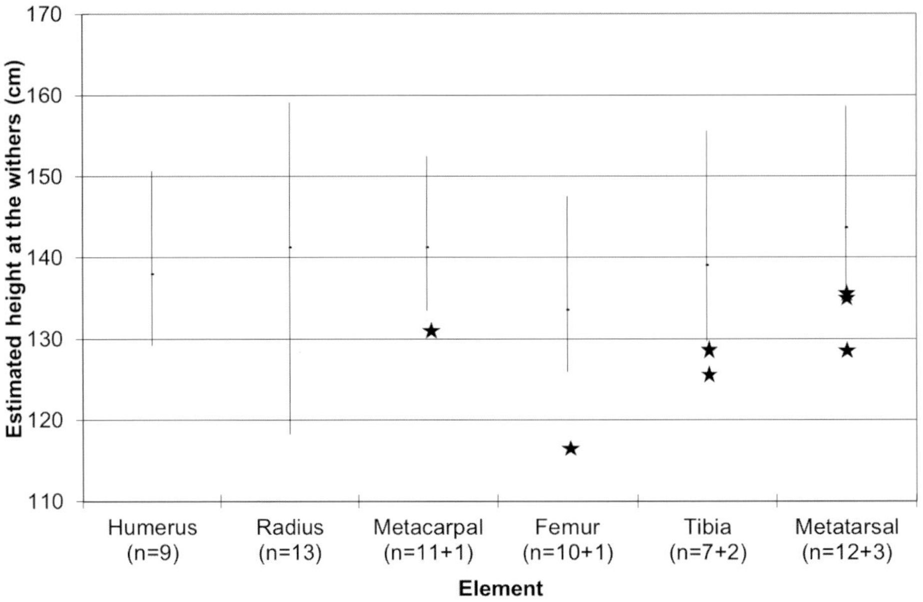

Fig. 25.6. Withers height reconstructions of the cattle found at Dayr al-Barshā. Calculations are made by multiplying the lengths of different long bones with the factors for castrates (complete animals) and females (additional animals) in von den Driesch and Boessneck (1974). Apart from a right femur of the additional animals, only left bones have been used (source measurements see Tables 25.7 & 25.8). The lines represent the range and average of the values by element for the complete animals, the stars indicate the values for the additional animals.

Pathologies and other traces

Several types of pathological deformations have been observed on the skeletal remains from the cattle pit at Dayr al-Barshā. Two skulls show evidence of a healing puncture of the frontal bone. Skull 9 is the most obvious example. It has a hole of almost 5cm wide in the right frontal bone, where part of the skull has been pushed down (Fig. 25.7, c–d). New bone tissue was already forming at the time of death of the animal, and the wound itself was therefore probably not the cause of death. In skull 5 a puncture of about 3 × 1.5cm occurs midway the left frontal bone. The nasal bone of this skull also shows abnormalities. Its cranial half is at least 12mm below the dorsal part of the premaxilla, while in normal circumstances it is level with it (Fig. 25.7, b). The right maxilla of skull 4 has an abnormally porous bone structure, and moreover shows light periodontitis (Fig. 25.7, a). Periodontitis can also be seen in skulls 7 and 10. Three mandible pairs show abnormal wear at the height of both the left and right first molar, and in one case (N9) also on adjacent parts of the P4 and M2, causing malocclusion (Fig. 25.8). On most of the tooth rows calculus has been formed, sometimes quite heavily, making observations on the eruption of the cementum-enamel-junction impossible (see above).

One right pelvic bone (U8) has an abnormal torsion at the height of the wing of its ilium, and in one sacrum (A11) the last two vertebrae are out of axis with the first three. A right metatarsal (V6) shows periostitis, midshaft on the medial side. One left (AD1) and one right (AF8) naviculocuboid from the cattle pit at Dayr al-Barshā have the 2nd and 3rd tarsal fused to it (Fig. 25.9). These can be considered as cases of spavin (Bartosiewicz *et al.* 1997, 71). Draught exploitation is in part responsible for the development of this pathology. However, deformations of metapodia and phalanges, often correlated with traction, have rarely been noted on the cattle from Dayr al-Barshā (Table 25.5). Several ribs with healed or healing fractures were found, sometimes involving the formation of a callus (Fig. 25.10, c). Other ribs show strong periostic reactions or bony outgrowths (Fig. 25.10, a, b & d).

Some of the cattle skulls from Dayr al-Barshā clearly have the tip of either one or both horns removed (Fig. 25.11). Only the very top of the horn is taken off, and therefore no traces are visible on the horn cores. This is in contrast to a bucranium found at Kerma in Soudan (Bucranium 234 from Tomb KN24; Middle Kerma, ca. 2050–1750 BC) where the ends of the actual horn cores were sawn off (Chaix 2001). It was postulated that this was done to equip the horns with protective tips. For some of the cattle skulls from Dayr al-Barshā the possibility was considered that the horn shape had been artificially modified. Especially for skull 5 this thought was raised (Fig. 25.12). Possibly, the horns of this skull were tied together in such a way that they grew more towards each other than they would do naturally, resulting in their almost ellipsoid shape. The present-day, East African Turkana deform the horns of their cattle, and it is thought that one type of deformation could give skull 5 as a result, although the Turkana apply it to

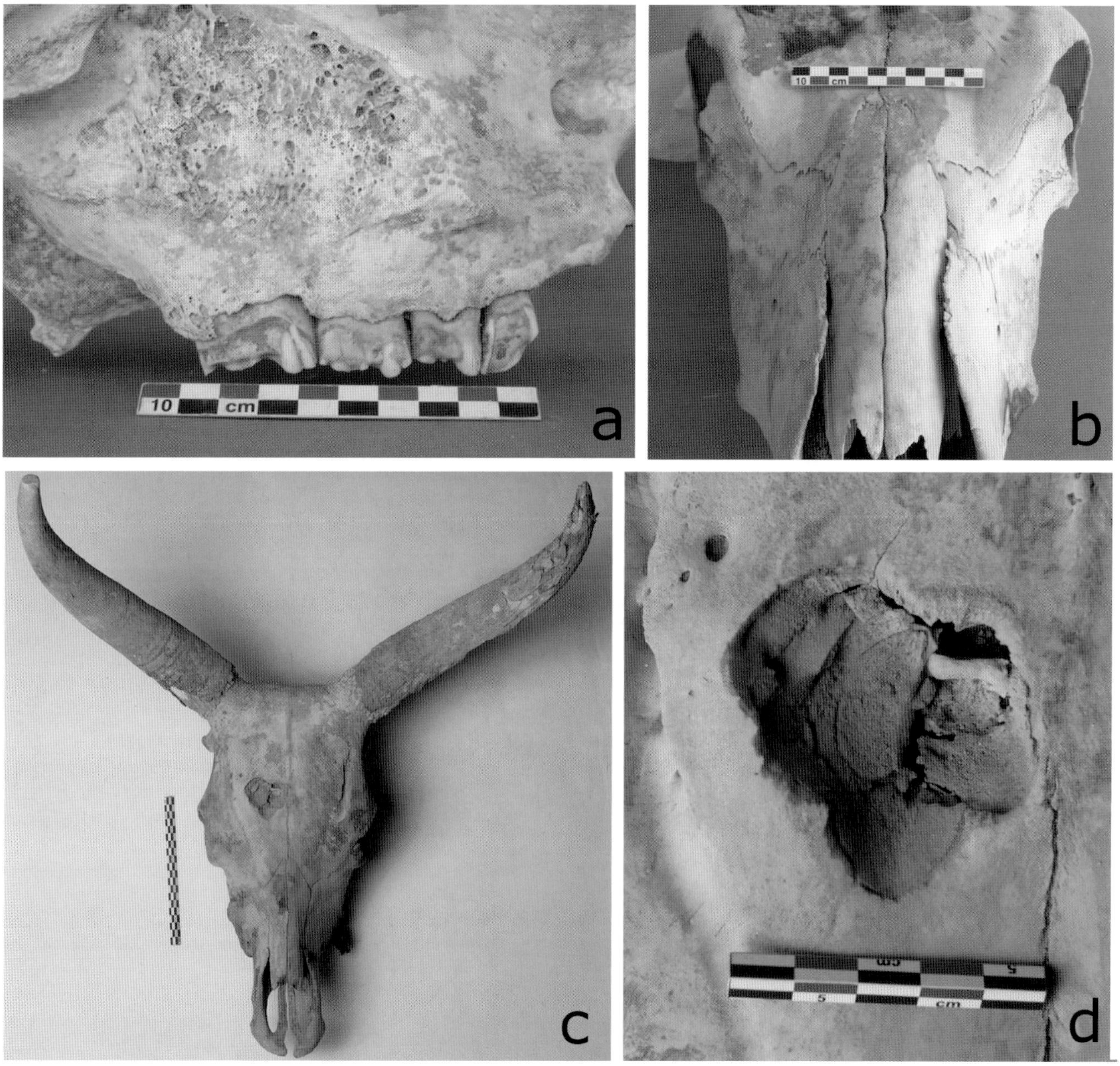

Fig. 25.7. Pathologies on skulls from Dayr al-Barshā. a. Right maxilla of Skull 4 with an abnormally porous bone structure and light periodontitis; b. Displaced nasal bone of skull 5; c.–d. Skull 9 with a puncture in the right frontal bone.

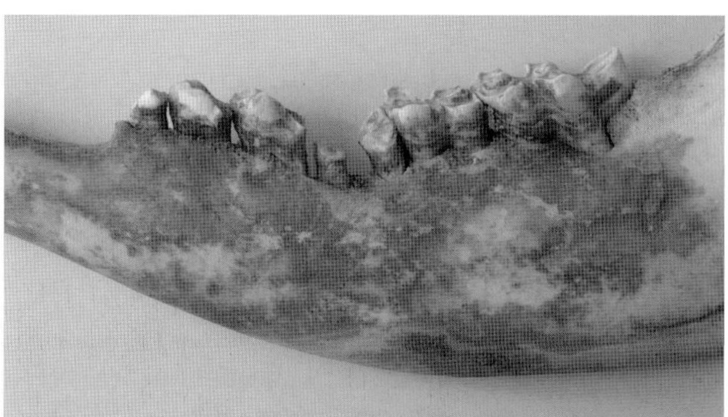

Fig. 25.8. Detail of a mandible (N9) with an abnormally worn M1.

Fig. 25.9. Normal left naviculocuboid (AF8) and left naviculocuboid showing spavin (R7).

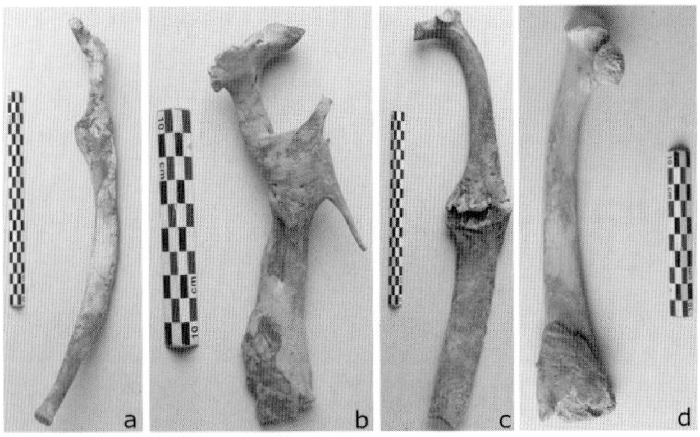

Fig. 25.10. Ribs with periostitis and bony outgrowths (a, b & d) and healing fracture (c) (From left to right: I1-AG19-I1-G8).

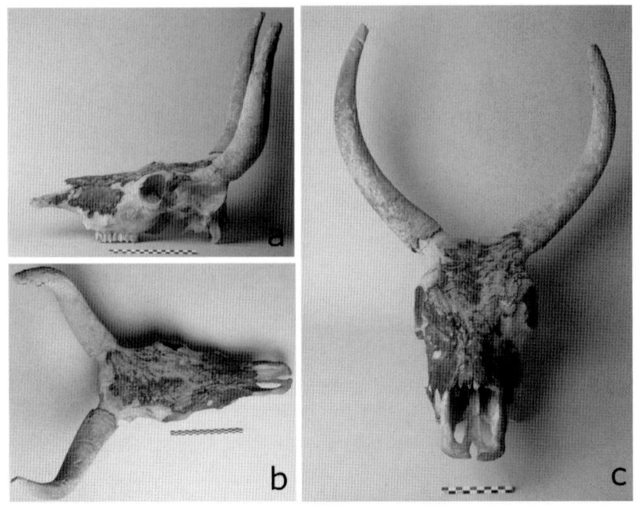

Fig. 25.11. Skull 6 with the left and right horn tips chopped off.

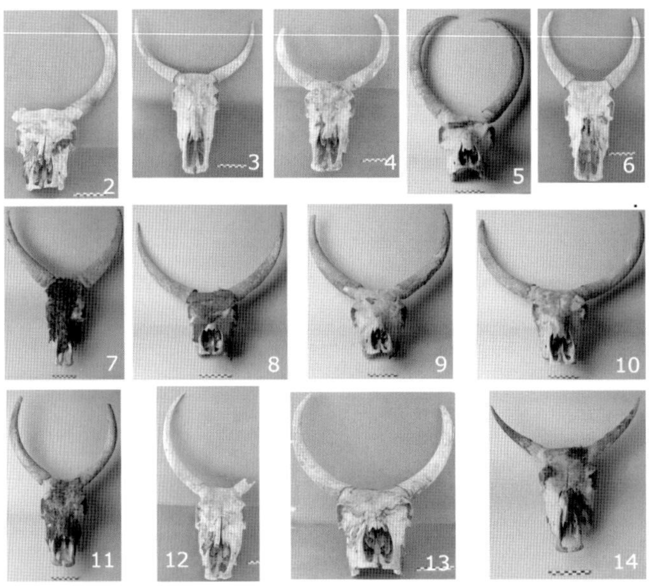

Fig. 25.12. The skulls of all adult cattle (Skulls 2–13) and of the individual of 2–2.5 years (Skull 14).

obtain parallel horns (Chaix 1996). The pressure exerted on the horn cores during this process can create a typical swelling on the intercornual crest that was observed at Kerma on rather severely modified specimens with forward pointing horns (Chaix & Hansen 2003). Such intercornual bulges have not been observed on the Dayr al-Barshā skulls. Other direct indications on how the deformations would have been achieved, i.e. flattening of the horns (cf. Chaix & Hansen 2003), rope imprints, cutmarks, etc., are also missing. The possibility therefore remains that the horns from the animals at Dayr al-Barshā are not deformed but rather part of the natural range of horn shapes of ancient Egyptian Longhorns. Nevertheless, horn deformations are known in ancient Egyptian iconography, where cattle are sometimes depicted with one horn trained downward and forward and the other in a more normal upward and backward direction and of which also ethnographic parallels exist among Nilotic groups (Schwabe 1984; Boessneck 1988, 70–71; Vercoutter 1980).

Despite careful observation, not a single butchery or slaughtering trace was found on the bones. Some hooves and horns seem to have been affected by insects, quite severely so in the case of the left horn of skull 9 and in the case of the right horn of skull 12. A few beetle remains have been collected from the inside of some of the skulls, but they could not be identified. It is likely, however, that they belong to the category of scavenging taxa (Elias 1994). One rib has a clear green stain, which may be due to the presence of copper or bronze, although no metal objects were associated with the cattle bones.

Bos africanus?

Ancient Egyptian Longhorns disappeared in the 19th

Table 25.5. Draught related pathologies on phalanges and metapodia from Dayr al-Barshā (cf. Bartosiewicz et al. 1997). Pex: proximal exostosis, plip: proximal lipping, dex: distal exostosis, brd: distal broadening, fus: fusion

No	*Element*	*pex*	*plip*	*dex*	*brd*	*fus*
Y4	Metacarpal left		2	2	3	2
AC2	Metacarpal right	2	2	2	3	2
E6	Metatarsal left			2		
B7	Metatarsal right		2	2		
I2	phalanx 1 anterior external		2			
O10	phalanx 1 anterior external	2		2		
P2	phalanx 1 anterior external		2	2		
Q1	phalanx 1 anterior external	2	2	2		
G8	phalanx 1 anterior internal		2			
T1	phalanx 1 anterior internal		2	2		
AB1	phalanx 1 anterior internal	2	2			
V6	phalanx 1 posterior external	2				
E6	phalanx 1 posterior internal		2			
U1	phalanx 1 posterior internal		2			
W1	phalanx 2 anterior			2		
Y1	phalanx 2 anterior		2	2		
AG1	phalanx 2 anterior	2		2		
T1	phalanx 2 posterior		2			
T1	phalanx 2 posterior		2			
U1	phalanx 3 anterior			2		
O10	phalanx 3 anterior	3				
P3	phalanx 3 anterior		2			

] 1 individual (Y4 + AC2; E6 + B7)

century AD due to rinderpest epidemics (Grigson 1991). Apart from some older descriptions of mummies, skulls and skeletons of these animals (Dürst 1899; Lortet & Gaillard 1903; 1905; 1909; Jackson 1934), and the description of a complete oxen skeleton by Boessneck and von den Driesch (1987a), there are very few osteological data on this cattle type. The complete skeletons from Dayr al-Barshā thus represent a unique source of information on the osteomorphology and osteometry of ancient Egyptian Longhorns (Tables 25.4, 25.6–25.8).

The ancient Egyptian Longhorns are thought to belong to an indigenous African cattle type, different from zebu and taurine cattle, and that has been referred to as *Bos africanus* (Grigson 1991; 2000), although others consider them as crossbreds between zebu and taurine cattle (MacHugh *et al.* 1997). Pia (1940) studied skulls of Egyptian archaeological cattle remains from different ages and came to the conclusion that they resemble taurines, but clearly not zebu cattle. When the skulls from Barsha are checked against the non-measurable criteria defined by Grigson (1980), to differentiate between zebu and taurine cattle, they fit usually better with zebu than with taurine cattle. All skulls have flat orbital rims for example. The persistence of flat orbital rims in older animals is typical for zebu, while in taurines they become sharp with age. Also the upward slope of the horns at the base is more zebu than taurine like. The shape of the sacrum of *Bos africanus* resembles that of taurines, but is said to be different from that in zebu type cattle (Grigson 1991; 2000). The cattle from Dayr al-Barshā also have sacral bones with a taurine shape. Six of the thoracic vertebrae from the cattle pit at the site have bifid dorsal spines (Fig. 25.13). Bifid dorsal spines in thoracic vertebrae are only rarely reported for taurine cattle, but are sometimes present in *Bos africanus* (Grigson 1991; 2000) and are common on some of the posterior thoracic vertebrae (6th–9th caudally) of the zebu type (Epstein 1971, 388). The long bones of *Bos africanus* are said to be slender, as in zebu, and unlike in taurine cattle. Logarithmic plots of maximal length and of distal breadth of metacarpals and metatarsals from Dayr al-Barshā, like those of other African

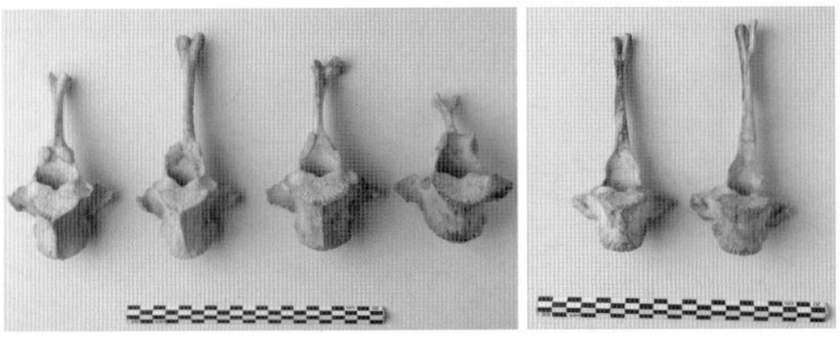

Fig. 25.13. Thoracic vertebrae from Dayr al-Barshā with bifurcated neural spine (from left to right: D4-D4-L1-AG1-Y1-AG1).

cattle, fit with zebu (Grigson 1991; 2000). Osteometrical data on the cattle from Dayr al-Barshā fall in the range of other Egyptian cattle (Laudien 2000) and of animals from Kerma in Sudan (Chaix 2007). However, the fact that the Dayr al-Barshā population is mostly composed of a single sex makes comparisons with these other mostly unsexed datasets, otherwise not very meaningful.

Discussion

Methodological issues

The study of the cattle from Dayr al-Barshā revealed some methodological problems. The results of the different ageing methods are not entirely in agreement, although this is probably partly due to the castration of the animals that especially must have affected the age of bone suture closure. The analysis was also hampered by the fact that there is inadequate literature on how to identify castrates. Most methodological studies on sex determination are based on numerous male and female reference animals, but include few castrates (e.g. Grigson 1982; Sykes & Symmons 2007). The withers height reconstructions for the Dayr al-Barshā animals showed slight variations depending on the element used. This is related to variations in body proportions between cattle of different types, and to the different effect of castration on the length of each bone, depending on the age its sutures close. The cattle from Dayr al-Barshā showed a wide range in horn shapes that seems to exceed the variation known up to now. It is believed that this can be a result of the increased sample size, although it could also be used to argue that some of the horns were artificially deformed.

Life and death of the Barshā cattle

Archaeological and osteological data from the cattle pit at Dayr al-Barshā, allow the deduction of some details about the life and death of the animals. The pathologies observed on the ribs, and the healing perforations in the skulls, can all be due to fights among the animals. Violent behaviour may have occurred rather frequently when animals were kept closely together in stables, as was done with meat animals in ancient Egypt (Boessneck 1988, 68). The preserved hooves from Dayr al-Barshā are not particularly long and therefore do not suggest inadequate wear due to a long stay in stables (cf. Boessneck 1988, 68). Although it is thought that the Egyptians dehorned some of their cattle (von den Driesch & Peters 2003, 14), the removal of horn tips has not previously been described. It may have been performed on the cattle from Dayr al-Barshā as a measure to reduce trauma in both cattle and the humans taking care of them. There exist some 19th century AD patent applications from the United States, US patent 373776 and 5237751, for machines designed to cut the top off cattle horns, exactly for this purpose (United States Patents and Trademark Office 2009).

Cattle, and especially oxen, were used in ancient Egypt for all types of heavy work (Boessneck 1988, 68). The cattle pit at Dayr al-Barshā dates to a period when the stone quarries near the site were intensively exploited and that animals potentially were put to work in the transport of stones. In fact, inside the quarries a cattle-drawn stone-cart seems to be depicted (Willems *et al.* 2009). However, the lack of traction related pathologies on the cattle bones from Dayr al-Barshā suggest that the animals did not serve such a purpose, or did so only for a short period.

It is known that, from the Old Kingdom onwards, castration was commonly practised on ancient Egypt cattle to optimise food and fat production (Boessneck 1988, 68). In addition, oxen were preferred for cattle offerings. However, there are no butchering or slaughtering marks on the Dayr al-Barshā cattle remains that would suggest the animals served as food or as food offerings. The presence of skin and hair also contradicts this.

Taphonomy

Before the cattle carcasses were placed inside the pit where they were eventually found, they were obviously disarticulated. Since no cut or chop marks occur, it is suspected that the carcasses were already disintegrated, falling apart naturally, before being buried. This initial rotting probably happened at some distance from the pit; at least the absence of some skeletal elements is suggestive of a loss during the transport between an unknown spot and the pit (see Table 25.1). Wherever the initial disintegration took place, carnivores had no access to the carcasses, as gnawing marks on the remains are absent.

Parallels

Burials of complete but articulated cattle are known in the Egyptian Nile Valley from the Predynastic period (ca. 4000–3050 BC) (Flores 2003; 2004) and even earlier, in the Western Desert of Egypt and beyond (Paris 2000; di Lernia 2006, Osypiński & Osypińska 2016). The site of Kerma in Sudan is famous for the numerous cattle bucrania that have been found associated with human graves from the Middle Kerma period (ca. 2050–1750 BC) (Chaix 1988;

1993; 2000; 2001). There are no exact parallels for the cattle pit at Dayr al-Barshā, where the animals were found disarticulated, but the closest comparison is a foundation deposit of an 18th Dynasty (15th century BC) temple in Tell el Dab'a (Bietak 2002), in the Nile Delta in the north of Egypt. The deposit was a round pit with a diameter of nearly 2m and which contained all remains of four bulls and one ox. Inside this pit were also four complete but broken pottery vessels. Butchery marks are not mentioned in the report, but such marks may have been rendered invisible due the poor preservation of the material. The bones are partly burned and are considered remains of an offering and a ritual meal. Breaking pots and the slaughtering and eating of cattle have been linked with the symbolic destruction of the enemy (Müller 2008). There are some clear differences between Tell el Dab'a and Dayr al-Barshā. At Dayr al-Barshā hardly any pottery was found, which, in addition, is not necessarily contemporaneous. The Dayr al-Barshā bones were not burned and the animals' meat does not seem to have been consumed. Moreover, there is a chronological difference of more than 1000 years between the deposit at Tell al Dab'a and the cattle pit from Dayr al-Barshā.

The cattle burial of Barshā in a larger context

An avenue to a more promising line of enquiry is offered by the Münchner Ochsenmumie (Boessneck *et al.* 1987), dated to second half of the 1st millennium BC. In fact, this is not an actual mummy, but the skeletal remains of one ox wrapped in linen in the shape of a bull. Other cattle mummies are known that are sculptured out of the bones of several individuals (Lortet & Gaillard 1903, 58–59). No cut marks have been found on the bones of the Münchner Ochsenmumie. This has been explained by Boessneck and von den Driesch (1987) by referring to a passage of the Greek historian Herodotus, who wrote in the 5th century BC about how Egyptians treat their dead cattle. In passage 41 of his book II he writes:

> 'the females they cast into the river, but the males they bury, each people in the suburb of their town … and when the bodies have rotted away and the appointed time comes on, then to each city comes a boat from that which is called the island of Prosopitis. … and when they have dug up the bones of the oxen they carry them off, and coming together they bury them in one single place.' (Macaulay 1852–1915)

Lloyd (1976, 186–189) has argued that the Island of Prosopitis was Kawm Abū Billū in the Western Nile Delta. There is much evidence that such long-distance displacements of animal remains were very common. For instance, in the Late Period (664–323 BC) the remains of ibises from all over Egypt seem to have been brought to the ibis galleries of Tūna al-Jabal (e.g., Kessler & Nur el-

Table 25.6. Cranial measurements (mm) according to von den Driesch (1976). Measurements for which special callipers are needed were not taken. When both sides were found associated (L&R), only the left side was measured. Note the marked difference in dimensions between the horns with horn sheath and the horn cores (cf. Grigson 1975).

Skull	*Sk. 2*	*Sk. 3*	*Sk. 4*	*Sk. 5*	*Sk. 6*	*Sk. 7*	*Sk. 8*	*Sk. 9*	*Sk. 10*	*Sk. 11*	*Sk. 12*	*Sk. 13*	*Sk. 14*
1 profile length	480	500	510	540	470	450	460	470	540	500	480	500	385
2 condylobasal length	465	490	480	520	470	460	460	470	500	500	470	490	390
3 basal length	430	460	450	490	440	430	420	435	460	465	440	460	365
4 short skull length	330	320	310	340	310	300	290	300	320	330	310	320	257
5 premolare-prosthion	140	140	150	150	130	130	120	110	140	140	130	140	113
7 viserocranium length	256	280	280	300	220	–	–	250	280	–	250	–	199
8 median frontal length	220	220	220	240	260	–	–	220	250	–	230	–	185
9 frontal length	263	260	260	290	230	–	–	260	300	–	270	–	219
10 short upper cranium length	390	415	410	420	–	350	390	370	430	410	390	400	(303)
11 akrokranion-infraorbitale of one side	360	380	380	410	360	340	370	350	430	390	360	390	273
12 greatest length of the nasals	170	190	190	180	–	–	–	160	180	–	150	–	121
14 lateral facial length	330	345	340	380	330	320	320	320	360	350	330	340	280
15 occipitale-infraorbitale	320	340	330	370	330	330	310	320	350	350	330	350	271
16 infraorbitale-prosthion	150	140	160	160	140	140	140	150	160	150	140	140	124
17 dental length	255	270	260	280	260	260	250	260	270	260	250	260	–
18 oral palatal length	202	196	210	230	200	190	190	200	200	190	190	200	164
19 lateral length of the premaxilla	155	150	140	160	130	130	130	140	150	–	125	170	128
20 cheektooth row length	126	137	129	133	138	130	130	137	140	140	135	130	–
21 molar row length	78	85	75	81	82	78	80	87	81	82	78	76	–
22 premolar row length	46	53	55	55	55	48	55	51	50	52	62	61	–
23 inner length orbit	52	70	68	73	65	62	66	65	70	67	70	60	45
24 inner height orbit	50	65	66	74	64	62	63	63	68	69	65	62	49

Skull	*Sk. 2*	*Sk. 3*	*Sk. 4*	*Sk. 5*	*Sk. 6*	*Sk. 7*	*Sk. 8*	*Sk. 9*	*Sk. 10*	*Sk. 11*	*Sk. 12*	*Sk. 13*	*Sk. 14*
25 mastoid breadth	198	207	214	232	210	196	191	219	222	216	210	216	152
26 breadth occipital condyles	105	107	102	117	100	105	106	114	105	111	105	117	89
27 breadth paraoccipital processes	150	160	158	170	152	145	146	257	155	165	162	162	118
28 breadth foramen magnum	30	28	37	43	29	40	40	30	33	38	39	31	23
29 height foramen magnum	41	37	37	40	35	41	36	40	41	38	42	35	24
30 least occipital breadth	118	132	126	131	134	117	141	144	118	131	138	141	99
31 least breadth between horncore bases	125	126	130	140	110	110	130	120	143	130	120	125	156
32 least frontal breadth	173	160	168	187	162	173	175	185	173	172	171	190	154
33 breadth across orbits	208	196	202	242	198	202	202	223	221	213	198	226	176
34 least breadth between orbits	151	148	150	175	153	156	147	161	180	161	169	170	130
35 facial breadth	135	135	141	166	142	140	138	160	158	159	146	152	123
36 breadth across nasals	44	51	63	55	–	–	–	53	60	–	49	–	44
37 breadth across premaxillae	78	77	87	93	85	82	80	93	90	88	75	89	67
38 palatal breadth	120	122	125	145	120	121	120	130	135	139	126	134	110
39 least inner height occipital groove	42	43	45	45	41	41	37	40	41	44	45	40	21
40 maximal occipital height	175	156	154	168	147	160	120	155	149	178	146	160	129
41 least occipital height	135	123	124	132	116	123	37	115	112	132	119	121	101
Horns without horn sheet													
42 least distance between horn tips	–	505	–	–	–	–	–	–	–	470	–	570	420
42a distance between horn tips - around curve	–	670	–	–	–	–	–	–	–	950	–	930	500
43 tangential distance between outer curves	–	505	–	–	–	–	–	–	–	500	–	570	420
44 basal circumference (BC)	190	180	210	225	200	210	200	215	215	205	180	200	135
45 greatest diameter base (DT-BA-GD)	63	60	73	73.5	66	70	65	72	70	70	59	68	46
46 least diameter base (DAP-BB-LD)	57	53	62	66	58	56	58	63	63	58	54	58	40
47 least outer curvature horn core	–	260	315	480	–	350	370	370	370	390	380	370	170
Horns with horn sheet													
42 least distance between horn tips	–	520	460	110	350	490	–	550	–	420	270	510	–
42a distance between horn tips: around curve	–	940	1060	1680	1000	1110	–	1130	740	1160	1320	1210	–
43 tangential distance between outer curves	–	530	530	550	450	530	–	590	1300	510	500	590	–
44 basal circumference (BC)	–	215	220	240	220	245	240	225	235	240	200	225	–
45 greatest diameter base (DT-BA-GD)	–	71	80	79	74	87	84	73	79	82	65	76	–
46 least diameter base (DAP-BB-LD)	–	61	61	72	63	66	58	64	63	68	56	62	–

Skull	*Sk. 2*	*Sk. 3*	*Sk. 4*	*Sk. 5*	*Sk. 6*	*Sk. 7*	*Sk. 8*	*Sk. 9*	*Sk. 10*	*Sk. 11*	*Sk. 12*	*Sk. 13*	*Sk. 14*
47 least outer curvature horn core	550	390	510	730	400	470	370	500	550	510	570	540	–

Mandibula	*H1*	*AG3*	*M3*	*Sk .9*	*Sk. 2*	*P10*	*Sk. 12*	*AG5*	*AC10*	*K5*	*Sk. 13*	*AC6*	*AG17*
Side	L	L&R	R	L&R	L&R	R	R	L	L	L&R	R	L	L&R
1 length gonion caudale-infradentale	363	365	366	370	370	375	380	380	385	390	390	390	400
2 length condyle process-infradentale	372	385	374	380	390	412	400	420	400	405	410	415	420
3 length gonion caudale: M3	111	109	107	97	116	110	111	111	118	123	127	124	124
4 length M3-infradentale	252	269	256	261	257	263	262	263	263	266	260	260	275
5 length gonion caudale-M2	242	245	239	244	256	250	255	267	256	257	264	264	265
7 length cheek tooth row	132	139	134	105	131	138	135	139	136	131	132	131	140
8 length molar row	85	86	86	91	83	90	89	90	84	85	87	84	90
9 length premolar row	46	49	47	50	45	50	45	50	50	47	46	45	51
10 length M3	36	36	36	41	36	37	39	37	39	37	40	38	39
10 breadth M3	19.1	17.0	13.4	16.8	15.2	14.0	16.4	14.4	14.8	16.7	16.6	16.5	14.9
11 lenght of the diastema	107	97	110	102	115	110	110	107	114	119	111	118	114
12 aboral height ramus	(137)	147	152	149	147	154	127	144	157	154	158	149	173
13 middle height vertial ramus	(141)	148	146	159	155	156	153	156	153	143	161	150	168
14 oral length vertical ramus	(210)	210	221	232	218	224	234	231	227	217	230	231	240
15a height mandibula behind M3	79.1	73.8	78.2	69.5	71.8	78.6	75.0	78.4	77.0	71.3	79.3	79.0	88.0
15b height mandibula in front of M1	52.3	49.2	52.4	54.3	52.9	49.0	49.1	53.7	51.7	52.9	57.8	52.0	64.7
15c height mandibula in front of P2	35.7	36.5	37.7	39.6	46.5	36.1	40.8	42.2	42.0	42.0	46.9	45.0	44.6

Mandibula	*M12*	*Q5*	*AC9*	*L3*	*R4*
Side	L	R	L&R	L	R
1 length gonion caudale-infradentale	400	403	415	–	–
2 length condyle process-infradentale	435	435	435	380	–
3 length gonion caudale - M3	128	122	126	118	110
4 length M3-infradentale	263	269	281	–	–
5 length gonion caudale-M2	267	268	264	250	250
7 length cheek tooth row	137	135	137	139	141
8 length molar row	91	90	89	90	88
9 length premolar row	46	–	48	50	53
10 length M3	40	40	42	37	38
10 breadth M3	16.0	18.9	17.7	13.8	13.8
11 length of the diastema	120	122	125	98	–
12 aboral height ramus	154	147	176	144	136
13 middle height vertical ramus	156	159	174	150	153
14 oral length vertical ramus	227	227	249	217	219
15a height mandibula behind M3	79	71.5	79.4	74.9	75.6
15b height mandibula in front of M1	50	51.4	51	50.3	61
15c height mandibula in front of P2	44	45.4	45.5	37.9	37

*Table 25.7. Measurements (mm) according to von den Driesch (1976) on postcranial bones of the complete cattle from Dayr al-Barshā. H1 on the pelvis according to Greenfield (2006). *unfused, **fusing; ( ): estimate.*

Scapula	*V9*	*Z2*	*AG16*	*U13*	*B6*	*U4*	*U9*	*Q4*	*Y2*	*T6*	*Y7*	*R10*	*Q13*
Side	*R*	*L*	*L*	*R*	*L*	*L*	*R*	*R*	*L*	*L*	*R*	*L*	*R*
HS	280	285	335	335	(352)	355	355	357	365	370	370	375	380
DHA	290	285	335	335	(260)	360	375	360	375	370	370	380	380
Ld	145	145	171	–	(195)	181	176	178	175	184	188	198	197
SLC	41.9	44.6	52.6	52.0	58.6	56.6	69.0	52.4	58.5	56.7	57.5	61.8	59.3
GLP	58.7	59.5	73.4	76.8	75.3	68.4	75.2	68.3	72.8	73.3	74.6	68.8	73.6
LG	49.9	40.5	61.7	62.4	58.8	58.7	60.8	58.3	59.3	62.4	62.4	59.7	60.9
BG	39.7	–	50.7	50.8	50.6	49.1	48.8	49.2	50.2	49.5	56.7	49.5	53.6

Scapula	*AB6*	*G3*	*A9*	*T8*	*E14*	*S5*	*A12*	*C1*	*D1*	*C14*	*F9*	*H6*	*O5*
Side	*R*	*L*	*R*	*R*	*L*	*L*	*L*	*L*	*L*	*R*	*R*	*R*	*R*
HS	380	390	(390)	395	400	400	420	–	–	–	–	–	–
DHA	–	–	(380)	395	395	–	420	–	–	–	–	–	–
Ld	–	194	200	189	190	–	200	–	–	–	–	–	–
SLC	55.8	50.2	54.2	53.6	56.5	61.7	64.4	55.3	55.3	54.5	58.8	65.5	60.2
GLP	78.8	67.2	–	73.2	72.5	70.5	87.5	77.7	76.0	77.9	73.5	85.8	72.1
LG	53.1	62.5	–	63.0	62.1	61.1	69.3	66.7	64.4	66.3	57.8	69.7	58.9
BG	–	52.8	–	49.3	48.7	56.7	58.7	52.5	50.8	46.4	49.2	56.0	55.9

Humerus	*M13*	*Q7*	*C4*	*C15*	*R2*	*Y14*	*P9*	*G1*	*G6*	*D7*	*AG1*	*R10*	*M10*
Side	*R*	*L*	*L*	*L*	*R*	*R*	*L*	*R*	*L*	*L*	*R*	*L*	*R*
GL	311	312	315	(320)	(320)	323	324	325	326	328	328	330	331
Bp	98.0*	99.5*	101*	96.9*	96.5*	104*	98*	100*	98.5*	96.1*	99*	103*	106.7*
SD	38.6	41.2	40.1	41.6	34.1	37.3	32.7	37.3	35.5	35.3	33.5	37.1	38.6
Bd	82.5	85.9	77.0	82.2	89.3	89.1	81.8	83.5	86.5	85.7	96.9	87.2	90.5
BT	71.6	80.7	74.0	74.6	76.6	75.0	77.1	78.4	76.8	76.4	79.2	78.4	77.5

Humerus	*T3*	*V7*	*E9*	*D7*	*D9*	*N9*	*Z3*
Side	*R*	*R*	*L*	*R*	*L*	*L*	*R*
GL	335	335	336	355	363	364	365
Bp	111	99.3*	106	(107)*	112*	120*	125*
SD	37.1	33.8	39.0	41.6	41.0	47.7	44.7
Bd	95.0	95.2	85.8	96.1	87.9	94.0	103.7
BT	78.3	77.7	78.8	83.8	83.2	87.4	86.9

Radius	*V11*	*Z7*	*G2*	*N5*	*C13*	*AG11*	*U5*	*AF9*	*AG15*	*R6*	*K2*	*AE3*	*V12*
Side	*L*	*L*	*R*	*R*	*L*	*R*	*L*	*L*	*R*	*L*	*L*	*R*	*R*
GL	275	310	310	315	318	320	321	323	326	327	328	330	330
Bp	74.6	81.0	81.3	79.3	79.9	84.0	89.8	82.8	85.1	86.0	85.0	84.7	87.0
SD	32.2	43.3	45.6	40.6	40.8	40.7	43.9	40.2	39.2	43.6	40.5	40.0	43.1
Bd	73.3*	80.4	80.3	77.7	79.3	77.1*	(83)*	77.6*	77.6	77.7	78.8	77. 3*	78.4

Radius	*L2*	*Q6*	*D11*	*AF4*	*V2*	*G1*	*E9*	*D9*	*S4*	*K6*	*U3*
Side	*L*	*R*	*L*	*L*	*R*	*R*	*L*	*L*	*R*	*L*	*R*
GL	331	335	336	340	340	340	344	346	348	370	370
Bp	84.3	85.3	86.5	85.9	87.7	86.3	87.3	93.0	93.7	90.6	91.7
SD	39.7	40.2	40.1	39.9	41.7	41.7	39.7	45.1	45.8	46.3	47.8
Bd	77.3*	74.3*	76.7*	79.7*	77.2*	81.3	84.9	85.1*	85.6*	90.3*	90.2

Ulna	*Y14*	*Z7*	*N1*	*U3*	*R6*	*K2*	*G1*	*S4*	*E9*
Side	*R*	*L*	*R*	*R*	*L*	*L*	*R*	*R*	*L*
LO	96.8	97.3	109**	110.0	110.9	114	114.2	119.6	120.4
DPA	66.3	68.4	70.2	72.2	71.7	67.9	54.4	64.6	67.5
SDO	56.3	57.8	58.5	55.7	56.8	55.2	69.9	51.3	53.7
BPC	48.5	48.9	47.6	50.2	50.1	49.5	50.7	50.6	–

Metacarpal	*G1*	*C13*	*AC5*	*K4*	*Y14*	*F2*	*AG12*	*V10*	*AG15*	*V12*	*D11*	*W3*	*N5*
Side	*L*	*L*	*L*	*R*	*R*	*R*	*L*	*R*	*R*	*R*	*L*	*L*	*R*
GL	217	218	218	218	218	221	223	223	225	225	227	227	228
Bp	62.0	61.8	64.1	61.4	64.3	62.4	62.1	60.4	64.0	65.1	63.5	63.4	62.3
SD	31.9	33.3	34.3	33.8	35.2	33.0	32.2	33.7	32.5	35.7	32.9	35.4	33.5
Bd	64.1	61.6	68.2	64.0	68.1	60.9	66.6	64.6	67.0	67.5	62.7	67.1	63.2

Metacarpal	*M16*	*M15*	*AE2*	*AG11*	*T5*	*AG1*	*Y4*	*AC2*	*L5*	*Y4*
Side	*L*	*L*	*R*	*R*	*L*	*R*	*L*	*R*	*L*	*R*
GL	234	237	237	237	238	238	239	240	248	248
Bp	63.6	63.4	65.1	65.3	63.6	65.2	69.2	69.5	67.2	69.2
SD	33.4	36.3	34.2	33.2	34.1	35.6	42.0	41.8	36.7	38.3
Bd	65.2	66.9	66.1	66.1	66.0	66.8	75.5	75.4	69.4	70.2

ph. 1 ant.	*AD1*	*AG1*	*Q1*	*Y9*	*U1*	*Y9*	*F2*	*W1*	*V1*	*C13*	*C6*	*N4*	*F2*
GLPe	55.9	56.3	57.0	57.4	58.1	61.6	61.7	61.8	61.9	62.9	63.1	63.2	63.2
Bp	27.0	26.9	25.9	27.7	26.6	33.6	31.2	30.9	33.5	32.0	30.8	32.6	32.4
SD	24.1	23.2	24.5	25.4	25.2	28.9	28.6	28.0	29.7	27.4	28.0	27.3	27.4
Bd	27.9	28.5	28.9	28.6	27.2	31.9	31.5	30.5	31.3	28.1	31.5	30.3	28.7

ph. 1 ant.	*Y9*	*AD1*	*G4*	*C6*	*G8*	*AB1*	*AG1*	*AG1*	*U1*	*N1*	*V1*	*10*	*AA1*
GLPe	63.3	63.8	63.8	64.3	64.3	64.4	64.5	64.6	64.7	64.8	64.9	65	65.0
Bp	33.7	28.1	33.0	30.8	33.3	33.2	30.6	34.1	34.4	33.0	31.9	29.8	33.2
SD	28.2	26.7	27.0	26.3	28.3	29.1	27.2	29.1	29.3	27.9	27.6	27.3	88.0
Bd	31.1	30.7	29.9	29.7	31.4	32.7	30.6	30.8	31.8	28.3	30.4	30.9	31.7

ph. 1 ant.	*D11*	*AE1*	*W1*	*AG1*	*X1*	*X1*	*I1*	*AG11*	*U1*	*M6*	*Q14*	*10*	*T1*
GLPe	65.0	65	65.1	65.5	65.5	65.5	65.7	65.7	65.8	65.9	66.3	66.4	66.5
Bp	32.6	33.7	31.5	33.5	33.7	34.0	34.2	33.3	33.8	33.0	32.6	30.7	32.7
SD	27.8	28.6	27.2	28.4	28.4	29.0	28.3	26.6	31.2	27.8	28.3	26.6	26.2
Bd	31.0	31.4	29.6	33.4	31.4	31.2	33.6	31.0	32.2	31.8	32.6	29.0	30.8

ph. 1 ant.	*AG1*	*AD1*	*AG1*	*AG1*	*D5*	*D5*	*AG11*	*AG1*	*P3*	*P3*	*T1*	*AB1*	*Q1*
GLPe	67.0	67.1	67.4	67.4	67.5	67.9	68.6	69.7	69.7	69.8	70.0	70.1	70.4
Bp	33.0	32.1	34.2	32.3	32.0	32.6	34.7	35.9	34.9	30.9	37.6	38.1	38.6
SD	27.6	27.6	27.1	28.3	28.0	28.2	27.5	29.8	28.0	30.2	33.2	34.2	34.2
Bd	32.3	32.2	30.3	31.8	31.7	31.8	31.9	32.6	33.0	32.5	36.4	36.5	35.2

ph. 1 ant.	*AG1*	*P2*
GLPe	70.5	71.6
Bp	35.9	33.5
SD	28.0	30.3
Bd	33.3	32.6

ph. 2 ant.	*V1*	*P1*	*AG1*	*U1*	*W1*	*AG1*	*AD1*	*AG1*	*AE1*	*W1*	*F2*	*C13*	*Z1*
GL	33.9	34.8	35.1	35.3	35.9	36.0	36.2	36.3	36.6	41.4	41.5	41.6	42.7
Bp	31.6	28.6	29.8	29.0	29.9	28.4	29.4	29.6	30.2	30.4	30.9	30.2	31.9
SD	26.9	26.8	23.3	20.9	23.8	22.1	22.9	22.4	24.5	25.4	26.6	25.6	25.7
Bd	27.9	24.3	25.3	23.6	25.1	24.8	26.0	24.5	25.3	27.0	26.9	27.3	27.5

ph. 2 ant.	*AE1*	*AF1*	*X1*	*8(1)*	*AF1*	*Q1*	*AG1*	*AB1*	*Y9*	*M9*	*AC1*	*Y9*	*AC1*
GL	42.9	43.0	43.2	43.4	43.5	43.6	43.6	43.9	43.9	44.0	44.0	44.0	44.4
Bp	31.5	33.0	32.7	30.1	32.0	31.2	32.5	32.8	32.2	32.0	32.6	32.0	32.2
SD	27.4	26.8	28.0	25.5	25.3	26.3	26.9	26.9	27.1	27.5	26.6	27.8	28.1
Bd	30.0	30.1	30.7	26.8	27.1	27.3	30.3	28.9	29.5	28.8	29.2	29.2	27.9

ph. 2 ant.	*AA1*	*D5*	*W1*	*G8*	*M6*	*AE1*	*O1*	*AG11*	*M6*	*V1*	*W1*	*Y1*	*G4*
GL	44.4	44.4	44.6	44.6	44.7	44.7	44.8	44.8	44.9	44.9	44.9	44.9	44.9
Bp	32.2	33.6	33.1	31.6	33.4	32.3	31.7	33.1	32.3	33.0	31.7	32.9	31.3
SD	26.3	28.4	29.1	25.4	27.4	27.8	26.7	26.8	26.6	27.5	27.6	27.2	25.4
Bd	29.0	28.4	30.8	28.5	30.4	29.7	27.6	29.4	29.3	29.1	28.8	29.1	29.5

ph. 2 ant.	*G4*	*AG1*	*AG1*	*Q14*	*AG1*	*10*	*U1*	*Y1*	*U1*	*P3*	*T1*	*AG11*	*U1*
GL	45.2	45.2	45.2	45.2	45.8	45.8	46.2	46.2	46.3	46.6	47.0	47.3	47.5
Bp	31.9	33.5	33.5	32.5	32.4	30.0	37.1	33.6	33.6	33.0	36.4	33.0	37.1
SD	26.3	28.3	30.4	26.9	26.8	30.9	33.5	26.5	27.2	26.7	30.7	27.3	30.6
Bd	27.2	28.2	33.6	27.9	30.5	27.2	36.1	30.7	30.5	30.7	33.0	30.9	32.1

ph. 3 ant.	*8(1)*	*X1*	*10*	*G4*	*AG11*	*M6*	*Y9*	*Q14*	*F2*	*Z1*	*AG11*	*M6*	*P3*
DLS	71.1	71.1	76.1	76.3	77.0	78.1	78.2	80.1	80.2	80.5	80.5	82.8	83.5
Ld	57.1	57.0	46.2	58.6	61.8	60.8	61.8	60.1	58.4	60.4	62.5	62.7	59.4

ph. 3 ant.	*M9*	*U1*
DLS	84.2	92.5
Ld	67.4	70.2

Pelvis	*P4*	*V4*	*U12*	*E13*	*N2*	*A3*	*AB5*	*O3*	*C10*	*U8*	*Y11*	*AG14*	*AA2*
Side	R	R	R	L	R	L	L	R	R	R	R	L	R
GL	480	–	–	–	–	–	–	–	–	–	–	–	–
LA	76.6	63.3	71	72.9	73	73.1	74.5	75.0	75.0	75	75	75.4	75.8
SH	41.3	30.8	41.0	39.7	43.7	42.5	37.1	42.9	39.2	38.5	45.9	40.4	46.7
SB	29.5	21.6	24.2	26.3	25.0	27.5	23.5	22.2	28.5	21.9	30.3	25.6	22.5
SC	110	86	110	110	115	113	106	106	115	105	123	110	118
LfO	–	71.4	77.9	89	90	–	83	95.8	83.0	80.4	78.6	93.7	118.8
H1	18.6	17.5	21.5	19.1	12.9	15.0	20.5	17.5	18.7	18.5	24.5	15.5	–

Pelvis	*O6*	*AF5*	*Y10*	*U11*	*AG4*	*AC12*	*C9*	*AG7*	*H3*	*Q8*	*V3*	*AD3*
Side	L	L	L	R	L	R	L	L	R	L	L	R
LA	75.9	76	76.8	77	77.5	78.0	78.8	79.9	85.6	86.0	86.6	88
SH	40.5	46.6	42.4	41.9	47.2	42.9	39.6	42.5	49	46.0	45.8	48.5
SB	22.6	29.0	23.6	24.0	23.5	25.3	30.1	26.3	27.6	32.3	34.0	29.1
SC	105	123	112	108	120	113	112	117	122	125.0	130	132
LfO	95.4	71.6	84.2	85.0	84.6	85	85.7	82.2	–	107.0	95.6	98
H1	14.4	19.6	15.9	21.0	16.0	17.9	18.3	12.4	16.1	12.0	20.2	22.5

Femur	*E2*	*F4*	*AC4*	*AG1*	*C8*	*X4*	*Q10*	*K3*	*AC11*	*AB2*	*AG12*	*AG1*	*AG6*
Side	L	R	L	L	R	L	R	R	L	R	R	L	R
GL	393	393	400	400	401	402	402	404	405	405	412	420	421
Bp	118*	117.3*	123*	127*	123.1*	124.6*	125.7*	123.3	118.7*	127*	120*	127*	129*
DC	47.0*	46.7*	48.5*	46.5*	48.3*	47.9*	48.3*	47.0	47.3*	46.1*	47.0*	49.3*	49.2*
SD	92.4	39.9	35.7	35.8	35.7	36.1	35.3	34.2	33.9	35.8	33.3	35.7	35.3
Bd	–*	90*	95*	97	93.3	96.3*	96.6*	102.7	100*	97.4*	100.5*	102.6*	98*

Femur	*AG13*	*O9*	*AD4*	*Z4*	*AD4*	*E12*
Side	L	R	R	L	R	L
GL	422	425	445	457	459	–
Bp	126.0	126.6*	132*	138*	140*	116*
DC	49.6	48.5*	51.2*	55.2*	55.5*	47.9*
SD	35.5	36.7	41.4	40.3	41.6	35.8
Bd	97.9	96.2	106*	114*	114*	98.8*

Patella	*AG1*	*C8*	*M1*	*8(2)*	*Q10*	*8*	*H8*	*U8*	*W1*	*AD1*	*AG1*	*AB2*	*Y1*
Side	R	R	L	L	R	R	L	R	R	R	L	R	L
GL	52.3	66.3	66.4	67	67.6	67.9	68.4	69.3	69.3	69.4	69.5	69.7	70.1
GB	53.4	52.9	53.6	54	56.4	52.2	56.4	52.7	56.2	66.6	62.7	58.9	56.0

Patella	*AG1*	*Z1*
Side	*R*	*L*
GL	71.2	73.5
GB	61.8	63.8

Tibia	*A2*	*B1*	*AC4*	*C8*	*AF2*	*AF3*	*AG12*	*B8*	*E2*	*Q10*	*AB3*	*AG1*	*X6*
Side	*R*	*L*	*L*	*R*	*L*	*R*	*R*	*L*	*R*	*R*	*L*	*L*	*R*
GL	375	376	377	378	393	395	400	401	402	403	405	418	420
Bp	95.9*	95.2*	95.8*	96.0*	97.8*	99.2	101.3*	102.8*	103.0*	98.3*	98.0*	101.0	100.3
SD	41.9	40.6	43.5	45.3	45.1	45.4	38.4	40.5	40.5	39.8	41.2	45.0	46.4
Bd	61.3	63.1	67.2	65.0	66.7	65.3	63.2	66.1	65.1	64.5	66.6	68.4	66.8

Tibia	*Y3*	*AD4*
Side	*L*	*R*
GL	451	452
Bp	110.3*	111.7*
SD	45.0	47.3
Bd	72.1	70.6

Os malleolare	*M1*	*H8*	*AD1*	*R3*	*AG1*	*M1*	*8(2)*	*AD1*	*R8*	*B7*	*Q10*	*C8*	*Y5*
GD	32.2	32.4	33.2	33.4	34.4	34.6	35.0	35.2	35.6	35.7	35.7	36.4	36.8

Os malleolare	*C16*	*AG1*	*R7*	*O1*	*L1*	*X1*	*AG1*	*V1*	*AF1*	*N1*
GD	37.1	37.7	38.2	38.3	39.1	39.4	39.4	39.5	40.7	41.5

Naviculocuboid	*M11*	*8(1)*	*R3*	*L4*	*AD1*	*Y5*	*Q1*	*B1*	*8(2)*	*Q10*	*R7*	*X1*	*AC1*
Side	*R*	*R*	*L*	*R*	*L*	*R*	*R*	*L*	*R*	*R*	*R*	*L*	*R*
GB	53.6	53.9	54.4	55.9	56.6	57.5	57.7	57.9	58.0	58.0	58.1	58.6	58.6

Naviculocuboid	*D5*	*AB1*	*H8*	*D12*	*AD1*	*N4*	*C8*	*Y13*	*AF8*	*T1*	*AD4*	*Z1*
Side	*L*	*L*	*L*	*L*	*R*	*L*	*R*	*L*	*R*	*L*	*R*	*L*
GB	58.6	58.7	58.9	58.9	59.1	60.1	60.8	61.2	63.2	66.2	66.5	66.7

Calcaneus	*Y13*	*C8*	*A7*	*A1*	*L4*	*P7*	*N1*	*AB1*	*Q10*	*AF1*	*E3*	*R7*	*AD1*
Side	*L*	*R*	*R*	*L*	*R*	*R*	*L*	*L*	*R*	*L*	*R*	*R*	*L*
GL	138.2	138.3	142.4	143.4	144.6*	146	146.5	149.3*	150	151.3*	151.7	152*	152.6*
GB	52	49	43	44	42	50	43	47	50	47	44	48	45

Calcaneus	*AC1*	*AG1*	*W1*	*AG1*	*Z1*	*L4*	*L4*
Side	*R*	*L*	*R*	*R*	*L*	*L*	*R*
GL	153*	155.9	156*	156.5	157*	161.1	162
GB	40	50	50	50	52	50	51

Talus	*8(1)*	*R3*	*AG1*	*C8*	*P7*	*Y13*	*P1*	*E3*	*L4*	*B7*	*Y5*	*Q10*	*D7*
Side	*R*	*L*	*L*	*R*	*L*	*L*	*R*	*R*	*R*	*L*	*R*	*R*	*L*
GLl	63.4	64.1	65.5	66.1	67.8	68.3	69.0	69.7	70.4	70.8	71.7	71.8	71.9
GLm	60.2	61.1	61.2	60.7	65.0	61.3	65.7	66.5	66.5	66.3	67.4	67.0	66.7
Dl	35.3	35.4	35.7	37.6	36.9	38.3	38.0	37.6	38.3	38.1	40.5	38.6	39.1
Bd	42.0	42.5	41.4	45.6	44.4	46.7	43.3	44.3	46.1	45.0	46.1	45.8	46.8

Talus	*P7*	*AG1*	*S1*	*AG1*	*T1*	*C16*	*AC1*	*AG1*	*T1*	*AF1*	*R7*	*AG1*
Side	*R*	*R*	*L*	*L*	*L*	*R*	*R*	*R*	*L*	*L*	*R*	*L*
GLl	73.2	73.4	73.6	74.3	74.7	74.8	77.3	77.6	77.7	78.1	78.2	79.0
GLm	67.6	67.8	66.8	67.1	70.9	70.0	70.4	73.7	69.9	68.3	68.5	72.3
Dl	40.2	40.8	39.9	40.5	40.9	41.4	41.5	43.0	40.5	41.2	41.4	43.0
Bd	44.2	43.3	45.2	43.2	48.5	49.5	47.3	47.8	47.6	45.5	46.3	49.4

Metatarsus	*V13*	*A7*	*Y13*	*C8*	*A1*	*E6*	*Y5*	*D12*	*B8*	*B7*	*D11*	*Z5*	*Q10*
Side	*L*	*R*	*L*	*R*	*L*	*R*	*R*	*L*	*R*	*L*	*L*	*L*	*R*
GL	249	254	254	254	255	256	256	256	256	258	260	260	262
Bp	51.4	49.6	52.4	52.6	51.2	51.1	51.4	52.1	52.2	50.7	52.0	52.6	52.9
SD	29.4	28.0	29.5	30.9	28.5	29.1	28.3	29.3	29.9	29.9	29.2	31.1	28.2
Bd	58.5	55.5	61.1	61.1	55.9	62.6	61.1	61.6	58.6	62.7	57.8	60.7	57.9

Metatarsus	*AF7*	*I5*	*P5*	*H4*	*Q12*	*V6*	*AG1*	*AD4*	*Y4*	*AF8*	*AD4*
Side	*R*	*L*	*L*	*R*	*R*	*R*	*R*	*L*	*L*	*R*	*L*
GL	263	264	270	270	271	272	272	272	273	289	291
Bp	53.4	51.3	50.3	57.6	50.9	51.4	52.7	57.1	51.6	56.1	55.2
SD	29.9	30.1	31.4	34.4	33.3	31.3	30.4	35.7	29.9	32.8	32.3
Bd	60.8	58.3	61.4	65.6	61.0	60.9	59.2	67.0	61.2	64.3	63.9

ph. 1 post.	*AG1*	*AG1*	*Q9*	*T1*	*AD4*	*A7*	*G8*	*A7*	*Y12*	*T1*	*Y12*	*Q10*	*S1*
GLpe	58.6	59.4	60.0	62.7	62.7	63.7	63.8	64.0	64.5	65.2	65.5	66.0	66.0
Bp	26.1	26.6	27.0	33	33.3	28.0	29.6	29.9	30.0	32.4	30.8	28.7	30.3
SD	22.2	23.3	23.8	29.7	28.2	25.9	25.4	24.8	26.4	27.6	26.0	25.0	26.5
Bd	27.7	27.4	27.0	32.5	30.4	28.4	26.9	27.4	30.4	30.9	29.2	29.1	29.2

ph. 1 post.	*I5*	*V6*	*V6*	*I5*	*D13*	*AA1*	*T1*	*P8*	*AG1*	*M1*	*AG1*	*Q14*	*P8*
GLpe	66.2	66.4	66.6	66.6	66.8	66.9	67	67.1	67.3	67.3	67.4	67.5	67.6
Bp	30.8	31	31.7	30.7	28.9	29.3	30	30.5	31.1	30.2	31.3	30.8	30.8
SD	25.3	27.2	27.3	24.5	25.2	25.0	26.8	27.8	30.9	24.8	26.0	27.6	25.5
Bd	29.0	30.6	28.8	28.0	29.5	29.0	29.4	34.5	28.6	27.6	29.6	31.6	29.9

ph. 1 post.	*D13*	*E6*	*Q10*	*E6*	*V1*	*M9*	*P2*	*Y1*	*AD1*	*AC1*	*E6*	*D12*	*X1*
GLpe	68.1	68.4	68.5	68.6	68.9	69	69.4	69.5	69.5	69.6	69.7	69.9	70.0
Bp	29.4	29.8	29.6	29.6	31.3	29.7	30.1	30.5	31.6	30.1	31.4	30.0	30.2
SD	26.0	25.0	25.8	24.9	24.4	25.6	26.2	24.4	25.2	25.9	24.7	27.1	25.8
Bd	28.5	30.9	29.0	29.7	30.9	–	30.1	27.2	28.6	30.0	29.4	32.4	28.5

ph. 1 post.	*AD4*	*AD4*	*U1*	*AF8*	*AF8*	*AG1*
GLpe	71.7	72.1	72.3	72.5	72.6	73.6
Bp	33.9	32.9	33.4	33.9	33.3	33.1
SD	25.0	27.8	31.1	28.0	28.4	28.6
Bd	30.9	32.7	32.5	30.9	31.8	31.8

ph. 2 post.	*R1*	*Z1*	*AG1*	*A7*	*8(2)*	*AD1*	*AC1*	*Y1*	*AD4*	*Q10*	*D13*	*N8*	*F7*
GL	40.4	40.7	42.2	42.8	43.1	43.1	43.8	43.9	44.1	44.8	44.9	45.0	45.1
Bp	27.6*	22.9*	28.5	28.5	29.6	28.8	29.7	30.9	32.1	29.4	29.1	29.7	29.4
SD	23.8	23.1	23.3	24.1	23.5	24.8	25.2	26.3	27.1	24.9	24.7	23.0	23.3
Bd	22.9	23.8	25.1	25.5	25.3	26.6	27.1	27.3	30.3	25.2	24.5	24.3	25.8

ph. 2 post.	*Q1*	*AE1*	*U1*	*D13*	*Q10*	*C6*	*F7*	*Y12*	*P8*	*V1*	*I5*	*U1*	*Y9*
GL	45.1	45.2	45.3	45.4	45.4	45.5	45.5	45.6	45.6	45.7	45.8	45.8	45.9
Bp	26.6*	28.9	29.0	30.1	29.1	30.3	29.9	30.8	30.4	30.6	30.1	29.9	29.1
SD	22.1	24.4	24.5	24.9	24.3	25.2	23.9	25.2	24.5	25.1	24.2	25.7	26.1
Bd	24.1	26.3	26.4	24.9	25.5	27.4	25.6	27.6	27.4	27.2	26.3	26.7	27.9

ph. 2 post.	*Y1*	*D7*	*X1*	*C16*	*M9*	*AG1*	*P8*	*Y1*	*N4*	*T1*	*T1*	*AG1*	*D12*
GL	46.0	46.1	46.2	46.3	46.5	46.6	46.6	46.7	46.8	46.8	46.9	47.2	47.2
Bp	29.6	31.9	29.7	31.6	30.2	31.5	33.5	30.3	30.5	30.8	30.2	30.5	32.3
SD	24.1	26.8	24.6	26.1	25.0	25.8	26.7	25.0	24.9	25.3	24.4	26.0	25.6
Bd	27.3	28.3	27.1	26.4	27.2	25.3	26.3	27.4	25.6	25.6	28.2	25.3	26.7

ph. 2 post.	*E6*	*T1*	*AF8*	*Y1*	*Y1*	*AD4*	*AD4*	*T1*	*AB1*
GL	47.4	48.2	48.6	48.7	48.9	49.1	49.2	49.2	50.0
Bp	29.8	35	31.1	34.2	34.6	31.9	30.7	34.4	31.8
SD	24.8	28.9	24.6	27.8	30.0	25.7	24.4	28.9	25.4
Bd	25.0	29.7	27.5	29.1	29.2	27.7	27.8	29.7	28.2

ph. 3 post.	*M9*	*N4*	*N8*	*P8*	*Y1*	*D13*	*Y12*	*A7*	*D13*	*D12*
DLS	69.9	81.1	66.7	70.2	65.2	75.2	74.5	74.8	73.4	68.1
Ld	58.9	61.3	53.4	57.5	56.8	60.5	62.8	58.6	62.5	57.8

ph. 3	*AG2*	*V1*	*U1*	*E15*	*AG1*	*AG2*	*V1*	*AD1*	*F7*	*Y1*	*AD1*	*X1*	*AC1*
DLS	56.4	56.6	56.7	57.7	58.0	58.6	58.8	58.8	58.9	61.8	62.9	63.0	63.5
Ld	44.7	62.8	45.0	48.6	46.0	49.3	47.1	43.9	47.0	47.4	49.2	49.4	49.0

ph. 3	*S1*	*H8*	*AC1*	*T1*	*AG1*	*AG1*	*AA1*	*A6*	*X1*	*AG1*	*U1*	*E15*	*E15*
DLS	66.6	67.5	67.6	68.7	69.5	69.6	69.8	69.8	70.1	70.3	70.3	70.5	70.6
Ld	56.1	55.6	56.3	55.0	56.6	55.8	55.8	56.6	55.1	58.6	56.8	55.0	54.4

ph. 3	*D10*	*Q10*	*T1*	*T1*	*AG1*	*P6*	*AB1*	*Y1*	*AC1*	*Q10*	*AB1*	*AG1*	*AG1*
DLS	71.4	71.6	71.7	72.0	72.2	72.2	72.9	73.1	73.1	73.3	73.5	73.5	73.5
Ld	55.8	56.2	57.5	56.3	57.4	56.8	56.1	56.3	56.5	61.5	56.8	56.6	61.4
ph. 3	Y1	AG1	M1	Q10	P6	D10	U1	P6	AA1	AG1	Q1	A6	AG1
DLS	73.9	74.4	74.4	74.4	74.9	75.0	75.5	75.8	76.9	77.0	77.7	78.0	78.1
Ld	62.0	61.8	60.8	60.6	60.7	58.3	58.3	60.7	61.1	61.0	60.3	59.5	59.4

ph. 3	*AG1*	*X1*	*AE1*	*X1*	*N8*	*U1*	*Z1*	*Y1*	*AG1*	*V1*	*W1*	*AB1*	*AD4*
DLS	79.1	79.3	79.3	79.5	79.7	80.0	80.2	80.7	80.7	81.0	82.0	82.5	83.0
Ld	60.5	61.6	58.7	59.7	57.7	60.6	62.4	61.9	60.1	60.4	60.3	62.5	58.8

ph. 3	*AG1*	*U1*	*AA1*	*U1*	*Y1*	*AG1*	*Y9*	*AB1*	*AA1*	*T1*
DLS	83.2	83.3	83.7	83.7	83.9	85.3	86.0	91.1	91.2	93.4
Ld	64.2	68.1	59.7	62.4	67.8	64.5	64.7	70.9	72.2	70.0

Atlas	*AE1*	*U1*	*Sk. 2*	*Q1*	*Sk. 12*	*T1*	*B3*	*Sk. 11*	*Sk. 13*	*Sk. 10*	*Sk. 9*	*D8*
GB	138	146	150	155	157	157	167	168	177	185	185	–
GL	100	95	90	105	105	105	113	110	106	109	112	110
BFcd	108	106	105	98	108	120	106	110	110	107	107	114
GLF	100	93	96	94	105	–	97	106	98	103	103	100
LAd	90	82	84	81	87	86	74	90	90	82	92	90
H	80	80	75	80	76	80	85	75	80	75	80	82

Axis	*B5*	*M1*	*M2*	*Q1*	*Sk. 2*	*X8*	*X8*	*Y1*	*AE1*	*AE1*	*Sk. 9*
LCDe	120	110	132	104	123	105	124	108	112	98	126
LAPa	95.4	104.4	103.7	90.1	88.4	95.1	96.0	94.7	88.8	86.1	91.3
BFcr	95.5	102.9	103.3	89.1	99.3	100.3	98.6	94.3	102.1	95.8	100.0
Bpacd	68.3	69.7	66.6	68.4	64.5	65.0	63.6	64.7	66.7	59.0	68.8
BPtr	98.5	108	111.6	90.0	83.6	102.2	103.9	94.4	89.6	82.0	110
SBV	50.8	58.3	55.9	50.5	54.6	58.1	58.4	49.1	53.2	59.5	57.0
BFcd	51.0	49.8	51.1	46.3	46.8	45.5	48.5	49.7	45.0	45.7	51.3
H	–	132	141	113	120	120	130	120	115	105	130

Sacrum	*C11*	*T4*	*H5*	*A11*	*O4*	*X5*	*AD2*	*Y6*	*AG10*	*S3*	*M7*
GL	234	238	274	–	–	–	–	–	–	–	–
PL	223	253	–	220	238	–	–	–	–	–	–
GB	193	214	217	205	203	162	(178)	182	187	(195)	–
BFcr	65.5	69.2	68.3	63	64.0	64	75	65.3	61.8	61.2	63.4
HFcr	31.7	32.2	34.1	30.5	30.7	28.2	28.8	27.4	30.3	30.5	27.7

Table 25.8. Measurements (mm) according to von den Driesch (1976) on postcranial bones of the additional cattle from Dayr al-Barshā.

Humerus	*W05*
Side	*R*
Bp	89
SD	33
Bd	88
BT	74

ph. 1 ant.	*AG02*	*AG02*	*AG02*
GLpe	63.7	64.3	61.5
Bp	30.0	29.7	27.2
SD	25.7	27.1	24.5
Bd	29.3	29.8	28.1

Patella	*G07*	*AG02*
Side	*L*	*L*
GL	65.1	58.1
GB	51.8	51.5

Talus	*AG02*	*AC03*	*AG02*	*AG02*
Side	*R*	*R*	*L*	*L*
GLl	64	65	66.0	70.8
GLm	58.2	63	61.5	64.9
Dl	35.8	38	38.3	37.8
Bd	38.4	43	43.3	43.8

Metatarsus	*O08*	*G07*	*AG02*	*G07*	*G07*	*AC07*
Side	*L*	*R*	*L*	*L*	*R*	*R*
GL	242	244	253	254	256	–
Bp	45.2	46	46.2	49	52.7	45
SD	26.4	25.8	27.3	26.6	26.9	26.4
Bd	52.4	52.7	–	56.8	56.5	–

ph. 2 post.	*AG02*	*AG02*	*AG02*	*AA01*	*AG02*	*AG02*	*AG02*
GL	40.2	40.2	41.4	41.7	44.1	44.4	44.5
Bp	27.9	27.1	28.1	28.3	30.3	28.8	28.9
SD	21.9	21.9	23.0	22.8	25.4	22.9	23.9
Bd	23.6	23.2	23.7	23.3	26.5	25.3	25.5

Phalanx 3	*AG02*	*AG02*	*AG02*	*AG02*	*AG02*	*AG02*	*AG02*
DLS	56.4	58.6	67.0	64.9	67.4	70.3	68.7
Ld	44.7	49.3	51.7	51.8	54.3	55.2	55.5

Metacarpus	*AG02*	*AC08*
Side	*R*	*L*
GL	217	–
Bp	57.6	–
SD	29.7	30
Bd	56.4	61

Femur	*AG02*	*G05*
Side	*R*	*R*
GL	362	–
Bp	113	43.1
DC	43	33.7
SD	31.9	92.7

Tibia	*AG02*	*AG02*	*N03*	*K07–G07*	*O02*
Side	*L*	*R*	*L*	*R*	*R*
GL	363	365	372	375	–
Bp	–	93	93	93	92.2
SD	37.7	39.7	38.8	39.5	–
Bd	60.1	–	62.0	–	–

Calcaneus	*O08*	*AC03*	*AG02*	*AG02*
Side	*R*	*R*	*L*	*R*
GL	126	133	134	140
GB	45	40	–	–

ph. 1 post.	*AG02*	*AG02*	*AC03*	*AC03*	*AC03*
GLpe	60.7	61.2	62	63	68
Bp	26.6	27.0	27	26	30
SD	23.5	23.2	24	25	26
Bd	26.7	25.8	28	26	29

Din 2005). It is clear however that Herodotus cannot be describing a practice of a similarly general application as, in his day, there were bull cemeteries in many places in Egypt, some of which must have been far more important than the one in the Delta he is referring to. Rather he is describing a very specific case, and one that cannot fully explain the situation in Dayr al-Barshā. The cattle pit found there suggests (just as Herodotus describes) that the cattle were first left to decay (maceration), and then dug up to be roughly cleaned. However, they were not taken to Kawm Abū Billū, but reburied in Dayr al-Barshā itself. Again in keeping with Herodotus' report, they were buried there together in one pit. The additional bones from Barsha are from female animals that seem to have been brought from an environment in or close to the Nile, as suggested by the loam-clay loam sediment associated to the bones. This fits remarkably with Herodotus' description of the treatment of female dead cattle; although he does not mention that they were eventually buried together with the males.

Although not much has yet been published about the practice, there is mounting evidence to show that, before the second Persian occupation of Egypt (343–332 BC), it was common to macerate the remains of dead animals by burial before they were committed to their last resting place. This could recently be shown, for instance, by a careful study of the chronological variations in body treatment in the ibis catacombs at Tūna al-Jabal (von den Driesch *et al.* 2005). It is important to note that the animal remains found not only belong to ibises, but also to various other animals including cattle. The cattle remains found in the catacombs of this period had also been macerated (D. Kessler pers. comm.).

While the treatment of the bones in the studied cattle pit thus strongly resembles that of the animals found in the catacombs at Tūna, it is still necessary to explain why this would have happened at Dayr al-Barshā, where there is no other evidence of animal burials linked to a temple cult. The explanation should moreover take into consideration the date of the deposit, 410–360 cal BC. After the New Kingdom, there is, for about a millennium, no evidence of human activity at the site, but in the reigns of pharaohs Nectanebo I (380–363 BC) and Nectanebo II (360–343 BC), huge limestone quarries were opened there. It seems more than likely that the cattle deposit should be linked to the activities of the former ruler. Textual evidence inside the quarries and on a stela from al-Ashmūnayn, dated year 8 of Nectanebo I, suggests that a major aim of the quarry activities was the construction of a huge pronaos in the temple of Thoth in al-Ashmūnayn (Willems *et al.* 2004; in press). Since ancient Egyptian temples were economic institutions as much as centres of religious cult, it is likely that the Thoth temple was directly involved in the organisation of the quarry exploitation in Dayr al-Barshā. Now Kessler (1995) has demonstrated that large cattle herds were linked to the Thoth cult in the region, which was concentrated in al-Ashmūnayn and Tūna al-Jabal. Both localities are just on the opposite side of the Nile at Dayr al-Barsā. From these herds, animals could be selected to serve as the cult animal or Thoth-bull, or as sacrificial animals, but most would be involved in more mundane tasks (as providers of meat or milk, or as draught-animals). It stands to reason that some herds were directed to the quarries. Possibly they were not deployed long for work there, since they had developed little pathologies related to draught activities. It is clear that, after their death, they were not transported to the animal catacombs at Tūna. Instead, they were committed to the earth close to the quarries where they may have been engaged. Their place of burial, suggests, however, that they were not unceremoniously discarded. The burial pit is located in what was, already in the 4th century BC, a very ancient cemetery, and in fact in an older tomb shaft that was slightly enlarged. Since Egyptian funerary symbolism revolved around the theme of the dead being transformed into a form of Osiris, it must have been a small step to envisage the cemetery as the realm of Osiris, precisely as the catacombs at Tūna were. If this is the correct interpretation, it seems that those responsible for the cattle burial may have transferred the ideology of the large, official animal cemeteries to this spontaneous deposit on the desert edge of Dayr al-Barshā.

Acknowledgements

The discovery of the cattle pit in desert plains at Dayr al-Barshā was made while David Depraetere (KU Leuven) was supervising the excavations of this sector in 2006. Véronique De Laet (Center for Archaeological Sciences, KU Leuven) and Bert Dusar (KU Leuven) are acknowledged for the analysis of the sediment associated with the additional cattle individuals. Elena Marinova (Center for Archaeological Sciences, KU Leuven) analyzed the plant remains, and Frederik Hendrickx (Royal Belgian Institute of Natural Sciences) investigated the beetles. Joris Peters and Angela von den Driesch (Institut für Paläoanatomie und Geschichte der Tiermedizin, Munich) commented on the photographs of the pelvic bones. We also thank Louis Chaix (Museum of Natural History of Geneva) and his doctoral student Jérôme Dubusson, for the fruitful discussions on cattle horn deformations in Africa. We are also very grateful to Dieter Kessler (Institute for Egyptology, Ludwig-Maximilians-University Munich), who helped with the interpretation of the cattle pit. Sheila Hamilton-Dyer (Southampton) corrected the English text. Finally we would like to thank Marleen De Meyer (KU Leuven) for spending a free day at Dayr al-Barshā on photographing cattle bones.

References

Aston, D. A. 1996, *Egyptian Pottery of the Late New Kingdom and Third Intermediate Period (Twelfth–Seventh Centuries BC): tentative footsteps in a forbidding terrain.* Heidelberger Orientverlag, Heidelberg.

Bartosiewicz, L., Van Neer, W. & Lentacker, A. 1997, 'Draught cattle: their osteological identification and history'. *Annales Sciences Zoologiques* 281, Musée royal de l'Afrique centrale, Tervuren.

Bietak, M., Forner, J. & Jánosi, P. 2002, 'Ausgrabungen in dem Palastbezirk von Avaris. Vorbericht Tell El–Dab'a/'Ezbet Helmi. 1993–2000'. *Ägypten und Levante* 11, 26–119.

Boessneck, J. 1988, *Die Tierwelt des Alten Ägypten, untersucht anhand kulturgeschichtlicher und zoologischer Quellen.* C. H. Beck, München.

Boessneck, J., Brunsch, W., von den Driesch, A., Grosser, D., Kessler, D., Kolta, K. S., Kudielka, R. A., Parsche, F., Schafer, H., Seifert, P., Storch, W., Vittmann, G. & Ziegelmayer, G. 1987, *Die Münchner Ochsenmumie.* Hildesheimer Archäologische Beiträgen 25, Hildesheim.

Boessneck, J. & von den Driesch, A. 1982, *Tierknochenfunde der Ausgrabungen des Deutschen Archäologischen Insitituts Kairo auf Elephantine. Studien an subfossilen Tierknochenfunden aus Ägypten.* Münchner Ägyptologische Studien 40, München/Berlin.

Boessneck, J. & von den Driesch, A. 1987a, 'Zoologisch-haustierkundliche Befunde an der Rindermummie', in J. Boessneck (ed.), *Die Münchner Ochsenmumie*, 55–71. Hildesheimer Archäologische Beiträgen 25, Hildesheim.

Boessneck, J. & von den Driesch, A. 1987b, 'Die Tierknochenfunde aus den Pavian- und Ibisgalerien von Tuna el-Gebel', in J. Boessneck (ed.), *Tuna el–Gebel I. Die Tiergalerien*, 39–221. Hildesheimer Ägyptologische Beiträge, 24, Hildesheim.

Boessneck, J. & von den Driesch, A. 1992, *Tell el Dab'a VII. Tiere und historische Umwelt im Nordost–Delta im 2. Jahrtausend v. Chr. anhand der Knochenfunde der Ausgrabungen 1975–1986.* Untersuchungen der Zweigstelle Kairo des Österreichischen Archäologischen Institutes 10.

Chaix, L. 1988, 'Le monde animal à Kerma (Soudan)'. *Sahara* 2, 77–84.

Chaix, L. 1993, 'The archaeozoology of Kerma (Sudan)', in W. V. Davies & R. Walker (eds), *Biological Anthropology and the Study of Ancient Egypt*, 175–185. British Museum Press, London,

Chaix, L. 1996, 'Les bœufs à cornes parallèles: archéologie et ethnographie'. *Sahara* 8, 95–97.

Chaix, L. 2000, 'Les animaux et les morts à Kerma (Soudan) entre 2500 et 1500 avant J.-C. Faits archéologiques et interprétations', in L. Bodson (ed.), *Ces animaux que l'homme choisit d'inhumer. Contribution à l'étude de la place et du rôle de l'animal dans les rites funéraires. Journée d'étude Liège 1999*, 15–39. Colloque d'histoire des connaissances zoologiques, Université de Liège 11, Liège.

Chaix, L. 2001, 'Animals as symbols. The bucrania of grave KN24 (Kerma Southern Sudan)', in H. Buitenhuis & W. Prummel (eds), *Animals and Man in the Past. Essays in Honour of Dr. A. T. Clason*, 364–370. ARC-publicatie 41, Groningen.

Chaix, L. 2007, 'Contribution to the knowledge of domestic cattle in Africa: The osteometry of fossil *Bos taurus* L. from Kerma, Sudan (2050–1750 BC)'. *Documenta Archaeobiologiae* 5, 169–248.

Chaix, L. & Hansen, J. W. 2003, 'Cattle with "forward–pointing horns": archaeozoological and cultural aspects. Cultural markers in the Later Prehistory of northeastern Africa and recent research'. *Studies in African Archaeology* 8, 269–281.

Davis, S. J. M. 2000, 'The effect of castration and age on the development of the Shetland sheep skeleton and a metric comparison between bones of males, females and castrates'. *Journal of Archaeological Science* 27(5), 373–390.

Depauw, M. 2008, 'A "verlan" scribe in Deir el-Bersha: some Demotic inscriptions on quarry ceilings'. *Journal of Egyptian Archaeology* 94, 293–295.

Dürst, J. U. 1899, *Die Rinder der Babylonien, Assyrian und Ägypten und ihr Zusammenhang mit den Rinder der alten Welt*. Georg Riemer, Berlin.

Driesch, A. von den. 1976, *A Guide to the Measurement of Animal Bones from Archaeological Sites*. Peabody Museum Bulletin 1, Harvard University, Cambridge MA.

Driesch, A. von den & Boessneck, J. 1974, 'Kritische Anmerkungen zur Widerristhöhenberechnung aus Längenmassen vor- und frühgeschichtlicher Tierknochen'. *Säugetierkundige Mitteilungen* 22(4), 325–348.

Driesch, A.von den, Kessler, D., Steinmann, F., Berteaux, V. & Peters, J. 2005, 'Mummified, deified and buried at Hermopolis Magna. The sacred birds from Tuna El–Gebel, Middle Egypt'. *Ägypten und Levante* 15, 203–244.

Driesch, A. von den & Peters, J. 2003, *Geschichte der Tiermedizin. 5000 Jahre Tierheilkunde*. Schattauer, Stuttgart.

Elias, S. A. 1994, *Quaternary Insects and their Environments*. Smithsonian Institution Press, Washington DC & London.

Epstein, H. 1971, *The Origin of the Domestic Animals of Africa*. Africana Publishing, New York.

Flores, D. V. 2003, *Funerary Sacrifice of Animals in the Egyptian Predynastic Period*. British Archaeological Report S1152, Archaeopress, Oxford.

Flores, D. V. 2004, 'Funerary sacrifice of animals in the Egyptian Predynastic Period', in S. Hendrickx, R. Friedman, K. M. Cialowicz & M. Chlodnicki (eds), *Egypt at its Origins*, 731–763. Orientalia Lovaniensia Analecta 138, Peeters, Leuven.

Gilbert, A. S. 1988, 'Zooarchaeological observations on the slaughterhouse of Meketre'. *Journal of Egyptian Archaeology* 74, 69–89.

Grant, A. 1982, 'The use of tooth wear as a guide to the age of domestic ungulates', in B. Wilson, C. Grigson & S. Payne (eds), *Ageing and Sexing Animal Bones from Archaeological Sites*, 91–108. British Archaeological Report S109, British Archaeological Reports, Oxford.

Greenfield, H. 2006, 'Sexing fragmentary ungulate acetabulae', in D. Ruscillo (ed.), *Recent Advances in Ageing and Sexing Animal Bones. Proceedings of the 9th Conference of the International Council of Archaeozoology, Durham, August 2002*, 68–86. Oxbow Books, Oxford.

Grigson, C. 1975, 'The craniology and relationships of four species of *Bos*. II. Basic craniology: *Bos taurus* L. proportions and angles'. *Journal of Archaeological Science* 2, 109–128.

Grigson, C. 1976, 'The craniology and relationships of four species of *Bos*. III. Basic craniology: *Bos taurus* L. sagittal profiles and other non-measurable characters'. *Journal of Archaeological Science* 3, 115–136.

Grigson, C. 1980, 'The craniology and relationships of four species of *Bos*. V. *Bos indicus* L'. *Journal of Archaeological Science* 7, 3–32.

Grigson, C. 1982, 'Sex and age determination of some bones and teeth of domestic cattle: a review of the literature', in B. Wilson, C. Grigson & P. Payne (eds), *Ageing and Sexing Animal Bones from Archaeological Sites*, 7–24. British Archaeological Report S109, British Archaeological Reports, Oxford.

Grigson, C. 1991 'An African origin for African cattle? – some archaeological evidence'. *African Archaeological Review* 9, 119–144.

Grigson, C. 2000, '*Bos africanus* (Brehm)? Notes on the archaeozoology of the native cattle of Africa', in R. M. Blench & K. C. MacDonald (eds), *The Origins and Development of African Livestock: archaeology, genetics, linguistics and ethnography*, 38–60. University College London Press, London.

Habermehl, K.-H. 1975, *Die Altersbestimmung bei Haus- und Labortieren*. Verlag Paul Parey, Hamburg and Berlin.

Jackson, J. W. 1934, 'The osteology. Report on the remains of sacred cattle from the Baqaria and Bucheum, Armant, Upper–Egypt', in R. Mond & O. Myers (eds), *The Bucheum*. London.

Kessler, D. 1995, 'Der Gott Thot–Stier', in D. Kessler & R. Schulz (eds), *Gedenkschrift Winfried Barta*, 229–45. Münchener Ägyptologische Untersuchungen 4, Frankfurt am Main.

Kessler, D. & Nur El-Din, A. 2005, 'Tuna el Gebel: millions of ibises and other animals', in S. Ikram (ed.), *Divine Creatures. Animal Mummies in Ancient Egypt*, 120–163. American University in Cairo Press, Cairo and New York.

Laudien, A. 2000, 'Osteometrische Untersuchungen zum altägyptischen Langhornrind', PhD thesis, Ludwig-Maximilians-Universität, München.

Lernia, S. di. 2006, 'Building monuments, creating identity: Cattle cult as a social response to rapid environmental changes in the Holocene Sahara'. *Quaternary International* 151(1), 50–62.

Lloyd, A. B. 1976, *Herodotus Book II. Commentary 1–98*. Brill, Leiden.

Lortet, A. & Gaillard, C. 1903, 'La faune momifiée de l'ancienne Egypte'. *Archive du Musée d'Histoire Naturelle, Lyon* 8, 1–205.

Lortet, A. & Gaillard, C. 1905, 'La faune momifiée de l'ancienne Egypte'. *Archive du Musée d'Histoire Naturelle, Lyon* 9, 201–330.

Lortet, A. & Gaillard, C. 1909, 'La faune momifiée de l'ancienne Egypte'. *Archive du Musée d'Histoire Naturelle, Lyon* 10, 1–325.

Macaulay, G. C. 1852–1915, 'The history of Herodotus volume', viewed 21 June 2009, http://onlinebooks.library.upenn.edu/webbin/gutbook/lookup?num=2456

MacHugh, D. E., Shriver, M. D., Loftus, R. T., Cunningham, P. & Bradley, D. G. 1997, 'Microsatellite DNA variation and the evolution, domestication and phylogeography of taurine and zebu cattle (*Bos taurus* and *B. indicus*)'. *Genetics* 146, 1071–1086.

Müller, V. 2008, *Tell el–Dab'a. XVII: Opferdeponierungen in der Hyksoshauptstadt Auaris (Tell el–Dab'a) vom späten mittleren Reich bis zum frühen neuen Reich.* Österreichische Akademie der Wissenschaften. Denkschriften der Gesamtakademie 45/ Untersuchungen der Zweigstelle Kairo des Österreichischen archäologischen Institutes 29, Ausgrabungen in Tell el-Dab'a 2. Verlag der Österreichischen Akademie der Wissenschaften, Wien.

Osypiński, P. & Osypińska, M. 2016. 'The Wadi Khashab ceremonial complex – a manifestation of cattle keepers in the Eastern Desert of Egypt before the end of the fifth millennium BC'. *Azania: Archaeological Research in Africa* 51, 257–281.

Paris, F. 2000, 'African livestock remains from Sahara mortuary contexts', in R. M. Blench & K. C. MacDonald (eds), *The Origins and Development of African Livestock: archaeology, genetics, linguistics and ethnography*, 111–126. University College London Press, London.

Pia, J. 1940, 'Rassenkundliche Untersuchungen an Schädelresten des altägyptischen Hausrindes'. *Zeitschrift für Tierzüchtung und Züchtungsbiologie* 48, 17–55.

Schwabe, C. W. 1984, 'A unique surgical operation on the horns of African bulls in ancient and modern times'. *Agricultural History* 58(2), 138–156.

Silver, I. A. 1963, 'The ageing of domestic animals', in D. Brothwell & E. Higgs (eds), *Science in Archaeology*, 250–268. Thames and Hudson, London.

Sykes, N. & Symmons, R. 2007, 'Sexing cattle horn-cores: problems and progress'. *International Journal of Osteoarchaeology* 17, 514–523.

United States Patents and Trademark Office, 2009, viewed 17 June 2009 http://patft.uspto.gov/netahtml/PTO/srchnum.htm

Vercoutter, J. 1980, *Livre du centenaire 1880–1980*. Institut français d'archéologie orientale du Caire, Cairo.

Willems, H., De Meyer, M., Depraetere, D., Peeters, C., Hendrickx, S., Herbich, T., Klemm, D., Op de Beeck, L. & De Pauw, M. 2004, 'Preliminary report of the 2002 campaign of the Belgian mission to Deir al-Barsha'. *Mitteilungen des Deutschen Archäologischen Instituts Abt. Kairo* 60, 281–282.

Willems, H., De Meyer, M., Depraetere, D., Peeters, C., Op de Beeck, L., Vereecken, S., Verrept, B. & De Pauw, M. 2006. 'Preliminary Report of the 2003 Campaign of the Belgian Mission to Deir al-Barsha'. *Mitteilungen des Deutschen Archäologischen Instituts Abt. Kairo* 62, 307–339.

Willems, H., De Meyer, M., Peeters, C., Vereecken, S., Depraetere, D., Dupras, T., Williams, L., Herbich, T., Verstraeten, G., Van Loon, G., Delattre, A. 2009. 'Report of the 2004–2005 Campaigns of the Belgian Mission to Deir al-Barsha'. *Mitteilungen des Deutschen Archäologischen Instituts Abt. Kairo* 65, 377–432.

26. The Opet Temple courtyard excavations: a new zooarchaeological study for Karnak (Luxor, Egypt)

Hervé Monchot & Guillaume Charloux

The archaeological excavations undertaken in 2006 and 2007 in the courtyard of the Opet Temple at Karnak (Luxor, Egypt) revealed, in 12 soundings, nine phases of occupation and/or construction dating from the end of the 3rd millennium BC to the present. In addition to an abundance of artifacts (ceramics, stone tools, seal impressions, shells, etc) more than 5000 faunal remains were recovered belonging for the most part to the classic domesticates (sheep/goat, cattle and pig) in addition to Nile fishes. Zooarchaeological analysis shows that the majority of bones are burned and that the Egyptians preferentially consumed young animals, lambs/kids, calves and piglets. This study offers a rare insight into the fauna of Karnak, and allows us to give a first glimpse on the life and the behavior of Thebans at the time of the emergence of the Amun cult.

Keywords archaeozoology, Egypt, Thebes, Middle Kingdom, New Kingdom, domestic animals, fish

Introduction

Located 3km north of Luxor (Egypt), the complex at Karnak (Fig. 26.1) was the heart of religious activity during the New Kingdom (1550–1070 BC), as well as an important economic centre. Three main temples dedicated to Mut, Montou and Amun constitute a 40ha religious complex, which was presumably erected at the beginning of the Middle Kingdom, ca. 2100 BC, if we take into consideration results of recent excavations in the central area of the Amun sanctuary (Charloux *et al.* 2004).

Although New Kingdom Karnak has been studied in detail, little is known of the periods preceding the monuments of the XVIIIth Dynasty kings. Obtaining information on the daily life of Thebans in particular has been neglected in favor of architectural and epigraphic research. Zooarchaeological investigations have always been considered as a secondary factor (McArdle 1989), which accounts for the absence of faunal reports for Middle and New Kingdom contexts. Most often the presence of bones is limited to a mere mention in publications (Debono 1982), although animal offerings, such as those found in foundation deposits (Azim 1982) or represented on tombs or temples reliefs (Beaux 1990), that is to say those associated with rituals, have been studied in somewhat greater detail.

The animal world deeply shaped Egyptian religion and imagery. Gods and goddesses are represented with the attributes of wild or sometimes tame animals, predatory birds, crocodiles, hippopotamus and others. Suffice to say that this animal world, like the whole environment, was a gift of the divine to the Thebans. Nevertheless, studies describing the economic behavior of local populations and their mode of exploitation of animal products in such a cultural sphere are still surprisingly missing. The lack of deeply stratified deposits at Karnak represents a major challenge in this matter. Two notable exceptions are current incomplete published works from a priests' residential area located South-east of the Sacred Lake at Karnak (Millet 2007) and from a domestic quarter at North Karnak (Jacquet 2001).

The Opet Temple

The temple of Opet is situated immediately to the west of the Khonsu temple (Fig. 26.2), in the southwest corner of the precinct of Amun at Karnak. It was dedicated to the

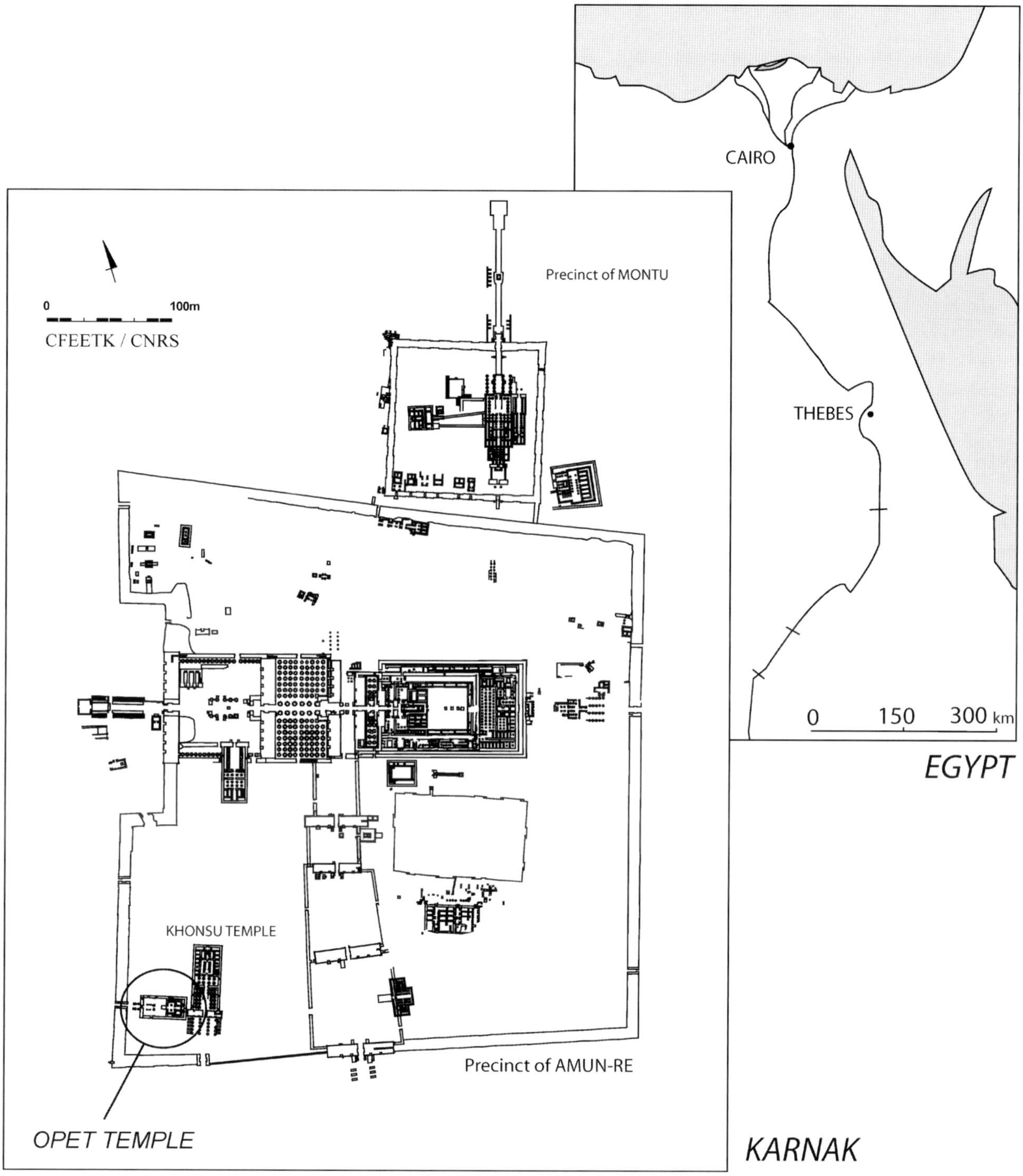

Fig. 26.1. Egypt map showing the Theban area on the Nile. Foreground, the location of temple sites within Karnak

hippopotamus goddess Opet by Ptolemy VIII Euergetes II (2nd century BC), and constitutes one of the last cult buildings erected in Egypt. Its decoration was carved until the reign of the Roman emperor Augustus Caesar (1st century BC–1st century AD), but the monument was left unfinished.

Although rather small in comparison to other sanctuaries, this temple presents a lot of unique features (location, plan, methods of construction, function, etc). It belongs to a very rare category called the 'mythological temples', like the *mammisi*. Originally, the concept of god as a divine entity was not the major goal of the cult, which focused more on the myths in which he was involved. This temple is dedicated to the goddess Opet in order to commemorate the main events of the life of her son Osiris: birth and resurrection, heir of God Amun, as well as the birth of Horus (Traunecker 2004).

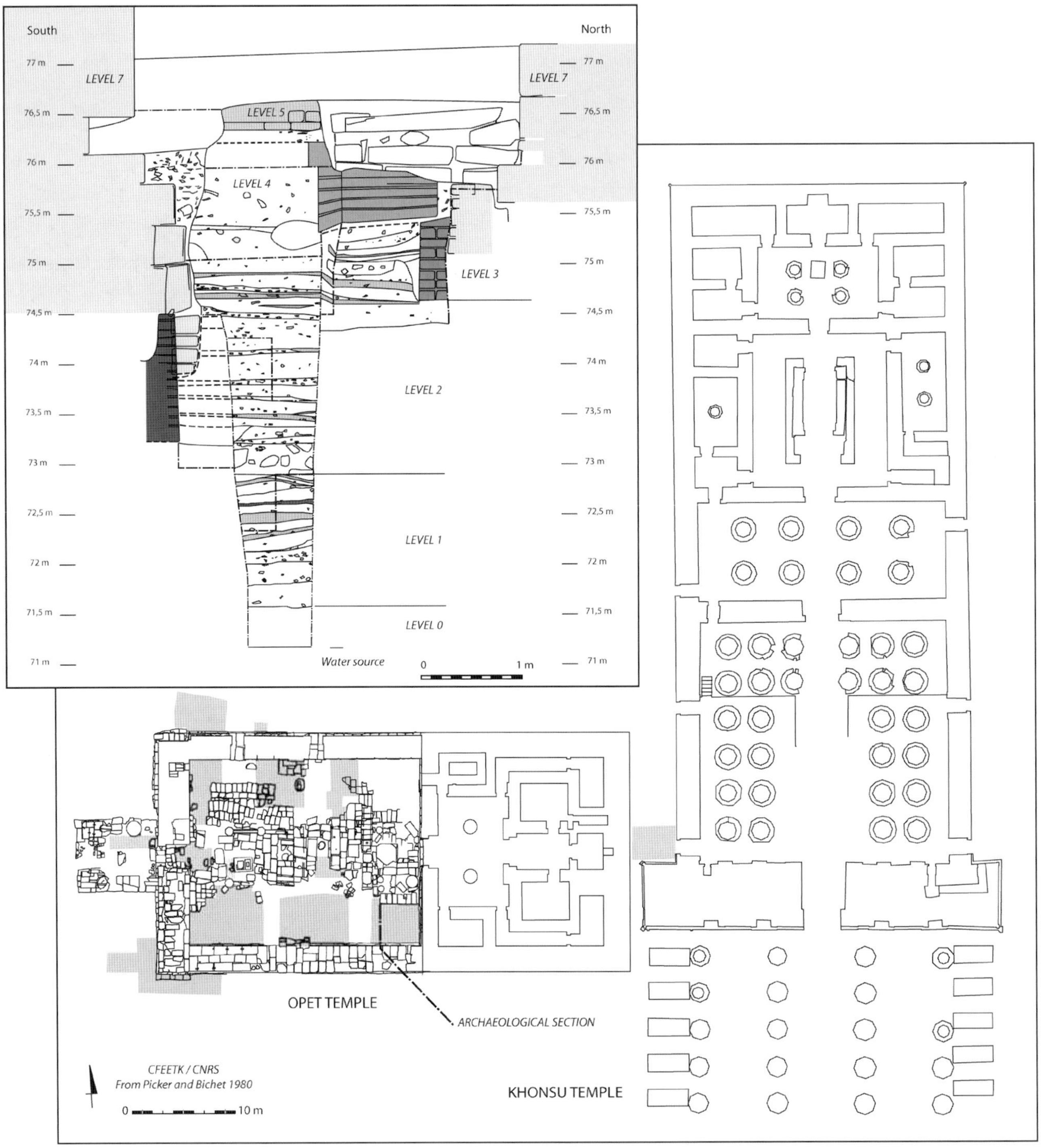

Fig. 26.2 Plan of the archaeological soundings in and around the Opet Temple courtyard and (inset) the deep section, with distinction of levels above the water table.

Stratigraphy, chronology and structural remains

Since 2005, a conservation and tourism management project has been undertaken under the direction of E. Laroze (architect, CNRS/CFEETK). This multi-disciplinary programme is a collaboration between the CNRS/MAE and the Egyptian Supreme Council of Antiquities. Two excavation campaigns in the courtyard of the temple were conducted by G. Charloux in 2006 and 2007 to answer specific architectural and archaeological questions, for instance the origin and the sequence of periods of occupation at the site (Charloux *et al.* 2009).

Nine archaeological levels were distinguished (Table 26.1), and a sounding reached the basal geological soil, revealing one of the deepest archaeological sequence uncovered at Thebes (five levels and 27 layers under the surface) (Fig. 26.2). Furthermore, the archaeological and architectural complexity found under the Opet temple seems to reflect a wide variety of contexts: domestic occupation (levels 1–3), production or storage facilities, monumental construction of unknown nature (level 4) and religious building (levels 5–9). As a consequence, this archaeological study offers a unique chance to understand the past of the whole area. No archaeological research undertaken in a 170m radius around the monument has as yet exposed structures older than the Ramessid period. With the exception of few superficial diggings, some of them incomplete or unpublished, the sector is clean of work. Concerning the archaeological levels we are especially interested in the XIth–XVIIIth dynasties (levels 1–4) according to ceramic studies – the only published data or ongoing excavations are located more than 350m away to the north or to the east (e.g. Charloux 2005; Millet 2007). For more archaeological information, the reader can refer to the review *Les cahiers de Karnak* published by the CFEETK (*Centre franco-égyptien d'étude des temples de Karnak*) and the CNRS (*Centre National de la Recherche Scientifique*) since 1968.

Table 26.1. Stratigraphical sequence of the Opet Temple courtyard excavations.

Levels	*Period*	*Preliminary relative ceramic dating*	*Preliminary relative chronology related to ceramic study*	*Archaeological context*
0	--	Virgin soil	--	--
1	Middle Kingdom to Early New Kingdom	XIth–XIIth dynasty	2100–1750 BC	Occupation floors
2				
3				
4		XIth–XVIIIth dynasty	2100–1400 BC	Foundation filling
5	Late Epoch ? to recent periods	XXVth dynasty according to stratigraphy (?)	7th century BC according to stratigraphy (?)	Foundation filling
6		XXVth dynasty	7th century BC	Temple foundation
7		until the Hellenistic period	until 2nd century BC	Temple foundation
8		until the Late Roman period	until 4th–7th century AD	Pit
9		medieval and modern periods according to stratigraphy	until 2008	Pit

During excavation of 12 soundings both within and just outside the Opet Temple, abundant cultural material – presently under study – was uncovered, including shells, lithics, stone beads, ceramics, seal impressions, as well as faunal remains. Except for surface fills and modern pits, all layers from the soundings were sieved – with an 8mm diameter mesh. A smaller sieve (5mm mesh size) was frequently used for sandy-silt and ashy layers.

The excavations of the courtyard Opet Temple yielded a total of 5234 animal remains and 91% of these (n=4761) belong to Middle Kingdom occupations to the early New Kingdom (henceforth MK–early NK; XIth–XVIIIth dynasties, ca. 2100–1400 BC),[1] which is the reason our

Table 26.2. Species list count recovered from the courtyard Opet temple excavations (Occ = occupations; Fil = filling).

AGE	*Middle Kingdom to early New Kingdom*						*Late epoch to modern period*					
Archaeological Period	*1*	*2*	*3*	*4*	*Total*		*5*	*6*	*7*	*8/9*	*Total*	*% total*
Context	*Occ*	*Occ*	*Occ*	*Fil*			*Fil*	*Temple*	*Temple*	*Pit*		
Consumed species					*N*	*%*						
Sheep/Goat *(Ovis aries/Capra hircus)*	16	35	29	123	203	4.3	1	20	11	12	247	4.72
Cattle (*Bos taurus*)		28	20	81	129	2.7	3	16	4	1	153	2.92
Pig *(Sus scrofa* dom.*)*	5	80	5	50	140	2.9	3	5	3	1	152	2.90
Pisces	2	39	17	66	124	2.6	1	7	1	2	135	2.58
Aves sp.	1	2	2	4	9	0.2	1	2	1		13	0.25
Various status												
Donkey *(Equus asinus)*		2			2	–				1	3	0.06
Dog *(Canis familiaris)*		2	2		4	–					4	0.08
Rodents (*Gerbillus* sp./*Arvicanthis niloticus*)		1	2	30	33	0,7	2	17			52	0.99
Unidentified mammallian remain												
Caprini size	34	72	13	183	302	6,3	20	22	1	6	351	6.71
Cattle size	2	31	4	41	78	1.6	1	20		1	100	1.91
Unknown size	144	1019	413	2161	3737	78.5	31	151	21	84	4024	76.88
Total	204	1311	507	2739	4761	–	63	260	42	108	5234	–
% Total	3.90	25.05	9.69	52.33	100	100	1.20	4.97	0.80	2.06	–	100

study will focus almost exclusively on these periods. Levels 1–3 (XIth–XIIth dynasties) comprisea succession of thin horizontal occupation layers – floors and/or household rubbish dump, although the fauna from level 4 comes mainly from the homogeneous filling of deep foundation pits with mixed MK–early NK material (Table 26.1). The upper levels (5–9) provide interesting insights into the later period, but their examination cannot be considered relevant for a comprehensive study due to mixing of deposits.

In this imperfect methodological context with small sample sizes, the amount of bones collected still constitutes an interesting first reference point for our knowledge of the ancient fauna of the late 3rd–mid-2nd millennium BC.

The faunal list

Among the totality of the collection of the faunal remains, 14.5% (759 elements) could be identified with some certainty to species (Table 26.2). The three main domesticates, caprids (sheep/goat), cattle and pig, made up 72.7% of the identified material. Except for the short bones, that are more robust and compact, such as the carpals, tarsals and phalanx, all bones are fragmented into many pieces, especially the long-bones. Consequently, 4471 bones could be identified only to body size group: (1) small size (caprid/young pig), (2) large size (large pig/cattle) and (3) unknown size (splinter). No worked bones and no wild ungulate species such as gazelle (*Gazella*) or Barbary sheep (*Ammotragus*) were recovered from the site. Beside the mammal bones, 135 remains belong to Nile fishes and 13 to wild birds were identified. This ratio is almost the same for MK–early NK material with 472 bones (73.3% of the identified material) for the main domesticate species (Table 26.2).

State of bone preservation

The bones show few signs of weathering processes suggesting that they were not exposed on the surface for very long. No connective tissue or articulated bones (complete joints) were found among the deposit. Most of the bones remains of levels 1 and 2, the older occupations not so far from the Nile water level, are covered by carbonated concretions, which affected the bones surface after their burial. This mineralisation results from water circulation through the deposits.

Around 25% of the identified bones and 40% of the unidentified material are burnt, sometimes intensely calcined. The examination of the colour and the macroscopic appearance shows that a great majority of bones are black (stages 2–3) while few are white (stage 6) the latter representing very high temperatures (Nicholson 1993; Stiner *et al.* 1995).

Consumed species: the classic domesticates (cattle, sheep/goat, pig)

Sheep/goat

Despite published criteria for separating *Ovis* and *Capra* (Boessneck *et al.* 1964; Payne 1985; Prummel & Frisch 1986; Helmer 2000; Fernandez 2001; Halstead *et al.* 2002), the anatomical distinction between them when studying bone fragments is difficult, particularly when they are butchered, cooked, and exposed to soil for millennia. The presence of some well-preserved elements like the talus, humerus or ulna, permitted clear identification of goats, suggesting that they were much more common at Karnak, than at most Egyptian sites where sheep predominate (e.g. Kom el-Hisn, Lehner, Ibrahim Awad, Merimde; Redding 1992) or as in the Middle East in general (Redding 1985). While sheep would have competed with cattle for floodplain grazing area, goats would have subsisted on poor-quality vegetation along the desert edge (Rossel 2006).

According to the diagrammatic representation of wear-state in the 3rd milk molar (dP4) (Payne 1973; 1985), the deciduous teeth (one lower dP3, nine lower dP4, one upper dP3, three upper dP4: 14/91=15.4%) clearly show predominance of young adults 1–2 years old. This result is confirmed by the presence of unworn teeth (three lower molars and two upper molars). Only one mandible belonging to the level 3 and one upper M^1 from level 1 attest to the presence of old individuals. Therefore, for the MK–early NK period, we observe approximately 50% lambs/kids, 20% young adults (teeth not worn), 23% mature and 7% old individuals. No yearlings were found in the assemblage. According to caprine-management within the present-day herding systems in the southeast of France (Blaise 2005; Helmer *et al.* 2007), such a choice corresponds with the consumption of young males for tender meat at the maximum weight. This profile was found notably during the Bronze Age in many Predynastic Egyptian sites (Van Neer 2002), or in the northern Near East, as at Sheik Hassan (middle Uruk period, Syria) and southeastern France (Vila 1998; Helmer *et al.* 2007).

For caprids, the whole skeleton is represented (Table 26.3) with a higher proportion of skull (isolated teeth) elements (n=124, 54.2%) and feet elements (carpals, tarsals, metapodials phalanx and sesamoids, n=70, 28.2%). The axial skeleton (ribs and vertebrae) are under-represented (4.5%), but if we take into account elements placed in the caprid size category (with respectively 30.5% of the elements), we obtain more coherent results (35%). The forelimb (6%) and the hindlimb (5.5%), rich meat elements, are scarce and the long-bones diaphyses – whether determined or not – are more plentiful than the extremities. Only ten bones exhibit unfused epiphysis (two distal radius, four calcanei, one femora distal and one 2nd phalanx) or juvenile aspect (two astragali), although this is not a large enough sample to construct a survivorship curve age distribution.

Table 26.3. Skeletal parts of domesticated mammals in NISP (Number of Identified Specimens) and % of NISP from the courtyard Opet temple excavations. (M.K. to N.K. = Middle Kingdom to early New Kingdom period; L.E. to R.P. = Late Epoch to recent period according to Table 26.1).

	CAPRID				CAPRID SIZE				PIG				CATTLE				CATTLE SIZE			
	M.K. to N.K.		L.E. to R.P.		M.K. to N.K.		L.E. to R.P.		M.K. to N.K.		L.E. to R.P.		M.K. to N.K.		L.E. to R.P.		M.K. to N.K.		L.E. to R.P.	
	NISP	%	NISP	%	NISP	%	NISP	%	NISP	%	NISP	%	NISP	%	NISP	%	NISP	%	NISP	%
Cranium	6	3.0	1	2.3	9	3.0	--	--	3	2.1	--	--	13	10.1	1	4,2	7	9,0	--	--
Mandible	13	6.4	--	--	4	1.3	--	--	9	6.4	--	--	13	10.1	1	4.2	--	--	--	--
Isolated Teeth	91	44.8	13	29.5	--	--	--	--	91	65.0	5	41.7	57	44.2	4	16.7	--	--	--	--
CRANIUM	110	54.2	14	31.8	13	4.3	--	--	103	73.5	5	41.7	83	64.4	6	25.1	7	9.0	--	--
Vertebra	6	3.0	--	--	32	10.6	8	16.3	--	--	--	--	8	6.2	5	20.8	12	15.4	2	9.1
Rib	3	1.5	2	4.5	60	19.9	16	32.7	--	--	--	--	11	8.5	--	--	16	20.5	7	31.8
TRUNK	9	4.5	2	4.5	92	30.5	24	49.0	--	--	--	--	19	14.7	5	20.8	28	35.9	9	40.9
Scapula	4	2.0	--	--	11	3.6	5	10.2	--	--	--	--	3	2.3	--	--	1	1.3	--	--
Humerus	3	1.5	2	4.5	10	3.3	2	4.1	2	1.4	1	8.3	--	--	--	--	--	--	--	--
Radio-ulna	5	2.5	3	6.8	2	0.7	1	2.0	2	1.4	--	--	2	1.6	1	4.2	1	1.3	--	--
FORELIMB	12	6.0	5	11.3	23	7.6	8	16.3	4	2.8	1	8.3	5	3.9	1	4.2	2	2,6	--	--
Pelvis	1	0.5	--	--	4	1.3	--	--	--	--	--	--	1	0.8	1	4.2	--	--	--	--
Femur	4	2.0	--	--	4	1.3	2	4.1	--	--	--	--	1	0.8	2	8.3	--	--	--	--
Patella	1	0.5	--	--	--	--	--	--	--	--	--	--	1	0.8	--	--	--	--	--	--
Tibia	4	2.0	--	--	1	0.3	1	2.0	--	--	1	8.3	--	--	2	8.3	1	1.3	--	--
Malleolus	1	0.5	--	--	--	--	--	--	--	--	--	--	--	--	--	--	--	--	--	--
HINDLIMB	11	5.5	--	--	9	2.9	3	6.1	--	--	1	8,3	3	2.4	5	20.8	1	1,3	--	--
Carpals	5	2,5	--	--	--	--	--	--	--	--	--	--	2	1.6	1	4.2	--	--	--	--
Metacarpal	3	1.5	--	--	--	--	--	--	1	0.7	1	8.3	--	--	--	--	--	--	--	--
Astragalus	10	4.9	--	--	--	--	--	--	1	0.7	1	8.3	--	--	--	--	--	--	--	--
Calcaneus	6	3.0	--	--	--	--	--	--	--	--	--	--	1	0.8	--	--	--	--	--	--
Others tarsal	1	0.5	--	--	--	--	--	--	8	5.7	--	--	1	0.8	--	--	--	--	--	--
Metatarsal	8	3.9	7	15.9	--	--	--	--	--	--	--	--	1	0.8	--	--	--	--	--	--
Metapodial	12	5.9	3	6.8	1	0.3	--	--	7	5.0	--	--	4	3.1	--	--	--	--	--	--
Phalanx 1	3	1.5	1	2.3	--	--	1	2.0	5	3.6	1	8.3	4	3.1	--	--	--	--	--	--
Phalanx 2	5	2.5	1	2.3	--	--	--	--	10	7.1	--	--	2	1.6	2	8.3	--	--	--	--
Phalanx 3	3	1.5	1	2.3	--	--	--	--	1	0.7	2	16,7	--	--	--	--	--	--	--	--
Sesamoids	1	0.5	--	--	--	--	--	--	--	--	--	--	2	1.6	--	--	--	--	--	--
FEET	57	28.2	13	29.6	1	0.3	1	2.0	33	23.5	5	41.6	17	13.4	3	12.5	--	--	--	--
Long bone	4	2.0	10	22.7	164	54.3	13	26.5	--	--	--	--	2	1.6	4	16.7	40	51.3	13	59.1
Burnt Bone	60	29.6	15	34.1	102	33.8	27	55.1	31	22.1	4	33.3	31	24.0	6	25.0	30	38.4	10	45.4
Total	203	100	44	100	302	100	49	100	140	100	12	100	129	100	24	100	78	100	22	100

Cattle

Maxilla, mandible and tooth fragments are the most numerous cattle remains (64.4% of all cattle skeletal elements) followed by the trunk (14.7%, and increase at 22.7% with the cattle sized elements) and the feet extremities (13.4%). In contrast for caprids, the rich meat parts of the skeleton, the forelimb and the hindlimb, are rare with 3.9% and 2.4 % respectively. The presence of two unerupted molars, of one unworn lower M_3, of one upper dP^4 and of one mandible with the M_2 in the crypt indicate the presence of calves or young adults. Only one complete mandible (level 5) testifies to the presence of a mature individual. Finally two bones had unfused epiphyses, a distal radius epiphysis (level 2) and a distal femur condyle (level 6). These two bones fuse around 3.5 years (Barone 1986).

Pig

With 152 elements, pig remains constitute a large proportion of the mammalian faunal remains. Like the caprid and the cattle, pig is represented mostly by skull remains with isolated teeth or enamel fragments (n=108, 71% of identified skeletal elements) and feet elements (n=38, 25%). Trunk and hindlimb elements are absent while the forelimb was represented only by four remains, two humeri and two radio-ulnae. As indicating the presence of burnt traces, on teeth, the abundance of enamel fragments is the result of the explosion of teeth when burnt. It is also very difficult to establish the kill-off pattern, because very few complete teeth are present. Nevertheless, we can note the presence of six fragments representing incompletely formed teeth, with open roots, and one mandible and one maxilla with deciduous teeth, evidence for piglets (Rowley-Conwy 1993). As such in Karnak like elsewhere, domestication is assumed in a context where there is large percentage of piglet bones, showing a systematic and regular culling, contrarily to hunters of wild boar which usually create bone assemblages comprising older animals (Lobban 1998). Abundant pig bones were mentioned in the excavations at Karnak to the east of the Sacred Lake beyond the outer wall of Tuthmose III (Debono 1982).

Before the appearance of the Judaic or Islamic taboo, one can see a long history of negative images and roles for pigs in Egypt. As A. J. Cagle says (Cagle 2002):

> 'The role of the pig (*Sus scrofa/Sus domesticus*) in the diet of ancient Egypt has been somewhat confusing for one simple reason: there is an almost total absence of pig remains from tomb provisions and a similar dearth of pictorial and textual representations from tomb and temple contexts. The main reason cited for this absence of evidence is mythological in nature: the male pig was often a manifestation of the evil god Seth and was thus considered a ritually unclean or impure animal much as it is today among various religions'. (see also Newberry 1928; Malaise 1988; Redding 1992; Lobban 1998)

The minor species: donkey, dog and rodents

Beside the domestic species, remains belonging to dog (n=4), donkey (n=3) and rodents (n=52) were identified.

Dog (*Canis familiaris*) is attested by the presence of teeth and may have been used for guarding, hunting or as a companion, but they could also have been used for cleaning the environment. In many rural communities dogs live essentially on garbage, and packs of pariah dogs too were a menace in towns and villages (Dixon 1989).

Three remains of donkey (*Equus cf. asinus*) were observed. The talus found in US 168159 (level 9) is clearly sub-recent, but the sesamoid and the phalanx III found in the level 2 are contemporaneous with the first phase occupation of the site. The donkey was the most important load carrier in Ancient Egypt, attested already in the Maadi Period (Predynastic period). It was used for ploughing the seed into the ground. Donkeys were also used to carry people (Blench 2000; Rossel 2006).

The rodents, essentially gerbils (cf *Gerbillus* sp.) and the Nile rat (cf *Arvicanthis niloticus)*, constitute intrusive species. The presence of these two species is attested in the majority of archaeological sites in Egypt where sieving was undertaken. The Nile rat lives near houses and is at present considered in Egypt as the most harmful rodent species because of the damage it causes to agriculture. Gerbils (the species represented at the Opet cannot be identified), live generally in sandy places of semidesert regions, containing a minimum of natural vegetation or cultures. Moreover the Theban necropolis was overrun by rodents which wriggled down through every crack and crevice to gain access to the food offerings placed with the dead; and in virtually every house in the Middle Kingdom workers' town at Kakhun the corners of the rooms had been tunnelled through by rodents (Dixon 1989, 194–195)

Table 26.4. Fish taxa present on the Opet Temple courtyard excavations (identification: Wim Van Neer, Natural History Museum, Brussels).

Family	*Species*	*Common Name*	*Anatomical Elements*
Clariidae	*Clarias* sp.	Catfish	Pectoral spine, cranium roof fragment, vertebrae, urohyal, articular
Mochokidae	*Synodontis* sp.	Catfish	Several pectoral spines
Mormyridae	*Hyperopisus* sp.	Elephantfish	Dentary
Latidae	*Lates niloticus*	Nile Perch	Anal pterygiophore
Bagridae	*Bagrus bajad*	Bayad (catfish)	Ceratohyal
Cichlidae	Tilapiini		Dorsal spine, 1st vertebra, 3rd vertebra
Cyprinidae	cf *Barbus-Labeo*		Precaudal vertebra

Nile fishes (identified by Wim Van Neer, Natural History Museum, Brussels)

One hundred and thirty-five fish remains were identified in the faunal assemblage, representing the major fish taxa that are encountered in the ichtyofaunal record of Egyptian sites (Van Neer 2004). In spite of the fragmentation of the bones and the presence of numerous vertebrae, for which the generic attribution is often difficult to establish, certain diagnostic pieces permitted the identification of several taxa (Table 26.4). The Clariidae – catfish– are well represented in the site. Two genera are found in the Egyptian Nile, *Clarias* and *Hetrobranchus*, but only the first one was determined with certainty based on the pectoral spine (Gayet & Van Neer 1990). Identification to species is only possible based on the entire vomer bone, an element absent in the Opet sample (Gautier & Van Neer 1989). Several pectoral spines allow us to identify *Synodontis*, but it is still difficult, without more diagnostic elements (i.e., the cleithra) to determine which species is present at the site. The other fishes identified in the studied area are *Bagrus bajad*, as well as Tilapia (tribe Tilapiini), Cyprinids (mainly *Barbus bynni*, sometimes *Labeo*) and the Nile perch (*Lates niloticus*).

The aforementioned taxa, which usually represent more than 90% of the fish bones in Egyptian archaeological sites, can be subdivided into two ecological groups, namely 'floodplain dwellers' and 'open water taxa' (Van Neer 2004). The floodplain dwellers are fish that prefer shallow environments and that can survive in adverse conditions. This group comprises the Clariidae that have accessory breathing organs, which enable them to use atmospheric oxygen. The Tilapiini and cyprinids also belong to this group; they do not have accessory respiratory organs, but their haemoglobin has a high affinity for dissolved oxygen (Fish 1956). The open water species typically spend most of their life in the main river and include Nile perch and the catfish genera *Bagrus* and *Synodontis*. As such the presence of bones from these larger fishes indicates harpoon, spear or net fishing from boats.

Birds

Thirteen wild/domesticate bird remains, of which nine belong to the MK–early NK period, were recovered in the bone assemblage and none of them could be identified to species as they represented fragments of long-bone diaphysis or of skull. If birds, such as geese, ducks and pigeons constituted an important source of the lists of funeral offering (Darby *et al.* 1977), it was not rare to find numerous breed fowls, captured or bought in villages as for instance in Deir El-Medineh (Malaise 1989).

Zooarchaeological interpretation

Few stratigraphic differences are evident between the different levels in the species and manner in which animals were consumed for the MK–early NK period. Caprines dominated, followed by pigs and cattle in equal proportions. The observed differences in some archaeological levels are essentially connected to the larger numbers of small easily recognizable dental fragments which have increased the NISP counts (for example for the pig in the levels 2 and 5) or connected simply to the size of the sample which is bound to the volume of deposit excavated in various soundings. Nevertheless, the body part representation (Table 26.3) is similar between the three domestic species: the rich parts in meat are even under-represented relative to the parts poor in meat such as the head or the extremities of the legs.

To explain this skeletal representation, we cannot call upon diagenesis or the presence of foundation deposits or offering for the reasons evoked above. Rather, the bones appear to represent cooking waste abandoned after treatment and consumption of carcasses. Indeed, the extremities of legs and crania are not preferentially consumed food, despite the fact that people like to eat brain or tongue.

Burned bones are present in all archaeological levels (Table 26.3). All the skeletal parts are affected by fire, but it is especially the fragments of long-bone diaphyses, vertebrae and ribs that increase the percentage of burned bones. The largest bone splinters are not completely burnt, but exhibit scorched areas due to fire, which, suggests that bones were in contact with fire during the preparation of the meat, such as during the roasting. Another explanation would be that the bones accidentally burned after being disposed of or even that they were intentionally burnt as garbage.

Lastly, one bone (a *Bos* size rib fragment) presents cut marks, but many pieces show clearly intentionally bone break (slicing marks, for meat preparation or marrow extraction) were present on the bone material, showing butchering damage. Gnaw marks made by rodents or carnivores were rarely seen in the sample: only four bones exhibited signs of carnivore (dog?) damage and an additional one may have been gnawed by a rodent. This suggests that the remains were rapidly buried.

Conclusion

Due to a complex sequence of occupation and to small sample sizes, the zooarchaeological analyses previously described allow only a partial reconstruction of the lifestyle of the inhabitants of Karnak during the Middle Kingdom up to the Early New Kingdom at the current temple of Opet. Sheep and goat were widely consumed, and in to a lesser degree cattle and pig. They were suppliers of meat, and the absence of an important number of adult means a limited exploitation of the secondary products (milk, wool, etc). The presence of numerous young adults among the domestic animals, associated with numerous remains of fishes, would show that the present occupations on the place of Opet, not so far from the divine domain of Amun-Ra, would be connected to an administrative zone with strong agricultural and craft connotation as the first results of the lithic industry suggests (Raphaël Angevin, pers. comm. 2008) and underlines the fact that the inhabitants were low-ranking consumers (city-dwellers) or granary staff, and not breeders. The absence of wild game common in some Predynastic sites (Boesneck1988; de Miroschedji *et al.* 2001, Van Neer 2002) among the Opet assemblage could have three main explanations: 1. with the intensification of agriculture on the floodplain and heavy hunting during the Predynastic period, wild animals disappeared with their habitat; 2. the urban context of the first occupations in Opet or; 3. the food remains in this sample represent low status food waste, hunted animals may have been luxury items at this time.

The low frequency of cattle would be explained by the fact that these animals were under the control of a central authority, with for instance the presence of cattle rearing in some places, such that they are commonly found on offerings lists and tables or in foundation deposits (Weinstein 1973; Redding 1992; Bartosiewicz 2000; Rossel 2006). The low number of pig bones indicates no reliance on pig. Redding (1992) argued that in the subsistence system of ancient Egypt the pig was not transported or manipulated by a central authority. Instead, the pig functioned as a locally maintained, inexpensive resource that an individual family could rear to supplement other sources of protein. But, the pig in Ancient Egypt is incompatible with intensive production of grains; hence at the site or in the area, as involvement in wheat/barley agriculture increases the use of pigs would decline (Redding 1992, 104). Finally, fish represented an important part of the Theban supplies, like a staple food, because the animals seasonally occur in large numbers on the floodplain and were intensively exploited. Moreover fish could be easily dried for future consumption (Van Neer 2004).

This zooarchaeological study provides another insight into Theban life and offers a deeper appreciation of the sacred domain of Amun during the late 3rd–mid-2nd millennium BC.

Acknowledgements

We wish to thank E. Laroze and the whole staff of the CFEETK for their daily help. We are also very much indebted to Mr. Boreik, Director of the Upper Egypt council of Antiquities, Mr. Suleiman, Director of the temples of Karnak, and the Egyptian staff of the SCA for their assistance. We also wish to thank Ms. Guichard and the American Foundation Michela Schiff Giorgini for their financial support. Thanks to Wim Van Neer for help with the fish identification and L.K. Horwitz for critical comments. The first author is grateful to the ASWA Conference organisers who invited him to Al Ain.

Note

1 XIth dynasty belongs to both late First Intermediate Period and early Middle Kingdom. It is here included in the Middle Kingdom period to simplify our discussion.

References

Azim, M. 1982, 'Découverte de dépôts de fondation d'Horemheb au IXe pylône de Karnak'. *Karnak* 7, 93–120.

Barone, R. 1986, *Anatomie comparée des mammifères domestiques. Tome 1. Ostéologie.* Vigot, Paris.

Bartosiewicz, L. 2000, 'Cattle offering from the Temple of Montuhotep Sankhkara (Thebes, Egypt)', in M. Mashkour, A. Choyke, H. Buitenhuis & F. Poplin (eds), *Archaeozoology of the Near East IV B*, 164–176. ARC Publicatie 32, Groningen.

Beaux, N. 1990, *Le cabinet de curiosités de Thoutmosis III.* Orientalia Lovaniensia analecta 36, Peeters, Leuven.

Blaise, E. 2005, 'L'élevage au Néolithique final dans le sud-est de la France: éléments de réflexion sur la gestion des troupeaux'. *Anthropozoologica* 40, 191–216.

Blench, R. M. 2000, 'The history and spread of donkeys in Africa', in P. Starkey & D. Fielding (eds), *Donkeys, People and Development*, 22–30. Animal Traction Network for Eastern and Southern Africa, Technical Centre for Agricultural and Rural Cooperation, Wageningen, Netherlands.

Boessneck, J., Müller, H.-H. & Teichert, M. 1964, 'Osteologische Unterscheidungsmerkmale zwischen Schaf (*Ovis aries* Linné) und Ziege (*Capra hircus* Linné)'. *Kühn Archiv* 78, 1–129.

Boessneck, J. 1988, *Die Tierwelt des Alten Ägypten.* C. H. Beck, Munich.

Cagle, A. J. 2002, 'Differential Consumption of Pig vs. Sheep/Goats at the Old Kingdom Site of Kom el-Hisn'. Unpublished paper presented at the 53rd Annual Meeting of the American Research Center in Egypt, Baltimore, Maryland.

Charloux, G. 2005, 'The Middle Kingdom Temple of Amun at Karnak'. *Egyptian Archaeology* 2, 20–24.

Charloux, G., Jet, J.-F., & Lanoë, E. 2004, 'Nouveaux vestiges des sanctuaires du Moyen Empire à Karnak. Les fouilles récentes des cours du VIe pylône'. *Bulletin de la Société française d'Egyptologie* 160, 26–46.

Charloux, G., Laroze, E., Angevin, R., Brouillet, A., Monchot, H., Roberson, J. & Virenque, H. 2009, 'Sous le temple d'Opet à Karnak'. *Archéologia* 463, 28–39.

Darby, W. J., Ghalioungui, P. & Grivetti, L. 1977, *Food: the Gift of Osiris.* Academic Press, London, New York.

Debono, F. 1982, 'Rapport préliminaire sur les résultats de l'étude des objets de la fouille des installations du Moyen Empire et "Hyksos" à l'est du lac sacré de Karnak'. *Karnak* 7, 377–383.

Dixon, D. M. 1989, 'A note on some scavengers of Ancient Egypt'. *World Archaeology* 21(2), 193–197.

Fernández, H. 2001, 'Ostéologie comparée des petits ruminants eurasiatiques sauvages et domestiques (genres Rupicapra, Ovis, Capra et Capreolus): diagnose différentielle du squelette appendiculaire'. Unpublished Ph.D dissertation, University of Geneva.

Fish, G. R. 1956, 'Some aspects of the respiration of six species of fish from Uganda'. *Journal of Experimental Biology* 33, 186–195.

Gayet, M. & Van Neer, W. 1990, 'Caractères diagnostiques des épines de quelques silures africains'. *Revue de zoologie africaine* 104, 241–252

Gautier, A. & Van Neer, W. 1989, 'The animal remains from the Late Paleolithic sequence in Wadi Kubbaniya', in F. Wendorf, R. Schild, & A. E. Close (eds), *The Prehistory of Wadi Kubbaniya 2, Stratigraphy, Paleoeconomy and Environment*, 119–161. Southern Methodist University Press, Dallas TX.

Halstead, P., Collins, P. & Isaakidou, V. 2002, 'Sorting the sheep from the goats: morphological distinctions between the mandibles and mandibular teeth of adult *Ovis* and *Capra'*. *Journal of Archaeological Science* 29, 545–553.

Helmer, D. 2000, 'Discrimination des genres *Ovis* et *Capra* à l'aide des prémolaires inférieures 3 et 4 et interpretation des ages d'abattage: l'exemple de Dikili Tash (Grèce)'. *Ibex, Journal of Mountain Ecology* 5, 29–38.

Helmer, D., Gourichon, L. & Vila E. 2007, 'The development of the exploitation of products from *Capra* and *Ovis* (meat, milk and fleece) from the PPNB to the Early Bronze in the northern Near East (8700 to 2000 BC cal.)'. *Anthropozoologica* 42, 41–69.

Jacquet, J. 2001, *Karnak-Nord IX: Le Trésor de Thoutmosis 1er. FIFAO* 44, Le Caire.

Lobban, R. A. 1998, 'Pigs in ancient Egypt'. *MASCA Research Papers in Science and Archeology* 15, 137–148.

Malaise, M. 1988, 'Les animaux dans l'alimentation des ouvriers égyptiens dans l'alimentation de Deir El-Médineh au nouvel Empire'. *Anthropozoologica* Special No. 2, 65–72

McArdle, J. 1989, 'Old bones, new ideas: zooarchaeology at Hierakonpolis, Part 1'. *Nekhen News* 5, 3–4.

Millet, M. 2007, 'Architecture civile antérieure au Nouvel Empire: rapport préliminaire des fouilles à l'est du lac Sacré, 2001–2003'. *Karnak* 12, 681–763.

Miroschedji, P. de, Sadeq, M., Faltings, D., Boulez, V., Naggiar-Moliner, L., Sykes, N. & Tengsberg, M. 2001, 'Les fouilles de Tell es-Sakan (Gaza): Nouvelles données sur les contacts égypto-cananéens aux IVe–IIIe millénaires'. *Paléorient* 27, 75–104.

Newberry, P. E. 1928, 'The pig and the cult-animal of Seth'. *Journal of Egyptian Archaeo*logy 14, 211–225.

Nicholson, R. A. 1993, 'A morphological investigation of burnt animal bone and evaluation of its utility in archaeology'. *Journal of Archaeological Science* 20, 411–428.

Payne, S. 1973, 'Kill-off patterns in sheep and goats: the mandibles from Asvan Kale'. *Anatolian Studies* 23, 281–303.

Payne, S. 1985, 'Morphological distinctions between the mandibular teeth of young sheep, *Ovis* and goats, *Capra*'. *Journal of Archaeological Science* 12, 139–147

Prummel, W. & Frisch, H.-H. 1986, 'A guide for distinction of species, sex and body size in bones of sheep and goat'. *Journal of Archaeological Science* 13, 567–577.

Redding, R. W. 1985, 'The role of faunal remains in the explanation of the development of complex societies in south-west Iran: potential, problems and the future'. *Paléorient* 11(2), 121–124.

Redding, R. W. 1992, 'Egyptian Old Kingdom patterns of animal use and the value of faunal data in modeling socioeconomic systems'. *Paléorient* 18(2), 99–107.

Rossel, S. 2006, 'A tale of the bones. Animal use in the Temple and Town of Wah-Sut'. *Expedition* 48, 41–43.

Rowley-Conwy, P. 1993, 'Season and reason: the case for a regional interpretation of Mesolithic settlement patterns', in G. L. Peterkin, H. Bricker & P. Mellars (eds), *Hunting and Animal Exploitation in the Later Palaeolithic and Mesolithic in Eurasia*, 179–188. Archaeological Papers of the American Anthropological Association 4.

Stiner, M. C., Kuhn, S. L., Weiner, S. & Bar-Yosef, O. 1995, 'Differential burning, recrystallisation, and fragmentation of archaeological bone'. *Journal of Archaeological Science* 22, 223–237.

Traunecker, C. 2004, 'Dimensions réelles et dimensions imaginaires des dieux d'Egypte: Les statues secrètes du temple d'Opet à Karnak'. *KTEMA* 29, 51–65.

Van Neer, W. 2002, 'Le matériel faunique', in B. Midant-Reynes & N. Buchez (eds), *Adaïma 1. Économie et habitat*, 521–565. Fouilles de l'Institut Français d'Archéologie Orientale 45, IFAO XI.

Van Neer, W. 2004, 'Evolution of prehistoric fishing in the Nile Valley'. *Journal of African Archaeology* 2, 251–269.

Vila, E. 1998, *L'exploitation des animaux en Mésopotamie aux IVe et IIe millénaires avant J.-C.* Monographie du CRA 21, CNRS Éditions, Paris.

Weinstein, J. M. 1973, 'Foundation deposits in Ancient Egypt'. Unpublished Ph.D dissertation, University of Pennsylvania, Ann Arbor: University of Michigan.

27. More animal burials from the Predynastic elite cemetery of Hierakonpolis (Upper Egypt): the 2008 season

Wim Van Neer, Veerle Linseele & Renée Friedman

Three animal burials from the Predynastic elite cemetery HK6 at Hierakonpolis are described. Recent excavations at this site have revealed large tomb compounds and a precinct of funerary temples dating to the Naqada IC–IIB (ca. 3800–3600 BC), all of which included animal burials. Shallow pits discovered by the enclosure wall of the cemetery and near the northeast corner of the temple precinct yielded the articulated skeletons of nine dogs, a baboon and six cats. A description of the finds is provided and special attention is given to the size reconstruction and possible breeds of the dogs, the pathology seen on the baboon skeleton and its interpretation in terms of conditions of keeping, and the status of the cats (wild, tamed or domestic).

Keywords offerings, ritual, dog, baboon, cat, domestication

Introduction

Hierakonpolis (Kom el Ahmar, ancient Nekhen) (25°06'N, 32°46'E) is a major archaeological site located on the west bank of the Nile, 17km north of the modern town of Edfu in Upper Egypt (Fig. 27.1). Covering nearly 3km^2, it is the largest site of the Predynastic period (4000–3100 BC) still extant and accessible. At its peak at ca. 3700–3400 BC, it must have been one of the largest centres anywhere along the Nile and its desert sands preserve domestic areas, industrial zones, ceremonial centers and discrete cemeteries for the different strata of society. Based on its size and other historical criteria, it is generally considered to have been the capital of a pre-unification southern kingdom in Upper Egypt (Hoffman 1982; Adams 1995; Friedman *et al.* 1999; 2002; Friedman 2008b).

Archaeological evidence to support its exalted status is currently being found in the cemetery of the elite segment of society called HK6, where recent excavations have uncovered two of the largest known tombs of the early Naqada II period (ca. 3800–3600 BC) and indications of the wooden superstructures that surrounded them. In addition, a distinct ritual precinct containing wood-built funerary temples, the earliest in Egypt, has also been discovered (Friedman 2008a; 2008b). Wide area clearance, with special attention to the recovery of architectural traces, has allowed the new and many of the older tombs excavated in previous campaigns in this cemetery (1979–1986 by Michael Hoffman; 1997–2000 by Barbara Adams) to be placed in a broader context. Results of the 2009 season of investigation indicate that large elite tombs of this period were placed at the center of an interconnected complex of fenced enclosures containing smaller subsidiary graves not only of (presumably) family members and court officials, but also a variety of animals, both wild and domestic (Friedman 2010).

Since the initial excavation in the 1980s (Adams 2000), the HK6 cemetery has been recognised as unique in the Predynastic period for the number and variety of animal taxa it contains. These take the form not only of butchered parts offered as food, but also, and more frequently, as complete carcasses often buried with care in tombs with or without human accompaniment (Fig. 27.2). Although the cemetery was mainly used during the Naqada IC–IIB (ca. 3800–3600 BC) and Naqada III periods (3300–3000 BC),

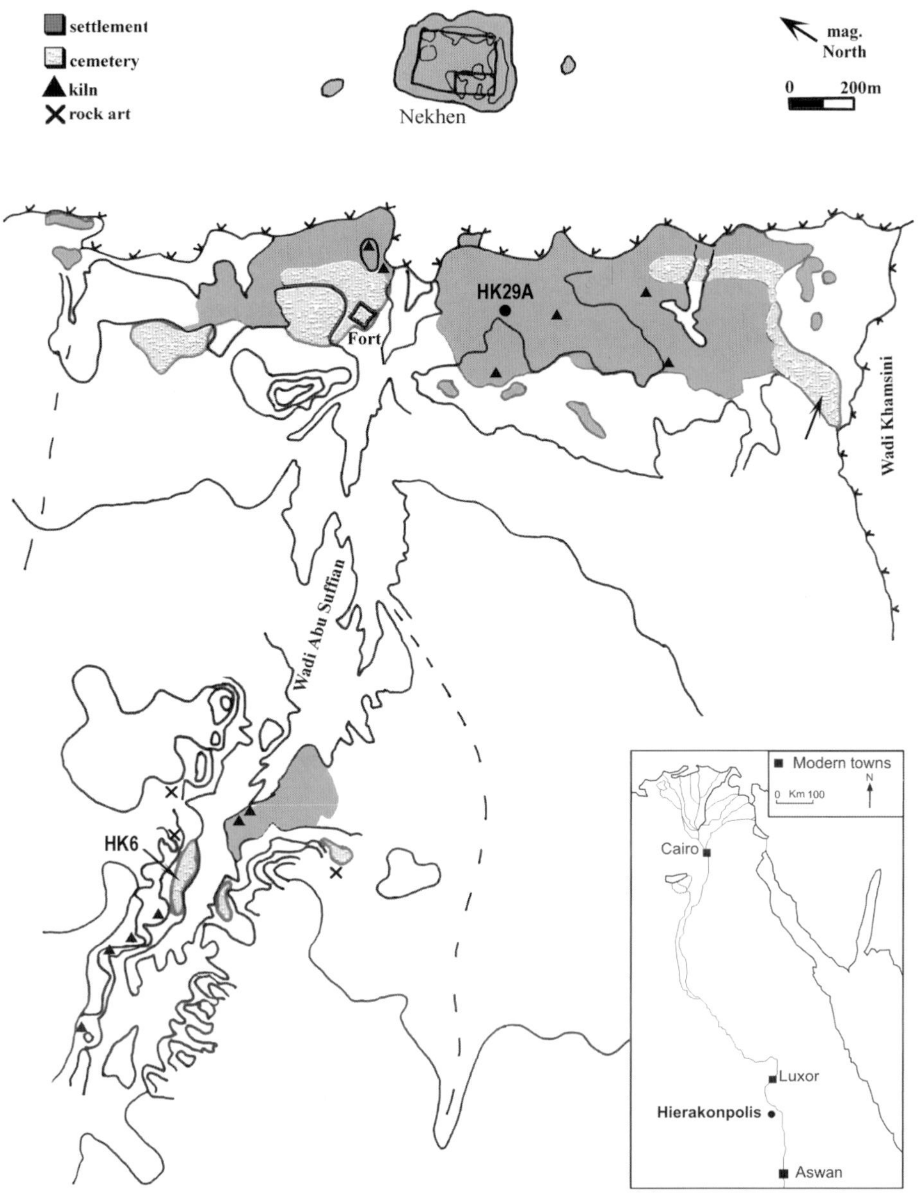

Fig. 27.1. Hierakonpolis and its localities mentioned in the text.

the independent burial of animals appears to be a feature of the earlier phases only. The faunal material found during the 1979–2003 campaigns has been described in Van Neer *et al.* (2004) and includes both domestic (cattle, sheep, goat, dog, donkey) and wild species (anubis baboon, aurochs, hartebeest, wild donkey, hippo, elephant, jungle cat). Almost all of the tombs contain some faunal remains, but due to considerable disturbance and plundering in ancient and modern times, the locus of primary deposition of some of this material cannot always be determined with certainty.

The present contribution deals with the animal burials discovered in March 2008 when investigations were focused on determining the extent of the funerary temple or 'pillared-hall' precinct (Fig. 27.2). To that end, excavation was made along the northern course of a wood–post wall (Wall B7) that runs along at the edge of the wadi terrace on which HK6 is located. As the purpose was specifically to determine whether this wall turned a corner and enclosed the ritual precinct, a wide area to the west of the wall was also investigated. The excavation along 25m of this wall revealed no corner; however, on the west side of this wall, in grid square B11, a deflated and rocky area otherwise almost entirely devoid of finds, three subsurface pit features were discovered which contained the first undisturbed burials of animals found in this cemetery. Feature B was the shallow grave of a juvenile baboon; feature C, an oval pit containing nine dogs; and feature E, a small circular pit with the articulated skeletons of six cats. All of the burials were intact, but none contained artefacts. Placed in close proximity to each other, the burials are almost certainly contemporary. Based on their position beside the wall, which can be dated to early Naqada II by materials in its

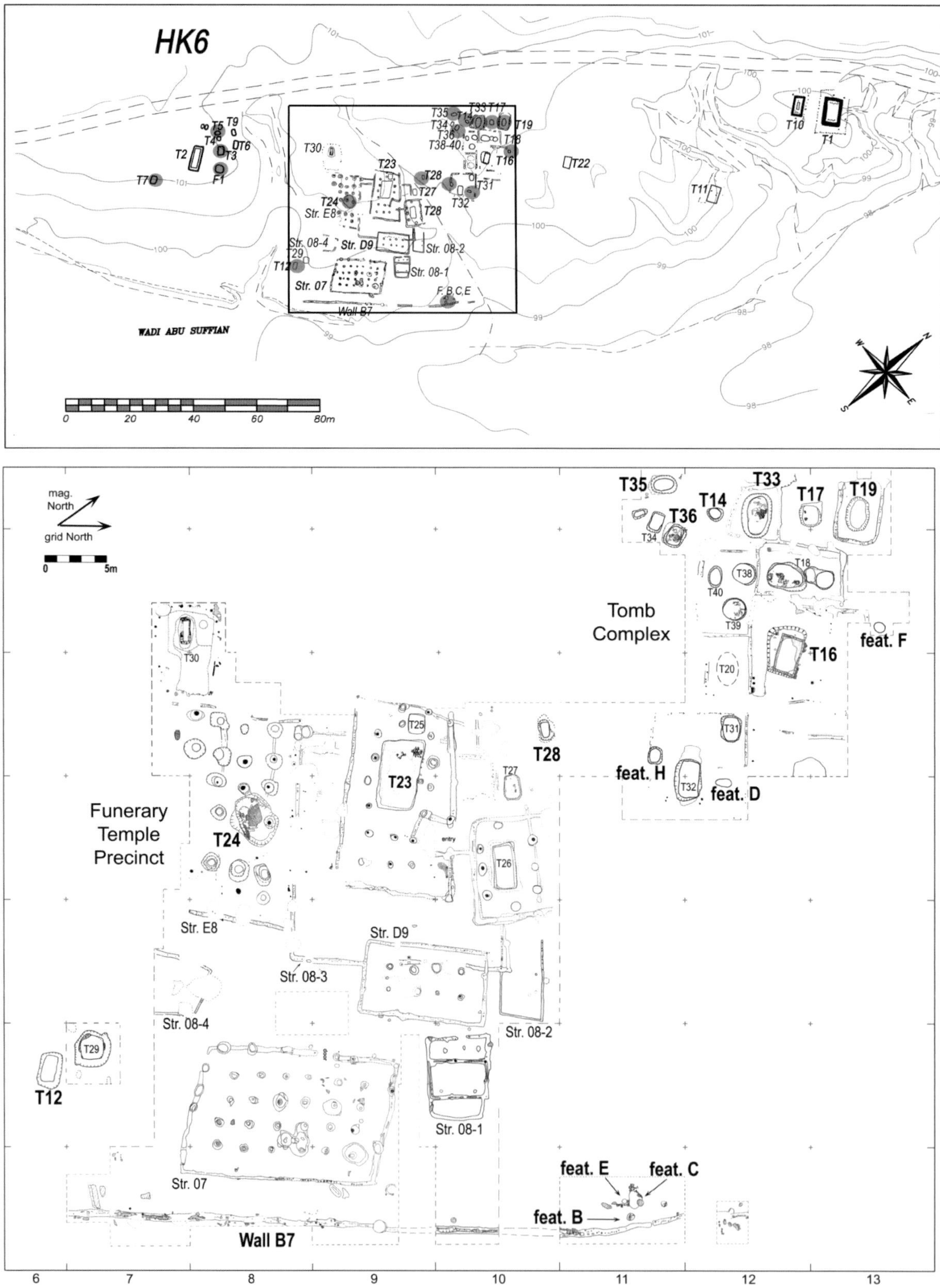

Fig. 27.2. Map of the cemetery at HK6 with tombs containing animal burials indicated in grey, and detailed plan of the funerary temple precinct and tomb complex investigated in 2000–2009.

foundation trench, it is assumed that the animal burials date to the same period. No material of later date was found in the immediate vicinity (Van Neer & Linseele 2008; Friedman 2010).

Description of the animal burials

The young baboon in feature B

In a shallow pit immediately below the present surface, and less than 30cm from the foundation trench for wall B7, a baboon was found, lying on its right side, facing north (Fig. 27.3). The animal was more or less in a foetal position, with its legs flexed upward and the lower arms resting on the knees.

The skull was poorly preserved, chalky and fragile, as a result of subsurface weathering and therefore a secure identification as either anubis (*Papio anubis*) or hamadryas baboon (*Papio hamadryas*) could not be achieved. However, it is likely that it is an anubis baboon since all individuals from HK6 that could be specifically identified thus far belong to this species (Van Neer & Linseele 2002; Van Neer *et al.* 2004). A secure sex identification of the individual was also not possible, since no observation of the canines could be carried out, not even during the excavation due to weathering.

Both the dentition and the fusion state of the long bones provide information allowing an age estimation of this individual. In the jaws, the second molars have attained the occlusal level, but they do not yet show any wear. The majority of the long bones of which the ends are preserved do not have their epiphyses fused to the shafts. This is the case for the proximal radius, the proximal and distal ulna, the proximal and distal femur, the proximal and distal tibia, the tuber of the calcaneus, the distal metapodials, and the proximal 1st and 2nd phalanges. The distal humerus, however, was already fused, a process that is terminated at about 3 years of age in females and 4 years of age in males (Bramblett 1971). Most of the other epiphyses mentioned above, that were not yet fused, do so between 6 and 7 years of age. Wear of the lower 2nd molar starts around 5–5.5 years of age. Taking these ageing criteria into account, it can be concluded that the baboon must have been older than 3 years and younger than 5.5 years, and most probably between about 4 and 5 years of age. Because of the young age of the animal and the poor preservation of its skeleton, the number of measurable bones was rather limited (Table 27.1).

Table 27.1: Measurements (mm) on the baboon from feature B.

lower M2, GL	14.0
atlas, GB	38.6
atlas, H	22.5
humerus, Bd	35.1
tibia, GL shaft	151
calcaneus, GL without tuber	44.2
astragalus, GL	29.4

Despite the brittle nature of the skeleton, it was possible to detect at least one pathology on the long bones. The lower third of the left ulnar shaft shows a transverse fracture, but the radius is intact over its whole length (Fig. 27.3). Above and below (i.e., proximal and distal of) the fracture, the shaft is swollen, but there is no evidence for the formation of a callus that unites both parts. This pathology is reminiscent of one observed previously in a baboon from Tomb 12, also in the HK6 cemetery. That burial contained seven anubis baboons exhibiting numerous pathologies related to traumas that the animals underwent (Van Neer & Linseele 2002; Van Neer *et al.* 2004). On the right ulna of one of these baboons was a fracture at the same level as that on the specimen from feature B (Fig. 27.4). In this case as well, a pseudoarticulation developed, probably as a result of the lack of immobilisation during callus formation. The young, female baboon of about 4 years of age found Tomb 17 in the HK6 cemetery also had a pathological forearm: the distal third of the shaft of the left radius and ulna are thickened over a length of about 4cm. As a result of this swelling the two bones are in contact and a synostosis was forming.

Dynastic depictions of constrained baboons or other cercopithecids show a rope tied either around their waist or neck (Osborn & Osbornová 1998), making it unlikely that the observed fractures of the forearms are related to the tethering of the animals. In humans, this type of forearm trauma has been called a 'parry fracture' and it is often seen as a result of interpersonal violence, although other causes, such as accident, stress, fatigue, or an underlying pathological condition are also possible factors (Smith 1996). These 'nightstick' or 'defense' fractures can occur when the forearm is used to parry or deflect blows to the head; however, it is unclear if this explanation also serves for the HK6 baboons. No cranial injury was observed on the feature B specimen or on the single more or less completely preserved skull that was found in Tomb 12. A mandible from that tomb had a healed fracture of the right horizontal ramus, but when this trauma occurred – during capture or later on in captivity – cannot be established. The same uncertainty applies to the ulnae of the baboons in feature B, Tomb 12 and Tomb 17: are their fractures related to accident or violence? There are numerous traumatic lesions on the feet and hand bones of the Tomb 12 baboons: at least four out of the seven baboons had a fractured hind foot and at least five out of seven had a fractured forefoot. These types of fractures are more likely to have occurred when the animals were held in captivity and thus show that the baboons were subjected to violence, possibly of a disciplinary nature.

Nine dogs in feature C

Less than half a metre from the baboon burial another pit was encountered beneath a loose pile of sandstone slabs.

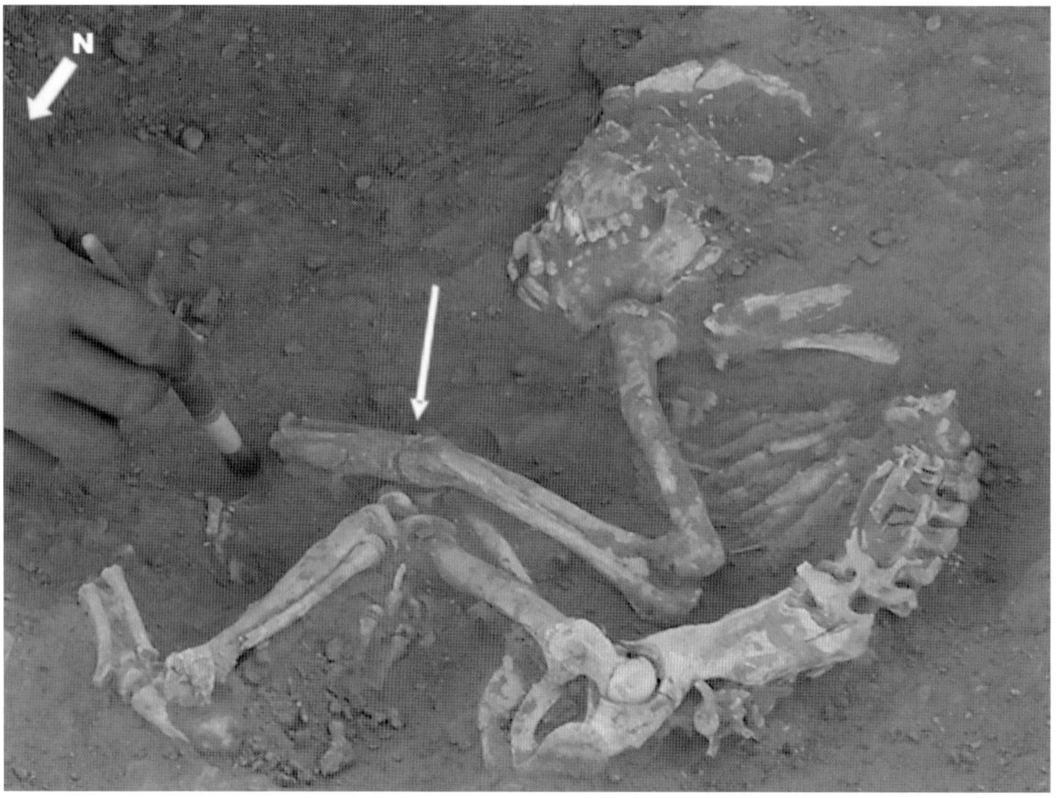

Fig. 27.3. Burial of the young baboon (Feature B). The fractured part of the left forearm is indicated with an arrow.

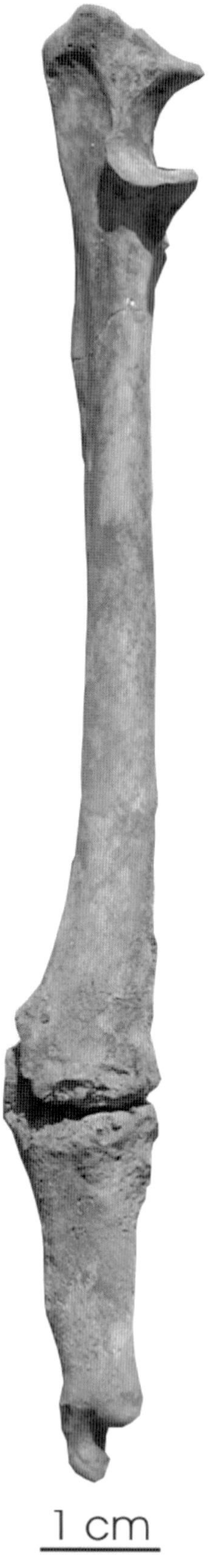

Fig. 27.4. Right ulna with a pseudoarticulation from a baboon of Tomb 12.

The oval pit, measuring roughly 1.4m east–west by 1m north–south and approximately 46cm deep, contained nine articulated dogs (Fig. 27.5). All the animals were on their left sides and oriented in the same direction with their heads to the east. The feet and legs were all directed southward, and many of these extremities were curved up along the southern side of the pit. This orientation of the skeletons shows that the dogs had been placed in the pit from the southern side. Observation during excavation of the relative positions and the superimpositions of the bones allowed the order in which the animals had been deposited to be reconstructed (Fig. 27.6).

Dog 9 was deposited first, on top of it came dogs 7 and 8, but in what order those two were cast in is unclear. Similarly it was not possible to establish in which order Dogs 6, 5 and 4 were deposited on top of dog 8. Dog 3 was placed in the pit after dog 6. Dog 2 was put on top of Dog 3 and the last animal deposited in the pit was Dog 1. During the excavations, when all of the individuals had been exposed, it appeared as if the animals had been laid along the bottom and around the sides of the pit since the upper, central part of the pit contained only sediment. It is likely, however, that the nine dogs initially filled the entire pit cavity, and that the 'empty space' above was only a result of the compaction of the corpses during and after their decomposition. Evidence for such movements is provided, for instance, in dog 1 by displaced ribs that were no longer articulating with their corresponding thoracic vertebrae.

Table 27.2: Measurements (mm) of the axial skeleton of the dogs. For the skull and mandible, numbers refer to those used by von den Driesch (1976). Measurements in brackets are approximate.

	Dog 1	*Dog 2*	*Dog 3*	*Dog 4*	*Dog 6 male*	*Dog 7*	*Dog 8*	*Dog 9*
Skull								
CBL	(164)	–	–	–	(159)	–	–	–
23	56.5	–	–	–	–	–	–	–
40	38.5	37.5	38.2	–	–	–	39.2	–
25	30.7	31.7	30.7	–	–	–	29.0	31
31	35.0	–	–	–	–	–	–	–
29	48.5	–	–	–	–	–	–	–
38	51.5	51.6	52.1	–	–	–	49.4	–
alv. L. P1–M3	–	–	–	–	62.5	–	–	–
alv. L. P1–4	–	–	–	–	33.6	–	–	–
alv. L. M1–3	–	–	–	–	30.2	–	–	–
Mandible								
1	119	–	119	–	119	–	115	118
2	119	–	118	–	119.5	–	115	119
18	47.2	–	45.3	–	51.8	–	48.7	–
alv. L. P1–M3	61.5	–	67.8	–	64.4	–	61.8	65.8
alv. L. P1–4	33.8	–	–	–	33.4	–	32.9	35.8
alv. L. M1–3	29.0	–	36.4	–	29.8	–	28.5	30.9
7	67.5	–	–	–	–	–	–	–
GL M1	18.9	–	19.0	–	20.0	20.0	19.0	20.6
GB M1	8.0	–	–	–	7.8	7.7	7.5	8.3
GL M2	9.8	–	–	–	–	–	–	–
GB M2	6.2	–	–	–	–	–	–	–
20	15.0	–	–	–	–	–	–	–
19	19.4	–	–	–	–	–	–	–
Atlas								
GL	32.1	37.7	–	–	–	–	31.1	32.0
GB	–	67.6	–	–	–	–	–	–
BFcr	–	–	–	–	–	–	31.6	3.8
LAd	14.8	14.2	–	–	–	–	13.6	12.5
Axis								
H	34.5	30.1	–	–	–	29.7	–	30.1
LCDe	44.0	44.4	–	–	–	41.9	–	41.8
LAPa	42.3	47.4	–	–	–	45.4	–	40.0
BFcr	24.6	25.3	–	–	–	23.0	–	24.7
Sacrum								
GL	–	–	–	27	–	–	32.2	36.4
GB	–	–	–	31.7	–	–	34.2	38.8
PL	–	–	–	25.1	–	–	28.2	29.9
BFcr	–	–	–	17.9	–	–	18.3	20.4
HFcr	–	–	–	8.1	–	–	8.3	9.0

Fig. 27.5. View of the dog burial prior to the lifting of the individuals.

Associated with the skeleton of Dog 6 was a baculum, indicating that this animal was a male. Early in the excavation process, when Dogs 1 and 2 were lifted, a second baculum was retrieved from the 2mm sieve. Despite the systematic sieving of all the sediment from the pit, no additional bacula were found. Because of the rather fragile nature of this skeletal element and the generally poor bone preservation, it would be speculative to conclude that seven of the dogs were females. Measurements of the various elements of the dog skeletons are given in Tables 27.2–4.

The age determination of the dogs can be carried out through combined use of dental and epiphyseal fusion data (Table 27.5). The mandibles of Dog 3 still have their deciduous canines, but these were in the process of being

Table 27.3. Measurements (mm) of the shouldergirdle and anterior extremities of the dogs. Measurements in brackets are approximate.

	Dog 1	*Dog 2*	*Dog 3*	*Dog 4*	*Dog 6 male*	*Dog 7*	*Dog 8*	*Dog 9*
Scapula								
HS	–	–	–	106	–	–	–	–
SLC	–	–	20.2	17.3	–	–	19.3	19.5
GLP	24.2	24.2	23.8	22.0	24.7	23.5	22.1	24.4
BG	15.0	15.2	15.2	13.5	14.5	14.1	13.6	14.4
Humerus				*NF prox.*	*NF prox.*			
GL	134	139	–	(133)	(142)	139	132	140
Bp	25.5	24.9	–	21.3	24.5	25.0	23.0	26.6
Dp	32.4	31.6	–	29.0	–	31.6	30.0	34.2
SD	11.7	10.9	–	9.9	–	10.2	10.3	11.2
Bd	26.0	25.3	–	22.8	25.3	25.8	24.2	26.2
Radius			*NF prox.; NF dist.*	*NF dist.*	*fusing prox.; NF dist.*			
GL	134	143	(130)	–	–	–	132	–
Bp	15.0	15.2	–	13.0	14.6	14.9	9.0	15.6
SD	9.9	10.3	–	–	–	–	10.1	9.7
Bd	19.4	19.5	–	–	–	–	17.7	20.2
Ulna				*NF dist.*	*NF dist.*			
GL	–	168.5	–	–	–	–	–	165
BPC		–	–	13.0	13.8	13.9	12.7	14.4
LO	24.2	–	–	23	26.3	23.9	22.6	24.7
SDO	16.9	18.1	–	15.5	17.1	16.7	16.2	17.3
DPA	20.2	21.1	–	18.1	20.5	19.4	17.8	20.0
McI								
GL	–	15.2	–	15.7	–	–	16.0	17.4
Bd	–	–	–	4.1	–	–	4.2	4.7
Mc II								
GL	48.0	47.2	–	45.4	47.3	–	50.8	47.5
Bd	6.6	7.7	–	7.0	7.7	–	6.4	7.2
McIII								
GL	54.5	53.5	–	51.6	55.0	–	44.1	53.6
Bd	6.9	6.8	–	6.7	6.6	–	6.6	7.0
Mc IV								
GL	53.8	52.2	–	50.5	55.6	–	49.8	53.5
Bd	6.8	7.6	–	6.3	6.5	–	6.2	6.8
Mc V								
GL	44.0	43.7	–	42.4	46.1	–	41.0	44.9
Bd	7.1	7.5	–	7.1	7.0	–	6.9	7.6

NF = non fused.

replaced by the permanent ones. This replacement takes place at around 5–7 months of age (Silver 1969). All of the other individuals have their complete permanent dentition in place, showing that these animals were at least seven months old. Dogs 4, 5 and 6 do not have all the epiphyses of their long bones fused and thus allow an age estimate on that basis. When the fusion state of the various bones within each of the individuals is considered, it appears that there are some slight contradictions within the scheme of Silver (1969). In Dog 4, for instance, the distal radius and ulna are not yet fused indicating that the animal was younger than 11–12 months, whereas the tuber calcis was already fused, a process that is terminated at 13–16 months of age. Nevertheless, it seems that Dog 4 and Dog 6 were around 1 year of age. Dog 5 was slightly younger, 8–9 months, judging from the distal humerus that is still fusing. Because the long bones of Dogs 1, 2, 7, 8 and 9 have all their epiphyses fused, these individuals must have been older than 1.5 years of age (Silver 1969).

The problems that dog remains pose in age determiation have been addressed by Horard-Herbin (2000), who developed a method for making age estimation on the basis of the occlusal wear of the lower 1st molar. When this scoring system is applied to the younger dogs, 3 and 6, it appears that the age estimations are less precise than (but in agreement with) the results obtained on the state of epiphyseal fusion. This method, however, was primarily devised to obtain more precise age estimates for dogs with

Table 27.4: Measurements (mm) of the pelvic girdle and posterior extremities of the dogs. Measurements in brackets are approximate.

	Dog 1	*Dog 2*	*Dog 3*	*Dog 4*	*Dog 5*	*Dog 6 male*	*Dog 7*	*Dog 8*	*Dog 9*
Pelvis									
LA	–	–	–	–	–	–	–	17.1	18.8
Femur			*NF prox.; NF dist.*			*NF prox.; NF dist.*			
GL	152	162	(145)	–	–	–	155	147	154
Bp	31.1	–	–	–	–	–	30.0	28.0	30.6
DC	15.7	–	–	–	–	–	14.8	14.3	16.0
SD	11.5	–	–	–	–	–	10.8	11.0	11.0
Bd	26.3	–	25.0	–	–	26.5	25.0	24.2	25.2
Patella									
GL	14.6	–	–	–	–	14.0	14.0	13.4	14.4
GB	9.0	–	–	–	–	–		–	–
Tibia				*NF prox.; fusing dist.*		*NF prox.; fusing dist.*			
GL	154	159	–	(155)	–	–	151	148	155
Bp	28.7	28.6	–	26.2	–	28.7	28.3	27.1	29.2
SD	10.5	10.4	–	9.1	–	–	10.3	17.0	11.0
Bd	(18)	18.7	–	16.5	–	17.8	18.1	17.0	18.0
Dd	13.3	13.9	–	–	–	14.2	15.7	12.7	14.2
Calcaneus									
GL	38	35.8	–	34.7	–	38.5	35.5	34.0	35.5
Astragalus									
GL	22.4	22.0	–	20.8	20.5	21.9	22.0	20.5	21.3
Mt II									
GL	53.2	54.0	–	–	–	53.9	–	49.8	52.4
Bd	6.4	5.9	–	–	–	7.1	–	6.1	7.1
Mt III									
GL	61.0	59.7	–	–	–	61.4	–	56.0	58.9
Bd	6.8	6.8	–	–	–	7.3	–	6.4	6.5
Mt IV									
GL	62.5	58.0	–	–	–	63.0	–	56.7	60.5
Bd	6.6	7.0	–	–	–	6.6	–	6.5	6.7
Mt V									
GL	55.6	50.7	–	–	–	56.2	–	52.4	53.0
Bd	5.9	6.4	–	–	–	6.9	–	5.4	6.0

NF = non fused.

Table 27.5. Age estimations of the dogs. The epiphyseal closures are according to Silver (1969), and the dental wear stages according to Horard-Herbin (2000).

Element	*Fusion at*	*Dog 1*	*Dog 2*	*Dog 3*	*Dog 4*	*Dog 5*	*Dog 6*	*Dog 7*	*Dog 8*	*Dog 9*
Humerus distal	8–9 months	F	F	NF	F	fusing	F	F	F	F
Humerus proximal	15 months	F	F	NF	NF	NF	NF	F	F	F
Radius proximal	11–12 months	F	F	NF	F	NF	fusing	F	F	F
Radius distal	11–12 months	F	F	NF	NF	NF	NF	F	F	F
Ulna distal	11–12 months	F	F	?	NF	?	NF	F	F	F
Femur proximal	18 months	F	F	NF	NF	NF	NF	F	F	F
Femur distal	18 months	F	F	NF	NF	NF	NF	F	F	F
Tibia proximal	18 months	F	F	NF	NF	NF	NF	F	F	F
Tibia distal	13–16 months	F	F	NF	fusing	NF	fusing	F	F	F
Calcaneus tuber	13–16 months	F	F	?	F	?	F	F	F	F
Metacarpal distal	8 months	F	F	?	F	?	F	–	F	F
Metatarsal distal	10 months	F	F	?	?	?	F	–	F	F
Age based on epiphyseal fusion		>18 mo	>18 mo	<8–9 mo	~12 mo	8–9 mo	~12 mo	>18 mo	>18 mo	<18 mo
Dentition		complete	complete	canines replacing	complete	complete	complete	complete	complete	complete
Wear stage of M1		E/F	–	A/B	–	–	A/B	A/B	F	Gd
Age based on tooth wear		24–48 mo	–	6–15 mo	–	–	6–15 mo	6–15 mo	36–48 mo	>72 mo

F = fused; NF = non fused.

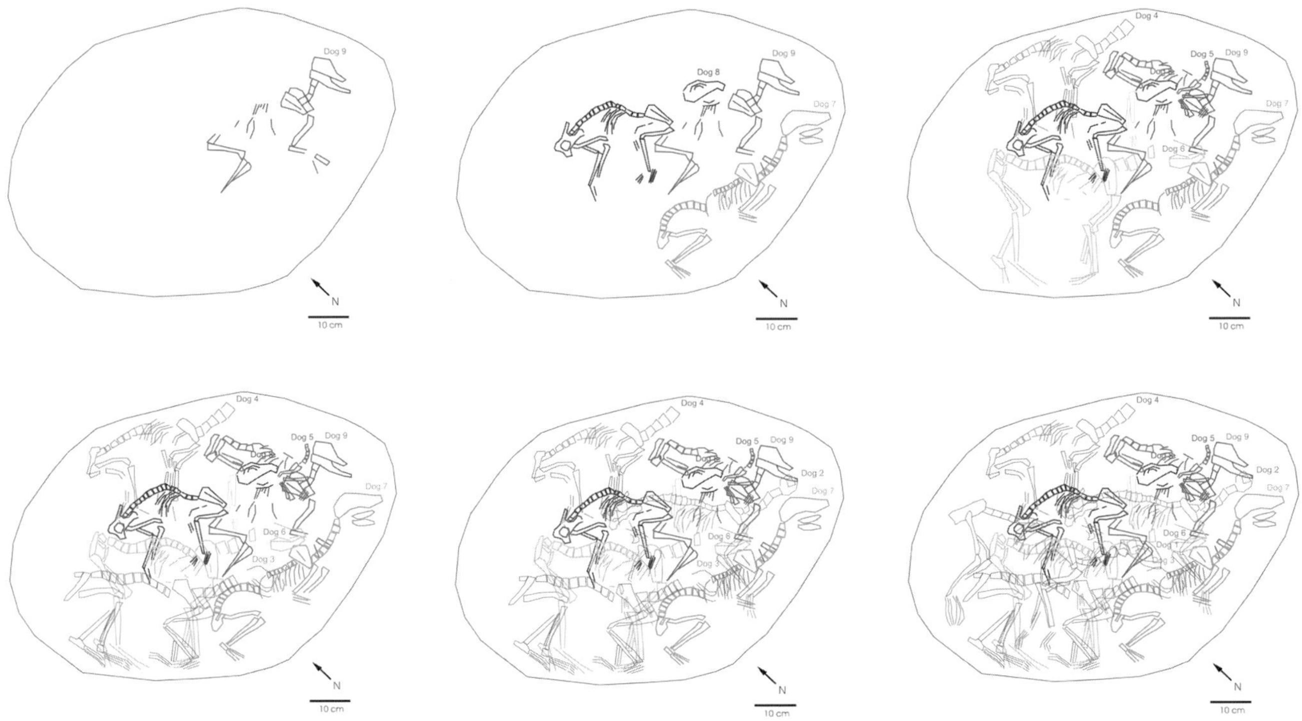

Fig. 27.6. A reconstruction of the various stages of the filling of the pit with indication of the numbers that were attributed to the dogs during the excavation.

Table 27.6: Size reconstruction of the dogs according to the different long bones.

	Dog 1	*Dog 2*	*Dog 3*	*Dog 4*	*Dog 5*	*Dog 6*	*Dog 7*	*Dog 8*	*Dog 9*
humerus	45	47	–	(45)	–	(48)	47	44	47
radius	43	46	(42)	–	–	–	–	43	–
ulna	–	45	–	–	–	–	–	–	44
femur	46	49	(44)	–	–	–	47	44	46
tibia	45	46	–	(45)	–	–	44	43	45

The shoulder heights are in cm; values in brackets were obtained on long bones of which the epiphyses were not yet all fused. NF = non fused; F = fused.

complete dentition and all their bones fused. The wear stages of the lower M1 indicate that Dog 1 was 2–4 years of age, Dog 8 was 3–4 years old and Dog 9 was a senile individual of more than 6 years of age. In the case of Dog 7, which should be older than 18 months because all of its long bones are fused, the wear stage of the lower carnassial suggests an animal of 6–15 months of age. This tooth had numerous fissures, which may be a possible reason why no wear facets could be detected and the dental age is underestimated.

Using the total lengths of the long bones and the indices compiled by von den Driesch and Boessneck (1974), it was possible to estimate the size of the dogs (Table 27.6). From the reconstructed shoulder heights, it appears that the dogs were rather uniform in size, and it is also clear that the young dogs, although they had not yet reached full maturity in terms of epiphyseal fusion, had already reached adult size. All of the animals are of 'medium' size, i.e., with heights at the shoulder of around 45cm or a little taller. This would correspond to the size of the average stray dogs that are still seen today in and around Egyptian villages.

The dogs in the feature C pit appear to have been in good physical condition, at least judging from their skeletons: not a single case of – the usually rather common – healed fractures or fused vertebrae was observed. The only pathology that was noticed was a deformed, slightly asymmetrical, axis in Dog 4.

No evidence was found for the way the animals were dispatched: the hyoid bones of Dog 1, which could be observed while still *in situ*, were unmodified, and none of the cervical vertebrae had any cut marks on the ventral part of the centrum which could be indicative of throat slitting. It should be kept in mind, however, that when throat cutting is carried out to bleed an animal to death, this need not produce any marks on the neck vertebrae or hyoid bones. This is especially the case when it is carried out by experienced persons who keep the animal immobilised properly, either by holding it firmly or by sedating it before the operation. No distorted cervical vertebrae that could be indicative of strangling were observed either.

Six cats in feature E

Directly adjacent to the southern edge of the feature C dog pit, a circular pit of about 50cm in diameter and only 25cm deep, held the bodies of six cats (Fig. 27.7). The animals were found fully articulated and draped along the bottom and around the sides of the pit. Unlike the situation in the dog pit, there was no preferential orientation in the position of the cats. Possibly their arrangement was to a large extent dictated by the small size of the pit.

A detailed description of these small felids can be found in Van Neer *et al.* (2014). Dental eruption (Habermehl 1980), fusion state of the long bones (Habermehl 1985), and size show that the remains are from a male, a young female, and four kittens from two different litters. The male had all its long bones fused and its dentition showed no wear suggesting that this individual was about 1 year of age when it was sacrificed. The female was a subadult that did not have all its epiphyses closed, but was near 1 year of age. The mandibles of the juvenile individuals still had their deciduous molars and in two of them the 1st molar can be seen almost piercing through the crypt. The mandibles of the two other young cats only have small openings in the ramus, meaning that these individuals were slightly younger and must be from another litter. This is also reflected in the overall size of the mandibles and the postcranial bones. Based on the eruption data of the lower molars, the age at death of these juveniles can be estimated at 4–5 months.

The measurements of the two older individuals show a clear difference in size that can be attributed to sexual dimorphism. Comparison of these measurements with the jungle cat (*Felis chaus*) previously found in Tomb 12 (Linseele *et al.* 2007; 2008) and with a female and male museum specimen of sand cat (*Felis margarita*) show that the felids from feature E can safely be attributed to *Felis silvestris*: they are clearly smaller than jungle cat and larger than sand cat.

Discussion

Status of the animals

In dynastic times, all three of the animal species found in the pit features under discussion were kept as house pets

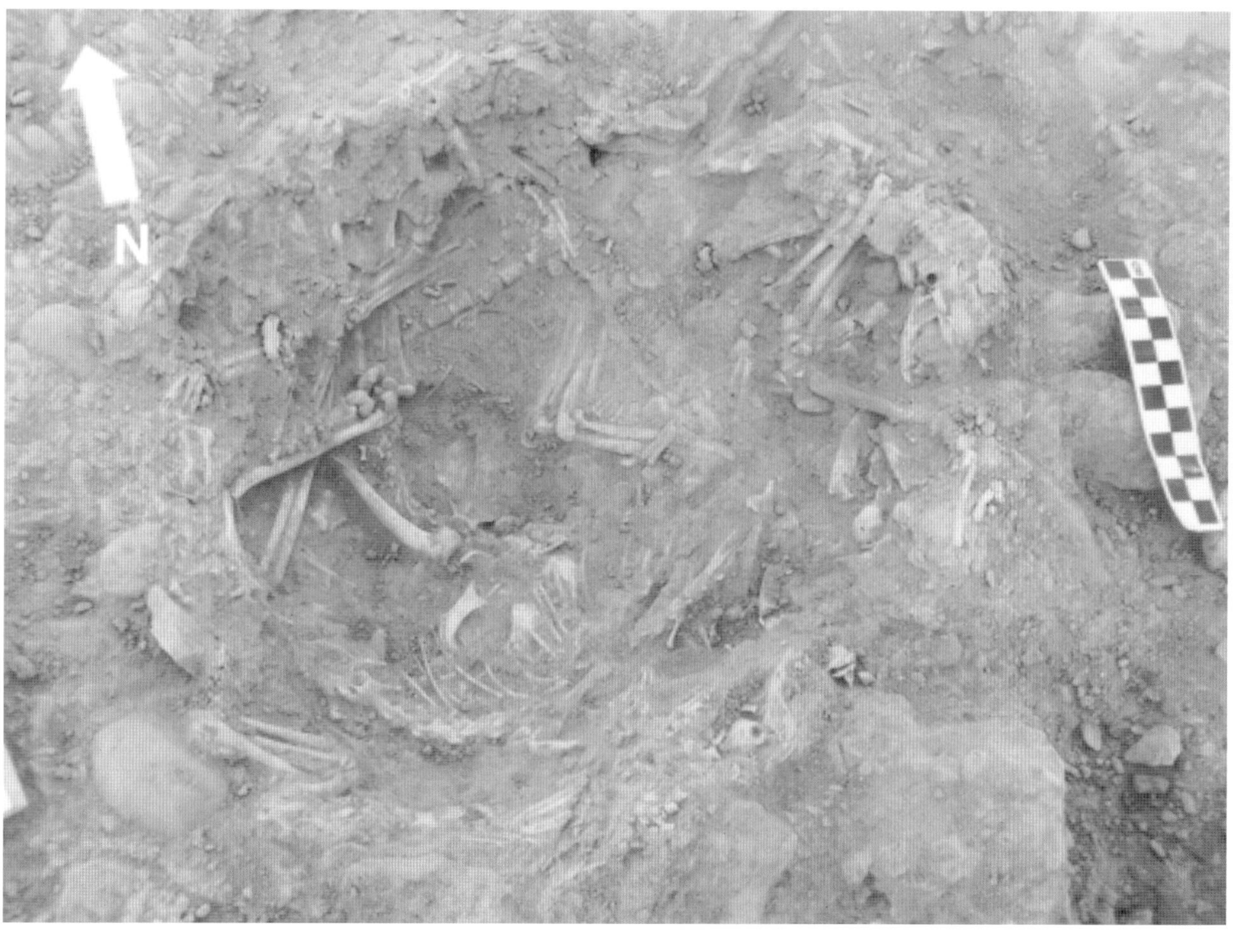

Fig. 27.7. View of the cat burial prior to the lifting of the individuals.

by the ancient Egyptians (Houlihan 1996, 75–108). Of the three, however, it is only the domestic dog that has a confirmed presence in the settlements and cemeteries at other sites of Predynastic age (4000–3100 BC; Flores 2003; 2004; Dreyer *et al.* 2000, 86–89). The dog is also the only one of these species to be depicted in the art of the period, where it is frequently shown involved in the hunting of wild animals, often wearing a collar or held on a leash by its human master (Hendrickx 2006). Prised for their skill in hunting, an elite activity of marginal significance in subsistence by this time, dogs no doubt were also valued for herding, guarding and companionship.

Dog burials are relatively frequent in the HK6 cemetery, but feature C is the only one thus far found undisturbed. It is also the first in which complete and articulated specimens have been recorded allowing for a reconstruction of their shoulder height. These can now be compared with the dogs recovered during the 2009 season of investigation, when excavations revealed one tomb (Tomb 14) and two pit features containing a total of at least ten dogs. Although all were disturbed, long bone preservation was such that six shoulder height reconstructions were possible: 46cm, 51cm, 52cm, 53cm, 54cm and 56 cm. From these it appears that the average size of these individuals seems somewhat larger than those in feature C. They also appear to have been buried with more care, i.e. in burials that were deeper and in two cases were provided with a mat lining. It would be tempting to see the differences in the size of the animals and the effort taken to bury them as evidence for different breeds or classes of dog, i.e., mongrels or the so-called pariah dogs in feature C and valued hunting dogs in the deeper and mat lined burials. Brewer (2001) suggests on the basis of pictorial and skeletal evidence of dynastic times that it may be possible to distinguish between the pariah dog, with a shoulder height of below 50cm, and the larger sight hounds, or the so-called tesem dogs. Further osteometric work should be carried out on this and other canid material in order to substantiate this idea. As suggested by Brewer (2001, 43) it cannot excluded that what has been observed thus far represents only a continuum. Boessneck (1988, 84) also warns that with the standard archaeozoological procedures it may be difficult to distinguish sight hounds from pariah dogs.

Outside of Hierakonpolis, there is no archaeozoological evidence for the presence of baboons in the predynastic period. Similarly, artistic representation is also completely absent until Dynasty 0 and early Dynasty 1 (3200–3100 BC), when statuettes of baboons become the most frequent among the small votive figures deposited in local temples throughout Early Dynastic Egypt (Bussmann 2010). The meaning of these figurines is unclear as is their relationship to the baboon god Hedjwer, who appears in the same period as the recipient of royal veneration. Hedjwer is generally believed to represent the royal ancestors, but the subject is still debated (Dreyer 1986, 68–73; Wilkinson 1999, 285; Winter 2006). Due to the stylised nature of many of the small statuettes, species attribution is difficult, but Osborn and Osbornová (1998, 32–33) identify them as hamadryas baboons. The first secure artistic representation of the anubis baboon does not occur until Dynasty 5 (ca. 2400 BC) of the Old Kingdom, when the animal is shown tamed, being led by a leash around its neck or waist through the market place (Houlihan 1996, 105–106). This species can also be identified in the god Baba (Bebon), known from Old Kingdom times, whose aggressive nature made him a potent protector against harmful creatures and inanimate dangers (Leitz 2002, 736–738; Wilkinson 2003, 196).

It is assumed that the baboon in feature B is an anubis baboon because all diagnostic specimens found thus far at HK6, in Tomb 12 and in the backdirt around Tomb 2 (Van Neer *et al.* 2004), belong to that species. Anubis baboons, like the elephants found in Tombs 24 and 33, probably did not live in the area, and most likely were imported from a region further to the south of Hierakonpolis. Even for the first half of the Holocene, when conditions were more humid, there is no evidence for anubis baboon in the Egyptian Nile Valley. Hamadryas baboons, however, probably inhabited the Red Sea mountains during predynastic times, but their presence in the Nile Valley would still represent an importation. From the traumatic lesions seen on the specimens from feature B, Tomb 12 and Tomb 17, it is clear that these animals were held in confinement for a long period of time prior to their sacrifice. In addition, results of dental microwear analysis conducted on the Tomb 12 specimens indicate the baboons were not allowed to feed freely. They were evidently hand fed on foods such as bread and river reeds, harder than the leaves and fruits they preferred (Merceron 2003).

While the baboon is shown as a tamed pet already in Dynasty 4 (ca. 2500 BC), the cat remains elusive. The identity of faunal remains from a Badarian period (before 4000 BC) grave at Mostagedda (Brunton 1937, 34, 57) is unconfirmed and depictions of what might even conceivably be considered as small felids (as opposed to lions) are extremely rare in predynastic art (see Payne 1993, cat. no. 143, 173; Dreyer 1986, 73, 177; taf, 36.192; taf, 59e). It is not until Dynasty 11 that the artistic record attests to a domesticated status, although some Old Kingdom depictions strongly suggest this was achieved much earlier (Malek 1993; Osborn & Osbornová 1998, 106–110).

It is also worth noting that as was the case for the baboons, a feline goddess named Mafdet makes a sudden appearance in Dynasty 1. Considered a protective power over the royal court, she was called 'the mistress of the mansion of life', a part of the palace that has been identified as the royal eating and food storage areas (Gardiner 1938). Although it is still unclear what specific feline is being represented (Osborn & Osbornová 1998, 117), it has been suggested that Mafdet was embodied in the cats which protected these areas from vermin and snakes (Wilkinson 1999, 290).There can be little doubt that the value of cats in this capacity was recognised at an early date.

Although cat bones occur in predynastic settlement deposits at several sites including Hierakonpolis, they attest only to the presence of the species in the vicinity of human habitation (Linseele *et al.* 2007). On the other hand, the cat burials in the elite cemetery at HK6 are the first confirmation of human interest in their presence.

In an attempt to establish whether the cats from feature E should be considered as wild, tamed or even domesticated different lines of possible evidence have been explored (Van Neer *et al.* 2014). Unlike the jungle cat from Tomb 12, the cats from feature E do not show any pathologies that could be indicative of their keeping in confinement. When the differences described in the literature to distinguish between wild and domestic cats, on the basis of the morphology of the mandible (Kratochvíl 1973), are applied to the feature E male and female, it appears that they best match the domestic form. A log-ratio analysis of the postcranial measurements was carried out using the standards calculated by O'Connor (2007) for the house cat. Comparison of these standards often allow wild and domestic cats to be distinguished in north European archaeofaunal assemblages. The values for the male and female from HK6 were calculated and the same has been done for the Roman cat from Quseir that was identified as domestic on the basis of the morphology of the skull and mandible (von den Driesch & Boessneck 1983). The Quseir cat is very large compared to the house cat standard, and the two Hierakonpolis individuals are smaller than the Quseir cat. The female cat from feature E has mainly negative values against the house cat standard, and the male has both slightly negative and slightly positive values. This means that the osteometric data also suggest that the cats are domestic.

Although both the morphology of the lower jaws and the postcranial measurements point towards the domestic form, some caution is needed. The criteria that were used are those established to distinguish wild and domestic cat in European sites. These may in fact reflect differences at the subspecies level (wild *F.s. silvestris* versus the domestic form derived from *F.s. lybica*). It remains to be verified, through the study of large series of wild *F.s.lybica*, if the criteria hold when applied to Egyptian material.

The six individuals clearly do not represent four kittens with their mother and father. As could be seen from the dental eruption and sizes of the juveniles, they belong to two different litters that were around 4–5 months of age. Wild cats in Egypt have a single litter of young per year (Le Berre 1990, 170). This means that when the juveniles were sacrificed, their mother had to be 16–17 months of age (or 28–29 months, etc.) had she been born in accordance with the same natural reproduction cycle. Since the female in feature E is slightly younger than 1 year of age, it is clear that the natural reproduction pattern has been disturbed. This discrepancy in the ages at death indicates that the cats at or near Hierakonpolis may have had more than one litter a year. Such deviation from the natural reproduction cycle has been observed in African free-ranging female wild cats that are hand-reared; they can have two to three litters per year (Estes 1991, 358–359). This departure from the birth pattern seen in wild cats that have experienced human interaction can be considered as circumstantial evidence for the cultural control of the cats in feature E.

Ritual meaning of the animals

In general, the burial of animals was a rare practice in Predynastic Egypt. Nevertheless, at most cemeteries, there are a few graves in which animals (dog, sheep, goat) have been placed most often with humans, perhaps to serve as pets, companions or food. On the other hand, the mass burial of animals without human accompaniment appears to be a very elite practice that can be documented at other elite cemeteries of the period (Naqada and Abydos) but never with the frequency observed at Hierakonpolis (Flores 2003; 2004).

As of 2009, more than 70 buried animals have been discovered in the elite cemetery at Hierakonpolis, albeit the graves are often severely disturbed and the skeletons not always complete. These burials appear in four basic forms:

1 Animals buried with humans, which tend to be domestic species like dogs or sheep/goats, although baboons have also been found.
2 Elaborate tombs for the large and exotic animals, like the elephants and aurochs, which were buried without human accompaniment but equipped with linen shrouds and matting as found in human interments. These tombs were also provided with above-ground wooden architectural markers.
3 Tombs containing multiple animals, both domestic and wild, buried in graves that have been dug deeply into the hard ground indicating significant effort, although associated above-ground architecture has not been observed.
4 The burial of animals in shallow pits never larger than that necessary for the purpose. This is the category into which the 2008 finds belong.

Recent investigations show that all of these categories are contemporary, thus the variety of taxa and the differential effort expended in their burial indicate the meaning and significance of these animal burials need not be the same. What that meaning might be, however, is difficult to determine.

Symbolically, the wide range of animals may well reflect the chaotic diversity of animal life, their captivity and eventual slaughter a way in which disorder was brought under control. The control of chaos, especially that inherent in nature, was one of the fundamental themes of predynastic iconography and religious beliefs with direct links to the political reality of developing kingship (cf. Kemp 1989). However, the spectrum in the cemetery does not include all of the animal species known and available to them, as demonstrated by the settlement remains. The cemetery record is notably different from the contemporary ceremonial center (HK29A) where gazelle, turtle, and fish play a significant role (Linseele & Van Neer 2003; Linseele *et al.* 2009). While it seems likely that the animals at the ceremonial center were gathered specifically for slaughter and consumption at that venue, it is unclear if this was also the case in the cemetery. In other words, it is unknown whether animals were collected for the expressed purpose of burial in the cemetery, perhaps as part of rituals of rebirth and/or displays of power over chaos, or whether they were selected for burial owing to the roles they had played in life and the desire for that to continue in the afterlife. As these purposes need not be mutually exclusive, the cemetery selection may have been influenced by all of the above.

In support of the latter suggestion are the results of the 2009 excavations in the HK6 cemetery. Investigations around a large and rich tomb of the Naqada IC–IIA period (Tomb 16), revealed a complex of interconnected fenced enclosures surrounding a variety of tombs and tomb groups containing both human and animal interments (Fig. 27.2). Examination of the owners of the surrounding tombs and their status suggest that tomb location was not arbitrary, but instead was specifically organised perhaps to reflect the layout of the royal/elite residence. In addition to at least nine tombs containing multiple human interments, the complex also includes the burial of an elephant (Tomb 33), aurochs (Tomb 19), hartebeest (Tomb 46), domestic cow and calf (Tomb 36), large male goats (Tomb 35), numerous dogs (Tomb 14, feature D) and a juvenile hippopotamus (feature H).

The elephant and aurochs were placed in elaborate tombs in proximity to the main burial of the ruler (Tomb 16). The ownership and maintenance of such beasts in both life and death would have been an eloquent statement on the power and wealth of their master, and the keeping of a royal menagerie would go on to become a means of legitimising the rule of later pharaohs (Müller-Wollermann 2003). In addition to ostentatious display, control of these animals also allowed their master to take on their powerful natural attributed for himself. The king as a wild bull is well known in late Predynastic art (Hendrickx 2002) and the close relationship the king had with the elephant is reflected on a carved ivory knife handle from Hierakonpolis, among other objects (Whitehouse 2002; Friedman 2004). The sacrifice of fecund and large domestic animals may be seen as an expression of wealth, power over chaos, as well as the desire for a continued food supply in eternity. The hartebeest, hippo and dogs probably allude to the thrill of the hunt, an activity that remained into Dynastic times an elite pastime and mark of physical prowess in addition to its significance in the symbol sphere (Decker & Herb 1994; Hendrickx 2010). Several of the human graves in the Tomb 16 complex included hunting gear as well as subsidiary pit graves containing the dogs of large size.

The excavations have not yet been able to incorporate the pit graves under discussion into a larger context. They remain isolated at the east edge of the cemetery. From their position they may be associated with the funerary temple precinct, potentially marking its limits in the northeast corner. In this they may mirror Tomb 12 in the southeast, which contained seven baboons, a cat, and a hippo. Marking the corner on the northwest may be Tomb 28, which held an adult dog and sheep; the southwest corner has not yet been investigated. On the other hand, they may be associated with a tomb complex yet to be revealed. Whatever the case, it seems most probable that a protective function can be attributed to the graves by virtue of their position and the animals they contained. The capacity of these animals to guard, protect and defend was certainly recognised by the kings of the first dynasties and this knowledge can no doubt be traced back to the observations and experimentations made by the elite society buried in the HK6 cemetery some 700 years earlier.

Acknowledgements

Bea De Cupere (RBINS, Brussels) helped with the osteometrical analysis of the cats. We also acknowledge the National Geographic Society for funding the excavations at HK6. The contribution of Wim Van Neer to this paper presents research results of the Interuniversity Attraction Poles Programme – Belgian Science Policy.

References

Adams, B. 1995, *Ancient Nekhen. Garstang in the City of Hierakonpolis.* Egyptian Studies Association 3, SIA Publishing, New Malden.

Adams, B. 2000, *Excavations in the Locality 6 Cemetery at Hierakonpolis 1979–1985.* British Archaeological Report S903, Archaeopress, Oxford.

Boessneck, J. 1988, *Die Tierwelt des Alten Ägypten.* C.H. Beck, München.

Bramblett, C. A. 1971, 'Non-metric skeletal age changes in the Darajani Baboon'. *American Journal of Physical Anthropology* 30, 161–72.

Brewer, D. J. 2001, 'Ancient Egyptian dogs', in D. J. Brewer, T. Clark & A. Phillips (eds), *Dogs in Antiquity. Anubis to Cerberus. The Origins of the Domestic Dogs*, 28–48, Aris & Phillips, Warminster.

Brunton, G. 1937, *Mostagedda and the Tasian Culture.* Bernard Quaritch, London.

Bussmann, R. 2010, *Die Provinztempel Ägyptens von der 0. bis zur 11. Dynastie. Archäologie und Geschichte einer gesellschaftlichen Institution zwischen Residenz und Provinz.* Probleme der Ägyptologie 30, Brill, Boston.

Decker, W. & Herb, M. 1994, *Bildatlas zum Sport in Alten Ägypten.* Brill, Leiden.

Dreyer, G. 1986, *Elephantine VIII. Der Tempel der Satet. Die Funde.* Archäologische Veröffentlichungen, Deutsches Archäologisches Institut, Abteilung Kairo 39, Philip von Zabern, Mainz.

Dreyer, G., von den Driesch, A., Engel E. M., Hartmann, R., Hartung, U., Hikade, T., Müller, V. & Peters, J. 2000, 'Umm el-Qaab. Nachuntersuchungen im frühzeitlichen Königsfriedhof 11./12. Vorbericht'. *Mitteilungen des Deutschen Archäologischen Instituts. Abteilung Kairo* 56, 43–129.

Estes, R. D. 1991, *The Behavior Guide to African Mammals*. Russel Friedman Books, Johannesburg.

Flores, D. V. 2003, *Funerary Sacrifice of Animals in the Egyptian Predynastic Period.* British Archaeological Report S1153, Archaeopress, Oxford.

Flores, D. V. 2004, 'Funerary sacrifice of animals in the Egyptian Predynastic period', in S. Hendrickx, R. F. Friedman, K. M. Cialowicz & M. Chlodnicki (eds), *Egypt at its Origins. Studies in Memory of Barbara Adams*, 731–763. Orientalia Lovaniensia Analecta 138, Peeters, Leuven.

Friedman, R. F. 2004, 'Elephants at Hierakonpolis', in S. Hendrickx, R. F. Friedman, K. M. Cialowicz & M. Chlodnicki (eds), *Egypt at its Origins. Studies in Memory of Barbara Adams*, 131–168. Orientalia Lovaniensia Analecta 138, Peeters, Leuven.

Friedman, R. 2008a, 'The cemeteries of Hierakonpolis'. *Archeo-Nil* 18, 8–29.

Friedman, R. 2008b, 'Excavating Egypt's early kings: Recent discoveries in the elite cemetery at Hierakonpolis', in B. Midant-Reynes, & Y. Tristant (eds), *Egypt at its Origins 2. Proceedings of the International Conference 'Origin of the State. Predynastic and Early Dynastic Egypt', Toulouse, 5–8 September 2005*, 1157–1194. Orientalia Lovaniensia Analecta 172, Peeters, Leuven.

Friedman, R. 2010, 'The Early Royal cemetery at Hierakonpolis: an overview', in F. Raffaele, M. Nuzzolo & I. Incordino (eds), *Recent Discoveries and Latest Researches in Egyptology. The Proceedings of the First Neapolitan Congress of Egyptology*, 67–86. Harrassowitz, Wiesbaden.

Friedman, R., Maish, A., Fahmy, A. G., Darnell, J. C. & Johnson, E. D. 1999, 'Preliminary report on field work at Hierakonpolis 1996–1998', *Journal of the American Research Center in Egypt* 36, 1–35.

Friedman, R. F., Watrall, E. , Jones, J., Fahmy, A. G., Van Neer, W. & Linseele, V. 2002, 'Excavations at Hierakonpolis'. *Archéo-Nil* 12, 55–68.

Gardiner, A. H. 1938, 'The Mansion of Life and the Master of the King's Largesse'. *Journal of Egyptian Archaeology* 24, 83–91.

Habermehl, K. H. 1980. *Altersbestimmung bei Versuchstieren.* Verlag Paul Parey, Berlin & Hamburg.

Habermehl, K. H. 1985. *Altersbestimmung bei Wild- und Pelztieren.* Verlag Paul Parey, Berlin & Hamburg.

Hendrickx, S. 2002. 'Bovines in Egyptian Predynastic and Early Dynastic iconography', in F. A. Hassan (ed.), *Droughts, Food and Culture. Ecological Change and Food Security in Africa's Later Prehistory,* 275–318. Kluwer Academic/Plenum Publishers, New York.

Hendrickx, S. 2006, 'The dog, the *Lycaon pictus* and order over chaos in Predynastic Egypt', in K. Kroeper, M. Chłodnicki & M. Kobusiewicz (eds), *Archaeology of Early Northeastern Africa. In Memory of Lech Krzyżaniak*, 723–749. Studies in African Archaeology 9, Poznań Archaeological Museum, Poznań.

Hendrickx, S. 2010, 'L'iconographie de la chasse dans le contexte social predynastique'. *Archeo-Nil* 20, 106–133.

Hoffman, M. A. (ed.) 1982, *The Predynastic of Hierakonpolis.* Egyptian Studies Association 1, Cairo and Illinois.

Horard-Herbin, M. P. 2000, 'Dog management and use in the Late Iron Age: the evidence from the gallic site of Levroux (France)', in S. J. Crockford (ed.), *Dogs through Time: an archaeological perspective.* British Archaeological Report S889, Archaeopress, Oxford.

Houlihan, P. F. 1996, *The Animal World of the Pharaohs*. AUC Press, Cairo.

Kemp, B. J. 1989, *Ancient Egypt: anatomy of a civilisation.* Routledge, London.

Kratochvíl, Z. 1973, 'Schädelkriterien der Wild- und Hauskatze (*Felis silvestris silvestris* Schreb. 1777 und *F. s.* f. *catus* L. 1758)'. *Acta Scientiarum Naturalium – Academiae Scientiarum Bohemoslovacae Brno* 7(10), 1–50.

Le Berre, M. 1990, *Faune du Sahara. 2. Mammifères.* Lechevalier-R. Chabaud, Paris.

Leitz, C. 2002. *Lexikon der ägyptischen Götter und Götterbezeichnungen.* Band II, Orientalia Lovaniensia Analecta 111, Peeters, Leuven.

Linseele, V. & Van Neer, W. 2003, 'Gourmets or priests? Fauna from the Predynastic temple'. *Nekhen News* 15, 6–7.

Linseele, V., Van Neer, W. & Hendrickx, S. 2007, 'Evidence for early cat taming in Egypt'. *Journal of Archaeological Science* 34(12), 2081–2090.

Linseele, V., Van Neer, W. & Hendrickx, S. 2008, 'Early cat taming in Egypt: a correction'. *Journal of Archaeological Science* 35(9), 2672–2673.

Linseele, V., Van Neer, W. & Friedman, R. 2009, 'Special animals from a special place? The fauna from HK29A at Predynastic Hierakonpolis'. *Journal of the American Research Center in Egypt* 45, 105–152.

Malek, J. 1993, *The Cat in Ancient Egypt.* British Museum Press, London.

Merceron, G. 2003, 'The diet of captive baboons: results of dental microwear analysis'. *Nekhen News* 15, 13–14.

Müller-Wollermann, R. 2003, '"Zoologische Gärten" als Mittel der Herrschaftslegitimation im Alten Ägypten'. *Die Welt des Orients* 33, 31–43.

O'Connor, T. P. 2007, 'Wild or domestic? Biometric variation in the cat *Felis silvestris* Schreber'. *International Journal of Osteoarchaeology* 17, 581–595.

Osborn, D. J. & Osbornová, J. 1998, *The Mammals of Ancient Egypt.* Aris & Phillips, Warminster.

Payne, J. C. 1993, *Catalogue of the Predynastic Egyptian Collection in the Ashmolean Museum.* Oxford University Press, Oxford.

Silver, I. A. 1969, 'The ageing of the domestic animals', in D. Brothwell & E. Higgs (eds), *Science in Archaeology*, 250–268. Thames and Hudson, London.

Smith, M. O. 1996, 'Parry fractures and female-directed interpersonal violence: implications from the Late Archaic period of West Tennessee'. *International Journal of Osteoarchaeology* 6, 84–91.

Van Neer, W. & Linseele, V. 2002, 'New analyses of old bones: the faunal remains from Hierakonpolis'. *Nekhen News* 14, 7–8.

Van Neer, W. & Linseele, V. 2008, 'More animal burials from the Elite Cemetery'. *Nekhen News* 20, 12–3.

Van Neer, W., Linseele, V. & Friedman, R. F. 2004, 'Animal burials and food offerings at the Elite Cemetery HK6 of Hierakonpolis', in S. Hendrickx, R. F. Friedman, K. M. Cialowicz & M. Chlodnicki (eds), *Egypt at its Origins. Studies in Memory of Barbara Adams*, 67–130. Orientalia Lovaniensia Analecta 138, Peeters, Leuven.

Van Neer, W., Linseele, V., Friedman R. F. & De Cupere, B. 2014, 'More evidence for cat taming at the Predynastic elite cemetery of Hierakonpolis (Upper Egypt)'. *Journal of Archaeological Science* 45, 103–111.

von den Driesch, A. 1976, *A Guide to the Measurement of Animal*

Bones from Archaeological Sites, Peabody Museum Bulletin 1, Harvard University Press, Cambridge, MA.

von den Driesch, A. & Boessneck, J. 1974, 'Kritische Anmerkungen zur Widerristhöhenberechnung aus Längenmassen vor- und frühgeschichtlicher Tierknochen'. *Säugetierkundliche Mitteilungen* 22, 325–348.

von den Driesch, A. & Boessneck, J. 1983, 'A Roman cat skeleton from Quseir on the Red Sea coast'. *Journal of Archaeological Science* 10, 205–211.

Whitehouse, H. 2002, 'A decorated knife handle form the "main deposit" at Hierakonpolis'. *Mitteilungen des Deutschen Archäologischen Instituts, Abteilung Kairo* 58, 425–446.

Wilkinson, R. H. 2003, *The Complete Gods and Goddesses of Ancient Egypt*. Thames and Hudson, London.

Wilkinson, T. A. H. 1999, *Early Dynastic Egypt*. Routledge, London.

Winter, E. 2006, 'Gedanken zum Pavian-Fund von Tell el-Dab'a', in E. Czerny, I. Hein, H. Hunger, D. Melman & A. Schwab (eds), *Timelines: studies in honour of Manfred Bietak*, 447–454, Peeters, Leuven.

Part 7

Animal exploitation during antiquity

28. Animal exploitation during the Classical/ Hellenistic period at Tepe Düzen (SW Turkey): preliminary results

Bea De Cupere, Wim Van Neer, Kim Vyncke & Hannelore Vanhaverbeke

Excavations at Tepe Düzen – located at less than 2km southwest of the classical site of Sagalassos (Burdur province, Turkey) – revealed the existence of a (proto-)urban centre that could be dated to the Classical/ Hellenistic period. The study of the faunal assemblage provides insights into animal exploitation during that period. The results were compared with data from Sagalassos (Roman to Mid-Byzantine period) in order to place Tepe Düzen within a broader perspective and to investigate how the relationship between humans and animals changed through time within the same area.

Keywords archaeozoology, Sagalassos, Classical/Hellenistic period, animal exploitation

Introduction

At about 1.8km to the southwest of the classical site of Sagalassos (Burdur province, SW-Anatolia), another settlement called Tepe Düzen (Fig. 28.1) was discovered on a broad flat area. Tepe Düzen extends over two promontories and covers an area of ca. 1.5km east–west and ca. 0.8km north–south, steeply sloping on the west, south and east sides that lead towards the green valley of the Ağlasun River. To the north it is dominated by Zencirli Tepe (1782m) (Fig. 28.2). Nowadays, the area is sparsely vegetated with low grasses and isolated stands of kermes oak (*Quercus coccifera*), while some wheat is cultivated on the colluvial fan located between Tepe Düzen and Zencirli Tepe. The area is mainly used for sheep and goat herding (Vanhaverbeke *et al*. 2010).

Fragmentary remains of structures at Tepe Düzen have been shown to derive from a (proto-)urban centre, surrounded by impressive defence walls. Excavations started in 2006 and have continued ever since. The excavations of the first 2 years revealed a complex archaeological sequence of at least four architectural building phases with associated floor levels, as well as some related refuse pits. Excavations during the 2008 campaign, on the other hand, yielded parts of what were most likely two different buildings consisting of several rooms, used as a dump area following occupation. A pit that was cut out of the bedrock was found as well. No water springs are available at the site and fresh water had therefore to be provided from cisterns. The remains of an at least 10 m deep cistern are preserved on the site, but most probably several more such structures must have been present at the site when it was occupied.

The pottery was preliminarily assigned to ca. 500–200 BC (Classical/Hellenistic period), with some

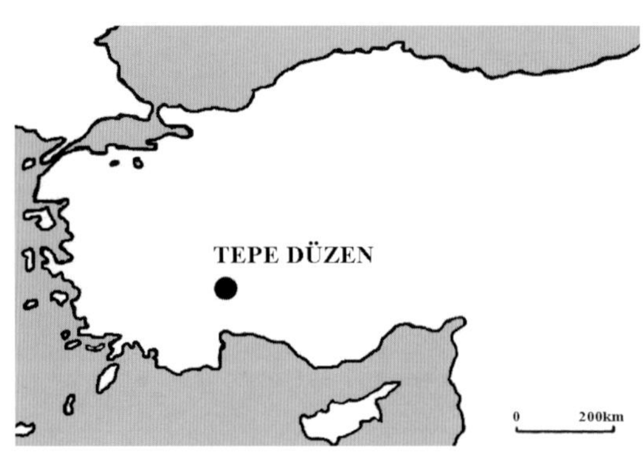

Fig. 28.1. Geographical location of Tepe Düzen.

Hellenistic and very rare Sagalassos red slip ware from the Imperial period. There was indeed only a small amount of chronologically diagnostic sherds, in general very weathered, and coming from an unknown local or regional production centre. Radiocarbon dates have been carried out on five bone samples and suggest a somewhat more recent date, up to the 2nd and possibly even the 1st century BC (Table 28.1).

This report presents the first results of the faunal analyses that have been carried out during the three first excavation campaigns (2006–2008) at Tepe Düzen. The study aims at providing a first impression of animal exploitation at this urban centre during the Classical/Hellenistic period. By comparing the data to those from Sagalassos, it should also be possible to draw some inferences concerning human–animal relationships through time within the same area.

Table 28.1. Radiocarbon ages of Tepe Düzen. Calibration was carried out using CALIB 5.0 (Stuiver & Reimer 1993).

Laboratory nr	*Context nr*	*Uncalibrated dates BP*	*Calibrated date (2 σ) BC*
Beta-244812	2006 TD 46-31	2150±40	360–280 & 260–60
Beta-244813	2006 TD 143-80	2350±40	510–380
Beta-244814	2007 TD 12-9	2220±40	390–180
Beta-244815	2007 TD 22-32	2210±40	390–170
Beta-244816	2007 TD 66-59	2150±40	360–280 & 260–60

The faunal material

The majority of the animal remains was obtained through hand-collection in the excavation trench. The material is very fragmented, resulting in a high number of unidentifiable bones. Soil samples were taken from interesting or promising contexts and wet-sieved using a series of 4mm, 2mm and 1mm meshes; a total volume of 538 litres was screened. However, the residues from these sieved samples yielded few identifiable remains. Until now no chronological phases could be distinguished and therefore, the faunal material has been treated as one large assemblage in this report. Some bone objects were found, mostly including worked astragali from sheep and goat, but these have not been included in this study.

The taxonomic composition of the animal remains from Tepe Düzen is provided in Table 28.2; hand-collected and sieved materials were kept separate from each other. Non-mammalian remains are low in number. Mollusc shells were found in small quantities and included terrestrial gastropods only; no shells from fresh water or marine molluscs have yet been found. The mollusc shells could be attributed to a large gastropod of the genus *Helix*, most probably *Helix cincta*, and to two small species, i.e. *Ceciloides tumulorum*

Fig. 28.2. View from Sagalassos onto the wide flat area of Tepe Düzen, and Zencirli Tepe.

and *Jaminia loewii*. These shells are considered to be the remains of intrusive species that lived on the site during or after the human occupation. Fish bones are very rare (n=10) and, due to the fragmentary nature of the material, only four pieces could be identified to taxon. These include two scales as well as an entopterygoid of Cyprinidae (carp family), that may come from the local Ağlasun River. Another scale derives from a tuna (*Thunnus* sp.) reflecting trade-links with the coast during the Classical/Hellenistic period. Reptiles are represented by the tortoise (*Testudo graeca*) although only carapace fragments have been found from this species. No anthropogenic traces were observed on this material and it is assumed that they are also the remains of intrusive animals. Wild birds are represented by the femur of a duck (Anatinae), by two posterior phalanges of stork (*Ciconia* sp.), by a scapula, a humerus, an ulna and two femora of chukar (*Alectoris chukar*), by the humerus of coot (*Fulica atra*) and by several skeletal elements from passerine birds of both medium (cf. *Turdus merula*) and large size (cf. *Pica pica* and *Garrulus glandarius*). Remains of domestic chicken represent a little more than half the total number of identified bird bones.

Small mammals are very poorly represented. There are remains of some small rodents, i.e. grey hamster (*Cricetulus migratorius*), house mouse (*Mus musculus*) and lesser mole rat (*Spalax leucodon*). Several bones could be assigned to the European hare (*Lepus europaeus*). Two skulls and a tibia have been identified as weasel (*Mustela nivalis*), while three other elements – a humerus, an ulna and a metapodial – belong to another, larger mustelid, i.e. the badger (*Meles meles*). Cut marks have been found on the humerus indicating that this individual was hunted, not only for its fur, but also perhaps for consumption. The brown bear (*Ursus arctos*) is represented by a single find, a distal radius that also displays cut marks, just above the distal articulation. Remains of cervids are more common and consist mainly of red deer (*Cervus elaphus*); some bones could be attributed to fallow deer (*Dama dama*), while roe deer (*Capreolus capreolus*) seems to be completely absent. One antler fragment has been positively identified as fallow deer, but the other antler fragments are more likely to be from red deer. The presence of cervids suggests that the landscape around the site was open with some woodland and shrubs.

The remains of domestic dog (*Canis lupus* f. familiaris) that were found in several instances, show a variation in size indicating that dogs of different sizes lived at Tepe Düzen. Both medium and large-sized individuals were retrieved from the excavation material. One fragment of a large ulna might rather be from a wolf (*Canis lupus*), but this could not be established with certainty. Equid remains were frequently identified and consist almost exclusively of donkey (*Equus africanus* f. asinus); two first phalanges have been identified with certainty as horse (*Equus ferus* f. caballus). The donkey remains all come from individuals of very small stature. A complete metacarpal permitted calculation of the withers height (von den Driesch & Boessneck 1974) yielding a value of approximately 1m. But, other, less completely preserved specimens show that even smaller individuals must have been present at Tepe Düzen. No traces of butchery or consumption have been found on the equid bones.

Table 28.2. Taxonomic composition of the faunal remains (expressed as NISP; + = present).

	Hand-collected	*Sieved total vol. 538 l*
Molluscs		
Helix cf. *cincta*	9	–
Ceciloides tumulorum	–	4
Jaminia loewii	1	–
Fish		
Cyprinidae sp.	–	3
Thunnus sp.	–	1
Reptiles		
Testudo graeca	11	–
Wild birds		
Ciconia sp.	2	–
Anatinae sp.	4	–
Alectoris chukar	2	–
Fulica atra	3	–
Passeriformes sp.	9	–
Wild mammals		
Lepus europaeus	19	–
Spalax leucodon	3	–
Mus musculus	–	2
Cricetulus migratorius	–	1
Rodentia sp.	–	77
Ursus arctos	1	–
Mustela nivalis	3	–
Meles meles	3	–
Canis lupus	1	–
Cervus elaphus	52	–
Dama dama	6	–
Cervus/Dama	2	–
Sus scrofa	+	–
Capra aegagrus	+	–
Domestic		
Gallus gallus f. domestica	22	–
Canis lupus f. familiaris	21	–
Equus ferus f. caballus	2	–
Equus africanus f. asinus	74	–
Equidae	9	–
Sus scrofa f. domestica	799	4
Capra aegagrus f. hircus	229	–
Ovis ammon f. aries	224	–
Capra/Ovis	5685	38
Bos primigenius f. taurus	2681	–
Total identified	9877	130
Unidentified mammal remains	19,051	5106
Unidentified bird remains	45	3
Unidentified fish remains	–	6

Most of the identified remains (99%), however, are from domestic species that were traditionally used as food animals, namely sheep (*Ovis ammon* f. aries), goat (*Capra aegagrus* f. hircus), cattle (*Bos primigenius* f. taurus) and pig (*Sus scrofa* f. domestica). Looking at the proportions (expressed in percentages of NISPs) of these food animals, it appears that sheep and goat are the best represented (64%) and that both species are present in about equal numbers. Cattle remains represent 28%, and only 8% can

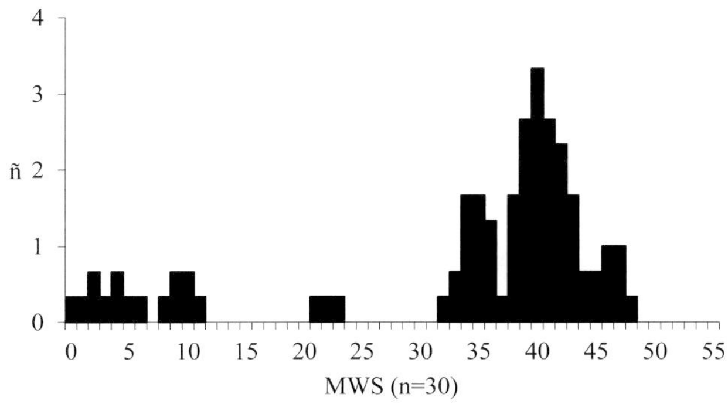

Fig. 28.3. Mandibular wear stages (MWS) of sheep/goat, according to Grant (1982); ñ: running mean.

be assigned to pig. Judging from the large size of some goat and pig bones, it seems likely that wild goat (*Capra aegagrus*) and wild boar (*Sus scrofa*) are present within the bone assemblage.

Sheep/goat

Age at slaughter of these domestic mammals has been established using the Grant's (1982) methodology. It is striking that a large proportion of the sheep/goats from Tepe Düzen are from older individuals (Fig. 28.3). The majority has a mandibular wear stage higher than 32, with a large peak around 40. The latter group comprises animals with a third molar displaying medium wear, or an age of over 4 years (Boessneck & von den Driesch 1975, 65). This indicates that they were not only kept for their meat, but that they also must have provided products such as milk. In the case of sheep, wool may have been important in textile manufacturing, something also suggested by several loom weights that were found on the site. Three complete metapodials and two radii permitted wither height calculations (von den Driesch & Boessneck 1974) of 57cm to be made for two goat individuals and 51cm, 59cm and 60cm for sheep. This corresponds to rather small animals. In one context, the almost complete skeleton of a foetus has been recovered; several other contexts yielded isolated bones of foetal animals.

Cattle

The number of mandibles of cattle was too low to establish slaughtering profiles. It was, however, clear that the remains of young individuals also occur in addition to the remains of older animals. A pelvic bone of cattle shows traces of the beginning of coxarthrosis, a pathological deformation that may be related to traction or other heavy-duty work (cf. Hüster 1990, 44–45, fig. 18). Other draught-related anomalies, as described by Bartosiewicz *et al*. (1997) on the phalanges and observed at Sagalassos (De Cupere *et al*. 2000), are present only to a slight degree. Therefore, it is believed that cattle were kept for both their meat and diary products, and – to a lesser extent – for draught power. The cattle remains appear to originate from rather small animals; calculation of wither height was only possible on one metacarpal from an animal of approximately 117cm high.

Table 28.3. Slaughter age of pigs, based on eruption and wear of the mandibular teeth; age classes according to Boessneck & von den Driesch (1975).

	Age	*Tepe Düzen*
M1 absent, Pd4 present	< 6 months	4
M2 erupted; M3 absent	1–1.5 years	2
M3 slightly worn	> 2 years	5

Pig

Mandibles of pig were also too few to establish slaughtering profiles, but data on the eruption and wear of the mandibular teeth have been summarised in Table 28.3. From these data it is clear that some of the pigs were slaughtered very young, before the eruption of the M1. Another part of the pig population reached an age of somewhat more than 2 years, so that their M3s appeared slightly worn (Boessneck & von den Driesch 1975, 100, table 23). There are, however, no indications of old animals within this assemblage.

Discussion

The relative abundances of the main domestic mammals at Tepe Düzen were compared to those observed at Sagalassos for different periods at the classical site (Fig. 28.4). For this comparison, data from Sagalassos were obtained from the large assemblages, related to large buildings and alterations of monumental buildings (De Cupere 2001, 137–141, fig. 99), and from the few contexts that have been dated to the Mid-Byzantine period. Some similarities do exist between forms of animal exploitation at Tepe Düzen and the Imperial period at Sagalassos. A heavy reliance on sheep and goat has indeed been observed for the first centuries AD at Sagalassos, although the proportion of the small livestock is even larger at Tepe Düzen. This may reflect a subsistence economy, favouring the herding of small livestock. This kind of herding economy is in sharp contrast with the increasing reliance on beef consumption – at the expense of sheep/goat – during the Late Roman period at Sagalassos. The economy at Salagassos in this period was flourishing and as a consequence, agricultural activities intensified (see also Vanhaverbeke *et al*. 2007). The need for draught animals may have increased, as suggested by the much older age at which cattle were slaughtered in Late Roman times (De Cupere 2001, 92–94) and the high incidence of pathological deformations on the phalanges that can be related to hard labour (De Cupere *et al*. 2000). The beginning of the Early Byzantine time

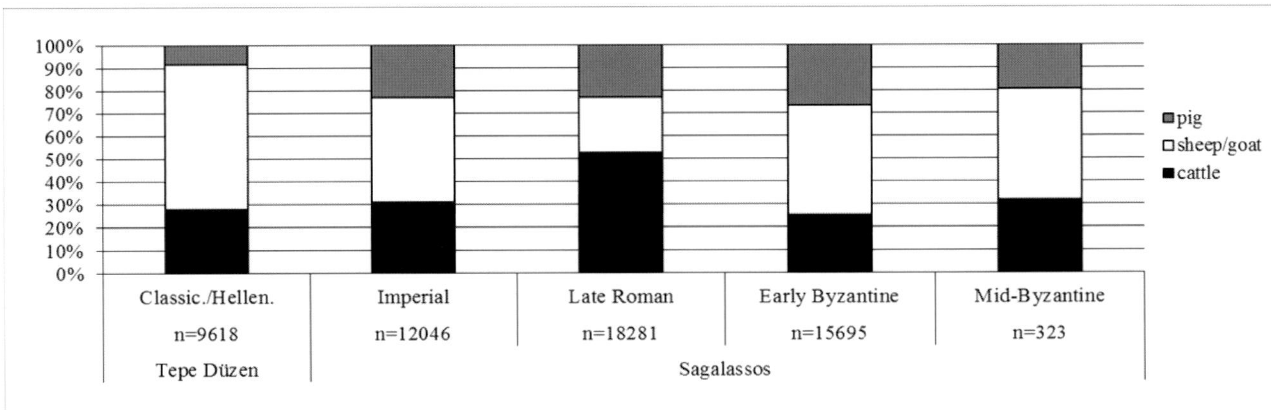

Fig. 28.4. Relative abundance of sheep/goat, cattle and pig at Tepe Düzen and Sagalassos (Classical/Hellenistic: 4th–1st century BC; Imperial: 1st–3rd centuries AD ; Late Roman: 4th–1st half 5th century; Early Byzantine: 2nd half 5th–7th centuries; Mid-Byzantine: 9th/10th–12/13th centuries).

was again characterised by a series of wars and natural disasters, which resulted in the destruction of Sagalassos' economy (Waelkens 1997). The inhabitants of the town and the territory may have turned again towards a more subsistence-based economy with an emphasis on the herding and breeding of sheep and goat (De Cupere 2001, 141). This situation continued into the Mid-Byzantine period.

Sheep and goat remains occur in, more or less, equal proportions at Tepe Düzen, in contrast to the situation at Sagalassos where goats are on average more numerous. Although goats thrive better than sheep in the rough environment of the region, the inhabitants of Tepe Düzen may have favoured sheep in view of the importance they attached to wool production. This is also suggested by the loom weights and spindle-whorls found at the site, although it is not excluded that fibres other than wool were also processed. Slightly higher numbers of sheep are observed for the Imperial period at Sagalassos and there is an inscription that indicates the presence of a guild of wool-dyers within the city during the 2nd century AD (De Cupere 2001, 141). Therefore, it is believed that this wool production must have retained its importance from the Classical/Hellenistic period through the Imperial period. In later periods, wool production probably diminished at Sagalassos as indicated by the increasing number of goats.

The site of Tepe Düzen – at an altitude of about 1400m a.s.l. – could be reached from the surrounding valleys by two rock-cut pathways that were not wide enough and too rough to be used by carts or wagons (Vanhaverbeke *et al.* 2010). Pack animals must have been ideal in such a situation. The rather large number of donkey remains that were found at Tepe Düzen therefore must be related to the topographical setting of the site and its accessibility. The absence of pronounced pathologies on the extremities of cattle, such as the distal metapodials and phalanges, suggests that these animals were of much less importance for transport or other heavy-duty work. This contrasts with Sagalassos, where the high incidences of draught-related pathologies found in cattle (De Cupere *et al.* 2000) were related to an intensification in agricultural production, especially during the Late Roman period (Vanhaverbeke *et al.* 2007).

As at Tepe Düzen, the overall share of wild mammals is very small within the mammalian assemblage of Sagalassos. However, when the cervids are considered, and red deer in particular, it is clear that their remains are relatively more common at Tepe Düzen compared to classical Sagalassos (Imperial, Late Roman and Early Byzantine periods). In the Mid-Byzantine contexts at Sagalassos, however, cervids occurred relatively frequently in the faunal assemblage. This higher number of deer bones in both the Classical/Hellenistic period of Tepe Düzen and the Mid-Byzantine period of Sagalassos might be explained in terms of changing environmental conditions. Considering the preferences of red deer and fallow deer for landscapes with interspersed forests, a possible explanation could be that the environment of the site was more wooded during these two periods than it had been in Roman-Early Byzantine times. The charcoal assemblages from Tepe Düzen are indeed indicative for better woodland resources in the Classical/Hellenistic period (E. Marinova pers. comm. 2009) and similar observations have been made from the pollen analyses (Vermoere *et al.* 2000). An alternative, or additional, explanation for the greater abundance of deer could be that by that time the area had become less densely inhabited by people. As a consequence, human pressure on the local environment may have been reduced, resulting in a considerably larger deer population.

Taphonomic reasons, such as bad preservation and recovery technique, could in principle be responsible for the very low number of fish remains. However, the identified pieces include fragile elements such as scales and the extensive sieving did not produce any significant amount of material. Among the fresh water fish only cyprinids have been identified and the presence of tuna bone is an indication for trade. It is, therefore, believed that fishing was not a common practice during the Classical/Hellenistic period in the area. Fish remains are much more common at Sagalassos and many different species are represented there, including Anatolian fresh water and marine fishes, as well as exotic fresh water fish (Van Neer *et al.* 1997; 2000). However, this taxonomic richness does not indicate intensification of local fisheries as all the fish, including the Anatolian freshwater fish, were imported (Van Neer *et al.* 2008).

Conclusion

The 2006–2008 excavations at Tepe Düzen yielded a faunal assemblage consisting of about 10,000 identifiable specimens. The data reveal that the economy was based mainly on the herding and breeding of sheep and goat, and to a lesser extent cattle and pig. Bones of sheep and goats are present in about equal proportions and both species provided products, such as milk and wool. Cattle, on the other hand, must rather have been raised for its meat and milk. There is no indication that cattle were regularly used for heavy duty work. The transport of goods was mainly carried out with donkeys. Comparison of faunal data from Tepe Düzen with the faunal material from Sagalassos, a Roman-Byzantine site less than 2km away, showed that sheep and goat herding, and the associated production of milk and wool, continued to be important during the Imperial period, although by this time these products seem to have been less emphasised.

Wild taxa make up a very small part of the assemblage at Tepe Düzen. Nevertheless, remains of cervids are not uncommon and are more frequently observed than at Sagalassos, suggesting environmental changes or greater human hunting pressure during the transition from the Classical/Hellenistic period to the Imperial period. The opposite trend is observed for the transition from the Early to Mid-Byzantine period where an increase in the numbers of cervid bones was observed in the faunal assemblage suggesting the presence of a larger deer population.

The overall picture of animal exploitation at the site of Tepe Düzen, as reflected in the faunal analysis, is that of a subsistence economy, mainly based on sheep and goat herding. The animals raised were rather small and the use of cattle in agriculture was probably quite limited. Such an animal husbandry strategy would correspond best to a rural community. However, in contrast, extensive town ruins have been brought to light at this site. In addition, the presence of tuna at the site provides evidence for sporadic (commercial) relationships with the coastal area during the Classical/Hellenistic period. Further excavations at Tepe Düzen will without any doubt provide much more information which will allow the verification of these preliminary results.

Acknowledgements

This text represents research results of the Belgian Programme on Interuniversity Poles of Attraction (IAP 6/22) initiated by the Belgian Federal Science Policy Office. Scientific responsibility is assumed by the authors.

References

Bartosiewicz, L., Van Neer, W. & Lentacker, A. 1997, *Draught Cattle: their osteological identification and history.* Annales du Musée Royal de l'Afrique Centrale, Sciences Zoologiques 281, Tervuren.

Boessneck, J. & von den Driesch, A. 1975, 'Tierknochenfunde vom Korucutepe bei Elaziğ in Ostanatolien', in van Loon, M. (ed.), *Korucutepe I, Studies in Ancient Civilization*, 1–220. North-Holland/American Elsevier, Amsterdam-Oxford-New York.

De Cupere, B., Lentacker, A., Van Neer, W., Waelkens, M. & Verslype, L. 2000, 'Osteological evidence for the draught exploitation of cattle: first applications of a new methodology'. *International Journal of Osteoarchaeology* 10, 254–267.

De Cupere, B. 2001, *Animal at Ancient Sagalassos. Evidence of the Faunal Remains.* Studies in Mediterranean Archaeology IV, Brepols, Turnhout.

Grant, A. 1982, 'The use of tooth wear as a guide to age of domestic ungulates', in B. Wilson, C. Grigson & S. Payne (eds), *Ageing and Sexing Animal Bones from Archaeological Sites*, 91–108. British Archaeological Report 109, British Archaeological Reports, Oxford.

Hüster, H. 1990, *Untersuchungen an Skelettresten vonRindern, Schafen, Ziegen und Schweinen aus dem mittelalterlichen Schleswig. Ausgrabung Schild 1971–1975.* Ausgrabungen in Schleswig. Berichte und Studien 8, Karl Wachtholz Verlag, Neumünster.

Stuiver, M. & Reimer P. J. 1993, 'Extended ^{14}C data base and revised Calib 3.0 ^{14}C age calibration program'. *Radiocarbon* 35, 215–230.

Vanhaverbeke, H., Martens, F. & Waelkens, M. 2007, 'Another view on late Antiquity: Sagalassos (SW Anatolia), its suburbium and its countryside in late Antiquity', in A. Poulter (ed.), *The Transition to Late Antiquity on the Danube and Beyond*, 611–648. Oxford University Press, Oxford.

Vanhaverbeke, H., Waelkens, M., Vyncke, K., De Laet, V., Aydal, S., Mušič, B., De Cupere, B., Poblome, J., Braekmans, D., Degryse, P., Marinova, E., Verstraeten, G., Van Neer, W., Šlapšak, B., Medarič, I., Ekinci, H. A. & Erbay, M. O. 2010, 'The "Pisidian" culture? The Classical-Hellenistic site at Düzen Tepesi near Sagalassus (SW Turkey)'. *Anatolian Studies* 60, 105–128.

Van Neer, W., De Cupere, B. & Waelkens, M. 1997, 'Remains of local and imported fish at the ancient site of Sagalassos (Burdur prov., Turkey)', in M. Waelkens & J. Poblome (eds), *Sagalassos IV. Report of the Fifth and Sixth Excavation Campaigns of 1994 and 1995*, 571–586. Acta Archaeologica Lovaniensia Monographiae 9, Leuven University Press, Leuven.

Van Neer, W., Wildekamp, R., Küçük, F., Ünlüsayin, M., Waelkens, M. & Paulissen, E. 2000, 'Results on the 1996 survey of the fish fauna of the Aksu river (Kestros) and some lakes in southwestern Anatolia, and the implications for trade at Sagalassos', in M. Waelkens & L. Loots (eds), *Sagalassos V. Report of the Survey and Excavation Campaigns of 1996 and 1997*, 828–842. Acta Archaeologica Lovaniensia Monographiae 11, Leuven University Press, Leuven.

Van Neer, W., Wildekamp, R., Küçük, F. & Ünlüsayin, M. 2008, 'The 1997–1999 surveys of the Anatolian fish fauna and their relevance to the interpretation of trade at Sagalassos', in P. Degryse & M. Waelkens (eds), *Sagalassos VI, Geo- and Bioarchaeology at Sagalassos and in its Territory*, 299–328. Leuven University Press, Leuven.

Vermoere, M., Waelkens, M., Vanhaverbeke, H., Librecht, I., Vanhecke, L., Paulissen, L. & Smets, E. 2000, 'Late Holocene environmental change and the record of human impact at Gravgaz near Sagalassos, southwest Turkey'. *Journal of Archaeological Science* 27, 571–595.

von den Driesch, A. & Boessneck, J. 1974, 'Kritische Anmerkungen zur Widerristhöhenberechnung aus Längenmassen vor- und frühgeschichtlicher Tierknochen'. *Säugetierkundliche Mitteilungen* 22(4), 325–348.

Waelkens, M. 1997, 'Interdiciplinarity in classical archaeology. A case study: the Sagalassos Archaelogical Research Project (southwest Turkey)', in M. Waelkens & J. Poblome (eds), *Sagalassos IV. Report of the Fifth and Sixth Excavation Campaigns of 1994 and 1995*, 225–252. Acta Archaeologica Lovaniensia Monographiae 9, Leuven University Press, Leuven.

29. Une accumulation d'équidés à *Berytus*: approche taxinomique et taphonomique

Yasha Hourani & Tarek Oueslati

L'étude porte sur un dépotoir riche en ossements animaux daté de la première moitié du Ier s. ap. J.-C. Cette accumulation est localisée à la frange Est de la ville romaine de Beyrouth. Les équidés dominent l'assemblage avec au moins 23 individus. Du point de vue taxinomique l'ensemble est très sélectif dans la mesure où tous les restes dentaires issus d'individus adultes ont été attribués à l'âne. Cela suggère sa forte implication dans l'économie de la ville. Du point de vue taphonomique, il s'avère que la viande de ces équidés a été au moins partiellement prélevée. De même certaines interventions sur leurs carcasses sont à l'origine du déficit du membre antérieur et des bas de pattes. En ce qui concerne les autres composantes de l'assemblage, des déchets de boucherie de production de masse de bœuf sont attestés. La technique de découpe est caractéristique de l'époque romaine et se retrouve dans les différentes provinces romaines de l'Empire. Ainsi, nous sommes en présence d'un dépotoir situé extra muros, *lieu d'élimination des déchets.*

Keywords Beyrouth, époque romaine, dépotoir, *extra muros*, âne, cheval, boucherie, équarrissage, gestion des déchets

An equine accumulation in Roman *Berytus*: a taxonomic and taphonomic approach

This study concerns a dump rich with faunal remains and dated to the first half of the 1st century AD. The accumulation is located at the eastern fringe of the Roman city of Beirut. The equids dominate the assemblage with 23 individuals. From a taxonomic point of view, the set is very selective as all the dental remains belonging to adult individuals were attributed to the donkey. This statement suggests that this animal was strongly involved in the economic activities of the city. The taphonomic approach reveals that the equine meat was partially defleshed. In addition, some interventions on the carcasses are the reason of the deficit in the representation of the fore and the lower limbs. Concerning the other components of the assemblage, bovine butchery waste is present. The butchery technique reveals a mass production activity, characteristic of the Roman period and found in the different provinces of the Empire. Thus we are in the presence of a dump located extra muros *in an area used for waste disposal.*

Keywords Beirut, Roman period, faunal remains, *extra muros*, donkey, horse, butchery, quartering, waste management

Introduction

Le site JEM002 se situe dans le quartier Jemmayzeh à Beyrouth. La fouille[1] a mis au jour une accumulation d'ossements animaux issue d'un niveau daté de la première moitié du Ier s. ap. J.-C. (Fig. 29.1). A l'échelle de la ville romaine, cette découverte se situe à la frange Est (Fig. 29.2), dans un secteur utilisé comme zone de décharge.

L'assemblage osseux comporte 2853 restes.[2] A

Fig. 29.1. Vue d'ensemble du dépotoir de JEM002, US 3346. L'échelle figurée est de 50cm.

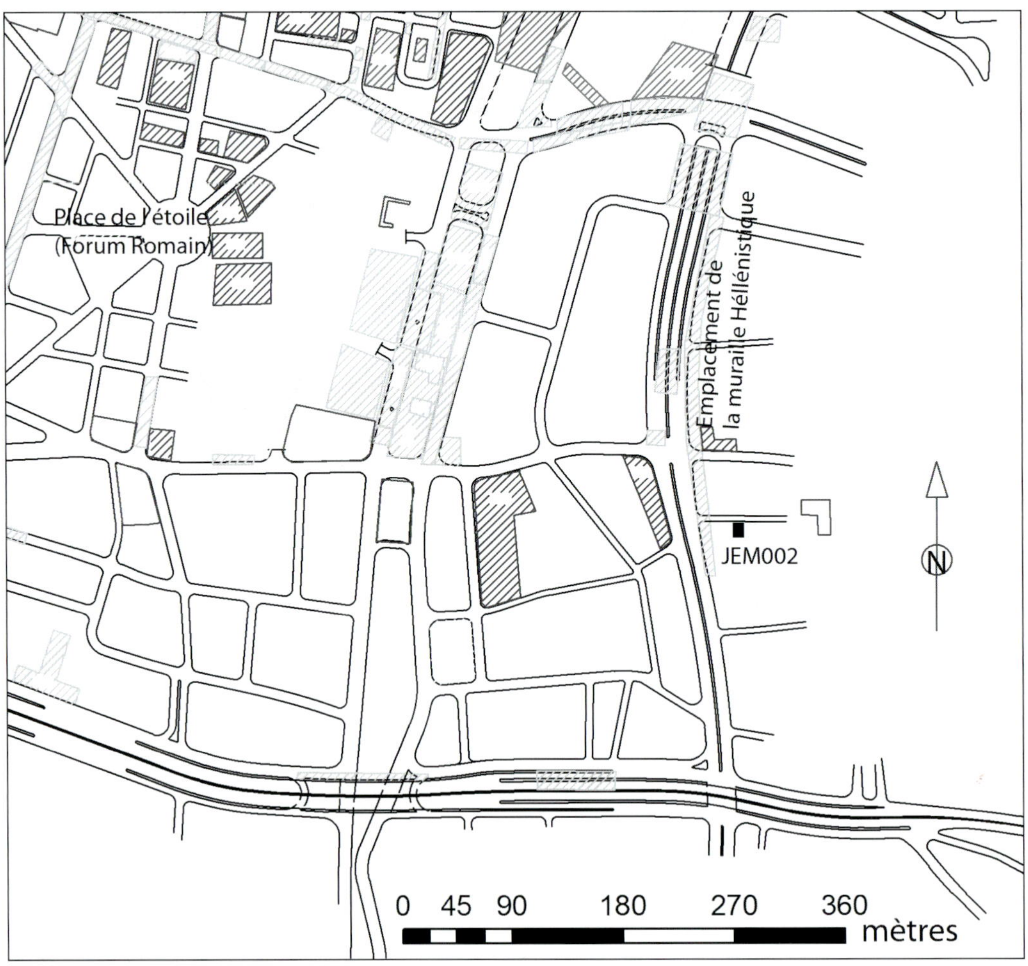

Fig. 29.2. Plan de Beyrouth (modifié d'après Curvers et Stuart 1998-1999). Localisation de la Place de l'étoile qui correspond à l'emplacement du forum romain, de la muraille hellénistique et du site JEM002. La muraille hellénistique correspond à la limite de la ville romaine jusqu'à la première moitié du Ier siècle après J.-C. A partir de la deuxième moitié de ce siècle, la ville s'étendra vers le Sud.

Table 29.1. Quantification des restes déterminés en Nombre de Restes (NR), Poids de Restes (PdR) et Nombre Minimal d'Individus (NMI).

	Taxa	*NR*	*PdR*	*NMI*
Equidés	*Equus africanus* f. *asinus*	42		21
	Equus ferus f. *caballus*	1		1
	Equus sp.	990		1
	Total équidés	1033	22,545	23
Bovinés	*Bos primigenius* f. *taurus*	191	9875	6
Caprinés	*Ovis orientalis* f. *aries*	4		2
	Capra aegagrus f. *hircus*	16		4
	Ovis/Capra	56		
	Total caprinés	76	817	6
Suinés	*Sus scrofa* f. *domesticus*	163	2134	7
Camélidés	*Camelus* sp.	9	1380	1
Canidés	*Canis* sp.	1	8	1
Hyénidés	*Hyaena hyaena*	5	97	1
	Total mammifères	1478	36,856	45
Oiseaux	*Gallus gallus* f. *domesticus*	7	6	1
Poissons	Carangidae	2	0.42	1
	Gadidae	1	0.44	1
	Moronidae	1	0.02	1
	Serranidae	5	0.71	1
	Sparidae	1	0.57	1
	Raie	1	0.13	1
	Requin	1	2.20	1
	Total poissons	12	4.49	12
	Total	1497	36,866.49	53

partir d'une analyse taxinomique et taphonomique, nous aborderons les activités qui sont à l'origine de cette accumulation importante de déchets. S'il est vraisemblable que nous sommes en présence d'un processus d'élimination de déchets générés par la ville, la prise en compte d'activités extra-urbaines reste néanmoins incontournable. Dans cette perspective, nous mettrons en évidence d'éventuelles remobilisations d'une partie de ces déchets dans le cadre de différentes chaînes opératoires du sous-système technique d'exploitation des ressources animales (Vigne 1998).

La détermination ostéologique révèle l'abondance des équidés. Le spectre est complété par le bœuf, les caprinés, le porc, les camélidés, les canidés, la hyène rayée, ainsi que des poissons de mer et le coq (Table 29.1).

La dominance des équidés a nécessité l'application des méthodes morphologiques et métriques pour la discrimination entre les différentes espèces du genre *Equus*. Pour notre cadre chronospatial, les données paléo-biogéographiques renseignent sur cinq taxons potentiels: le cheval et l'âne domestiques, l'hémione hémippe, la mule et le bardot (Ducos 1968 ; Burleigh 1986 ; Groves 1986 ; Uerpmann 1987).

Ainsi, dans un premier temps nous exposerons les méthodes et les résultats de la détermination des équidés. Nous décrirons ensuite la répartition de l'âge et du sexe des individus, leur stature et les pathologies osseuses.

Caractérisation des équidés

Détermination taxinomique

Les méthodes de discrimination entre les différentes espèces du genre *Equus* sont fondées sur des critères morphologiques et morphométriques de plusieurs éléments crâniens et post-crâniens. L'application de ces méthodes à notre échantillon a permis d'éliminer la probabilité de l'occurrence de l'équidé sauvage potentiel, *Equus hemionus* (Uerpmann 1986; 1991; Boessneck 1987; Vila 1998; Eisenmann 1980; 1981). Les résultats métriques que nous présentons dans cette étude excluent donc la comparaison avec l'hémione hémippe. Le référentiel métrique de comparaison est issu des données mises en ligne par V. Eisenmann (http://vera-eisenmann.com). Les mesures relevées sur le crâne suivent le système de mesures établi par Eisenmann (1980) et sur les dents jugales et les os des membres celui établi par Eisenmann et Mashkour (2000).

La représentation graphique des données craniométriques à l'aide d'un diagramme de rapports (Simpson 1941) permet la discrimination entre les espèces (Eisenmann 1980; 1986). Notre assemblage contient huit crânes. Toutefois les conditions de conservation et les fortes déformations limitent le nombre de spécimens étudiables et les seules mesures discriminantes ont été obtenues sur le crâne 31. La conformation du museau (la longueur, la largeur entre les incisives et la largeur minimale, respectivement les mesures 5, 17 et 17 bis; Fig. 29.3) de ce spécimen se rapproche de celle du cheval. Toutefois, ce crâne est issu d'un individu sub-adulte, ce qui impose une certaine prudence dans l'interprétation des résultats.

La morphologie des dents jugales, précisément le dessin de l'émail à la surface occlusale, permet également de caractériser les différentes espèces (Churcher et Richardson 1978; Armitage et Chapman 1979; Eisenmann 1980; 1981; 1988; Uerpmann et Uerpmann 1994). Les dents jugales sont les éléments les mieux conservés dans notre assemblage. La discrimination entre les espèces et les formes hybrides du genre *Equus* se base notamment sur ces éléments dans notre étude.

La plupart des dents jugales supérieures présente une morphologie asinienne (NMI=15; Fig. 29.4).[3] Le protocône est court et symétrique, le sillon postprotocônal est court, le parastyle et le mésostyle sont simples, le pli caballin est absent et le nombre de pli des fossettes est réduit. Cinq spécimens (dont le crâne 31) présentent de nombreux plis au niveau des fossettes et un pli caballin. Les fossettes complexes, le pli caballin et le protocône symétrique sont des caractères qui se retrouvent chez la mule (Uerpmann

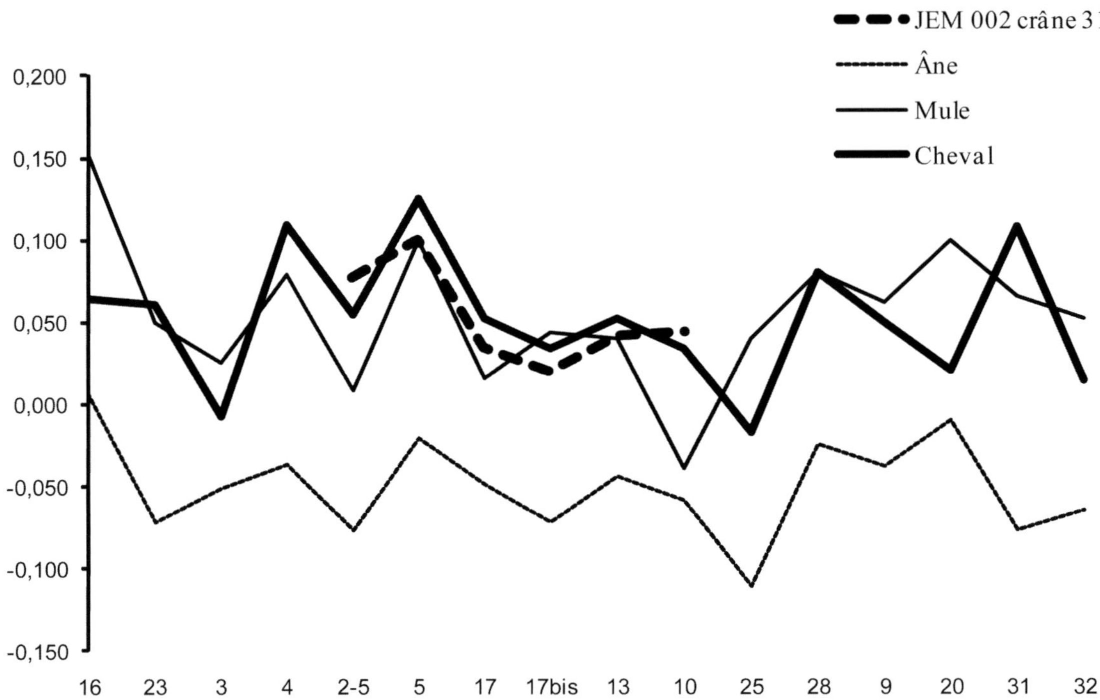

Fig. 29.3. Diagramme de différence en logarithme des mesures crâniennes du cheval (C117), de la mule (UM3), de l'âne et du crâne 31 de JEM002. L'espèce de référence est l'hémione (données métriques: http://vera-eisenmann.com).

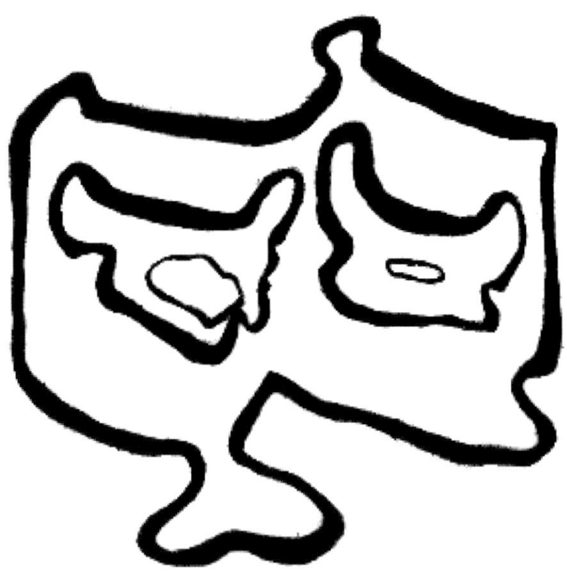

Fig. 29.4. Dessin en vue occlusale de la P3 gauche du maxillaire 39 (Longueur occlusale = 24.7mm; largeur occlusale = 23.7mm; Longueur du protocône = 9.17mm).

2002 in Johnstone 2004, tab. 4.1, 165 et fig. 4.3, 166). Nous avons donc comparé nos données métriques à des spécimens de référence de l'âne, du cheval et de la mule en rapportant la longueur du protocône à la surface occlusale[4] de la dent (Fig. 29.5, a–b; Mashkour 2001). Notre échantillon se trouve dans l'aire de répartition de l'âne. Seul le crâne 31 se distingue du lot. Par rapport aux autres spécimens, la P3 a une surface occlusale plus développée et le protocône est de longueur relativement moyenne; les molaires ont une surface occlusale plus grande et le protocône plus long. La P3 et la M1 se trouvent dans la zone de recouvrement entre le cheval et la mule. La M2 est incluse dans la variabilité du cheval. Comme nous l'avons déjà mentionné, ces dents appartenant à un individu sub-adulte sont peu usées et nous resterons prudents sur leur attribution spécifique.

Les dents jugales inférieures (NMI=23) présentent deux variabilités morphologiques. La première se caractérise par une double boucle de type sténonien; le sillon lingual est en 'V', le métaconide et le métastylide sont arrondis avec des bords linguaux convexes (Fig. 29.6, a). La deuxième variabilité se caractérise par un sillon lingual en 'V' peu profond à branche déjetée (*tick shaped*) et une double boucle asymétrique: le métastylide est arrondi et le métaconide est allongé (Fig. 29.6, b). Dans ces deux variabilités, le sillon vestibulaire est court ; il ne pénètre pas le col du métastylide et du métaconide. Ces deux morphologies caractérisent les asiniens (Armitage et Chapman 1979). Les dents jugales inférieures sont de ce fait issues d'ânes domestiques.

Nous notons que des dents de morphologie différente se retrouvent dans la même série dentaire. Par ailleurs, le pli caballinide est plus développé sur les dents appartenant à la deuxième variabilité morphologique.

La discrimination de l'espèce à partir des incisives est plus problématique. Dans notre échantillon, plusieurs troisièmes incisives inférieures ne présentent pas de cornet. Généralement, l'absence du cornet est plus fréquente chez les asiniens que chez les caballins et les hémioniens[5].

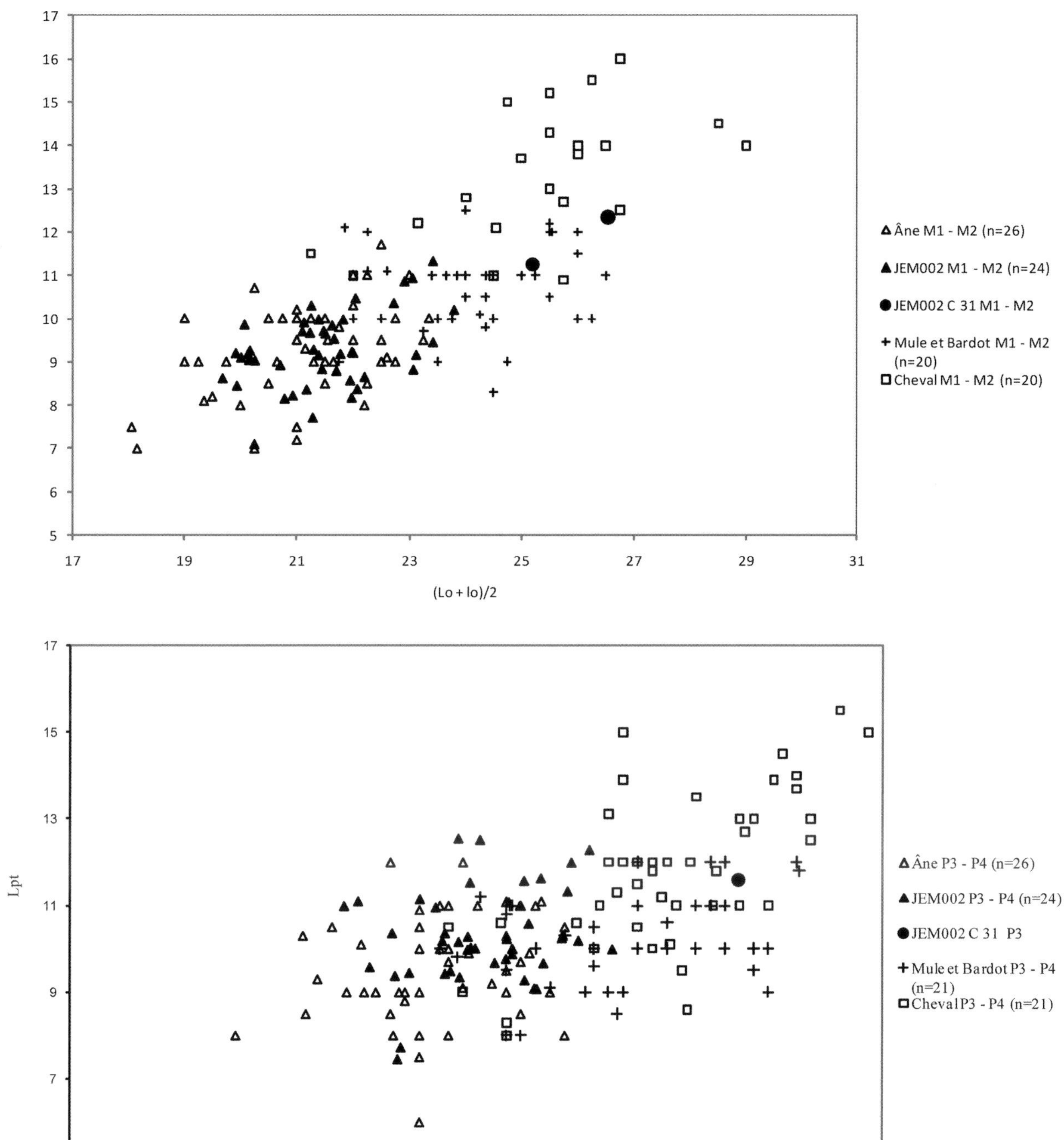

Fig. 29.5. Diagramme de dispersion des mesures des dents jugales supérieures des équidés de JEM002 comparées à celles de l'âne, du cheval, de la mule et du bardot. Le crâne 31 de JEM002 (C 31) est représenté individuellement (données métriques: http://vera-eisenmann.com): a: les molaires (M1 – M2); b: les prémolaires (P3–P4).

Les os des membres sont moins représentés que la tête osseuse dans notre assemblage (voir Fig. 29.21). Parmi ceux-ci, le métacarpe et la phalange I sont bien appropriés à la discrimination entre les espèces. Ces éléments peuvent être comparés avec les différentes espèces potentielles à l'aide d'un diagramme de rapports (Eisenmann et Beckouche 1986; Dive et Eisenmann 1991). Notre échantillon contient trois métacarpes ayant donné des mesures. La Figure 29.7 illustre les proportions d'un métacarpe robuste (voir le rapport entre la largeur et la longueur de l'os, respectivement les mesures 3 et 1). Sa largeur articulaire distale est plus développée que sa largeur supra-articulaire (mesures 11 et 10). La profondeur interne du condyle médial est moins développée que sa

Fig. 29.6. Dessins en vue occlusale des deux morphologies observées sur les dents jugales inférieures des équidés de JEM002. Les mesures de la surface occlusale sont en (mm): A: M2 droite de la mandibule 54 (Longueur occlusale = 24.1; largeur occlusale = 12.6; Longueur du postflexide = 8.7; Longueur de la double boucle = 12.6); B: P4 gauche de la mandibule 53 (Longueur occlusale = 23.7; largeur occlusale = 14.3; Longueur du postflexide = 10.7; Longueur de la double boucle = 14.7).

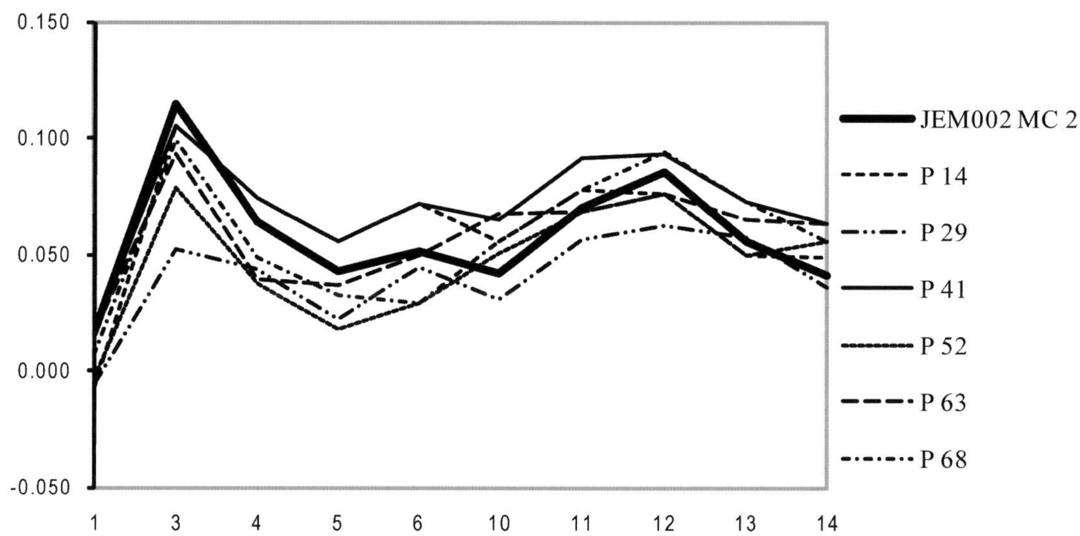

Fig. 29.7. Diagramme de différence en logarithme des mesures du métacarpe 2 de JEM002 et du cheval de Przewalskii. L'espèce de référence est l'hémione (données métriques: http://vera-eisenmann.com).

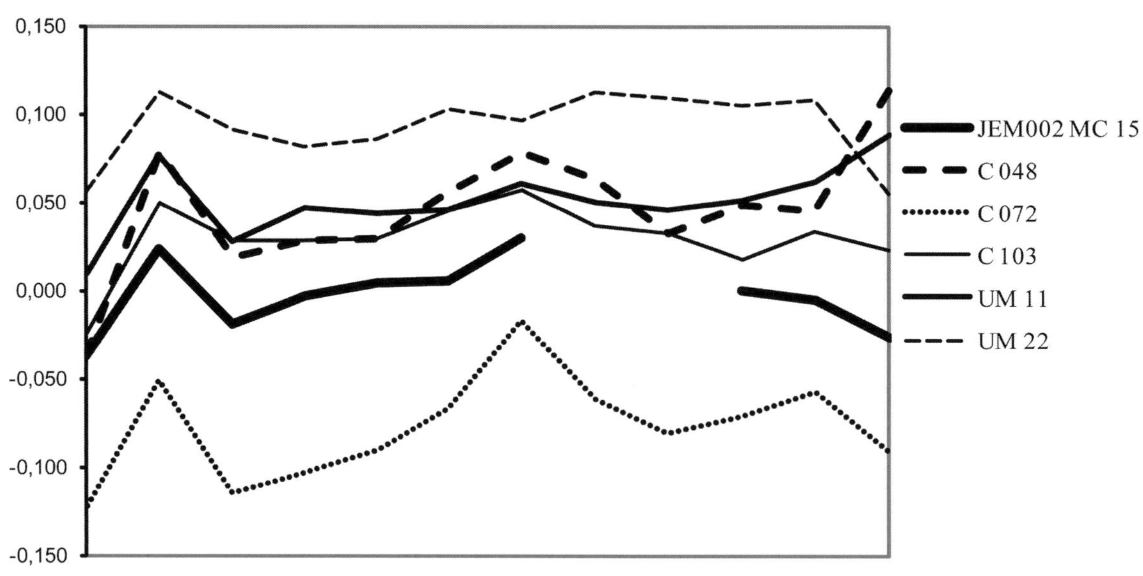

Fig. 29.8. Diagramme de différence en logarithme des mesures du métacarpe 15 de JEM002, du cheval (C048, C072, C103) et de la mule (UM11, UM22). L'espèce de référence est l'hémione (données métriques: http://vera-eisenmann.com).

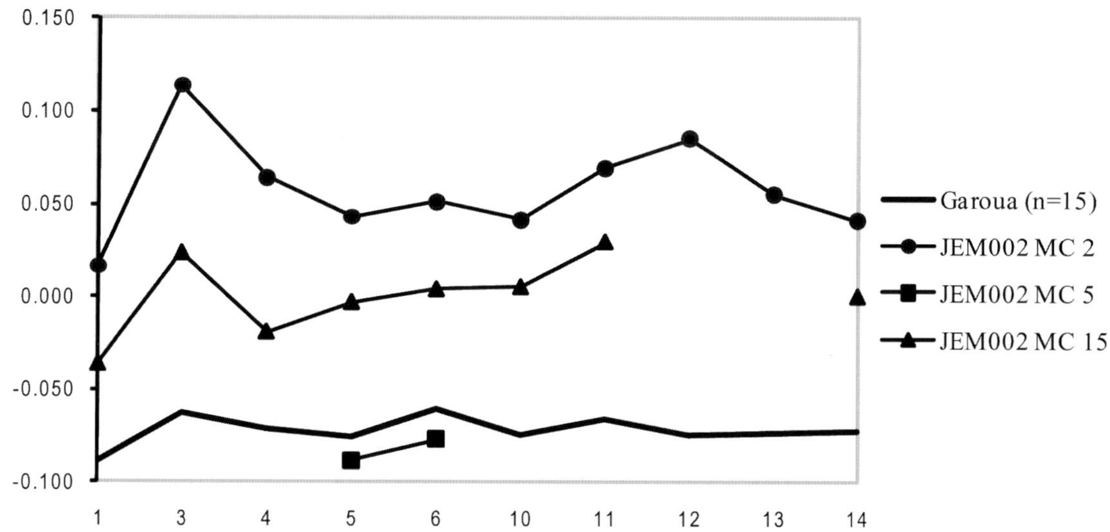

Fig. 29.9. Diagramme de différence en logarithme des mesures des métacarpes de JEM002 et de l'âne domestique de Garoua – Cameroun. L'espèce de référence est l'hémione (données métriques: http://vera-eisenmann.com).

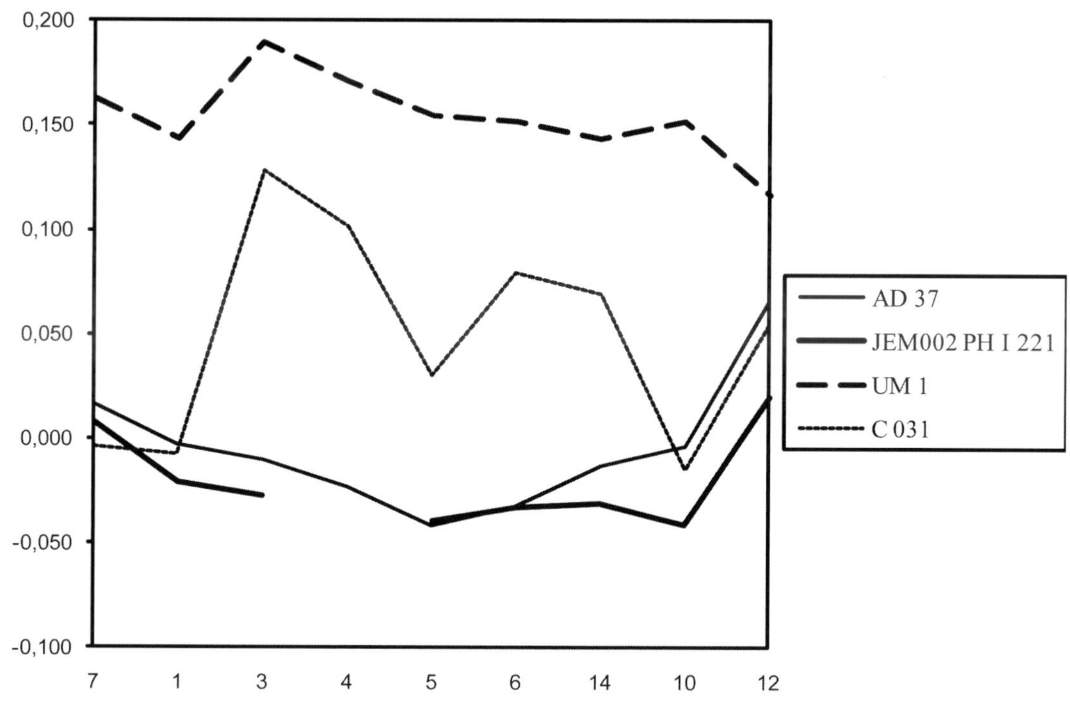

Fig. 29.10. Diagramme de différence en logarithme des mesures de la phalange I (221) de JEM002 et de la phalange I antérieure de l'âne (AD37), de la mule (UM1) et du cheval (C031). L'espèce de référence est l'hémione (données métriques: http://vera-eisenmann.com).

profondeur externe (mesures 13 et 14). Ce métacarpe est ainsi de morphologie caballine. La Figure 29.8 montre un métacarpe plus gracile. La diaphyse de cet os est plus large que profonde; et la largeur articulaire distale est plus importante que la largeur supra-articulaire. Ce métacarpe pourrait appartenir à un cheval de petite taille ou à une mule. La Figure 29.9 présente une comparaison entre les trois métacarpes de JEM002 et l'âne. Le métacarpe 5 qui a livré uniquement deux mesures est le plus petit spécimen, il se rapproche de la taille de l'âne.

Deux phalanges I ont fourni des mesures discriminantes. L'une est une phalange antérieure qui appartiendrait à un âne (Fig. 29.10) et l'autre est une phalange postérieure qui pourrait appartenir à un âne de grande taille ou à une mule (Fig. 29.11).

D'autres éléments des membres ont été identifiés

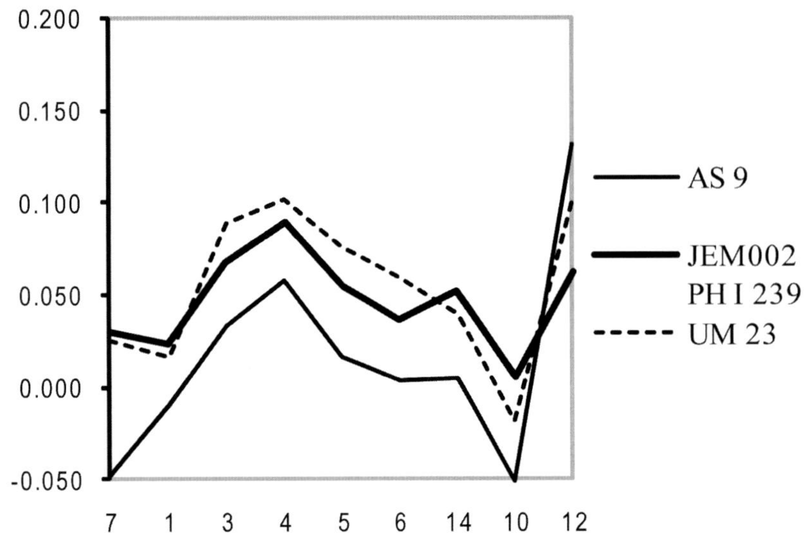

Fig. 29.11. Diagramme de différence en logarithme des mesures de la phalange I (239) de JEM002 et de la phalange I postérieure de l'âne (AS9) et de la mule (UM23). L'espèce de référence est l'hémione (données métriques: http://vera-eisenmann.com).

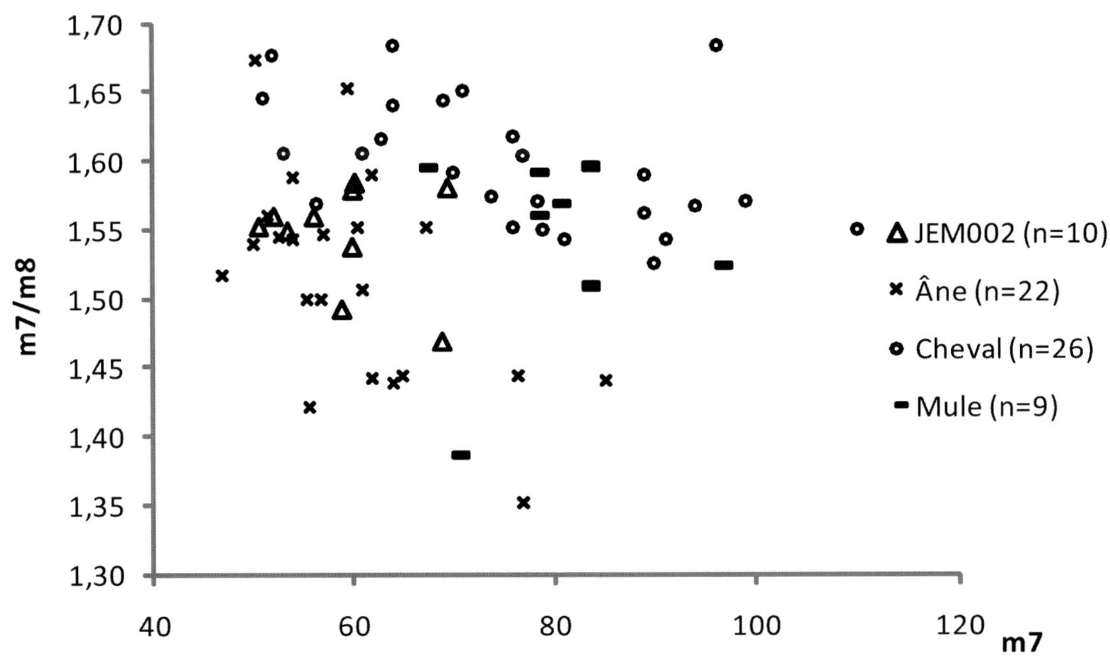

Fig. 29.12. Diagramme bivarié de la largeur (mesure 7) et de l'indice d'aplatissement du tibia distal (mesure 7/mesure 8) des équidés de JEM002, comparés à l'âne, au cheval et à la mule (données métriques: http://vera-eisenmann.com).

au rang de l'espèce à partir de critères morphologiques (Peters 1998). Le tibia distal (NR=10 ; NMI=7)[6] a une forme trapézoïdale. Contrairement à la forme rectangulaire caballine, la forme trapézoïdale du distum est asinienne. La Figure 29.12 montre que pour la largeur et l'aplatissement (la largeur divisée par la profondeur: mesure 7/mesure 8) de l'articulation distale, la majorité de nos spécimens se trouve dans la variabilité de l'âne. Le distum qui présente les plus grandes valeurs pour la largeur et pour le rapport entre la largeur et la profondeur de l'articulation, se trouve à des valeurs proches de celles du cheval et de la mule. Nous ne pouvons exclure son appartenance à l'un de ces deux équins.

La discrimination spécifique a ainsi montré que la majorité des restes d'équidés appartiennent à l'âne, notamment les dents jugales. Un crâne pourrait provenir d'un cheval ou d'une mule. Le squelette postcrânien provient aussi d'individus asiniens et d'un cheval. L'appartenance d'un métacarpe et d'une phalange à la mule n'est pas exclue.

Estimation de l'âge

Les incisives étant pour la plupart isolées, nous avons établi le profil de mortalité des équidés à partir du stade de développement des racines et de l'attrition des dents

Table 29.2. Répartition des classes d'âge en fonction de la hauteur de la P2 et de la M2 supérieures des équidés de JEM002. Les intervalles de hauteur de la couronne définissent les classes d'âge. La hauteur maximale correspond à la classe d'âge de 2.5/3–5 ans. Les trois quarts de la hauteur maximale correspondent à la classe d'âge de 5–9 ans. La moitié de la hauteur maximale correspond à la classe d'âge de 9–13 ans et le quart correspond à la classe d'âge de 13 ans et plus. Les effectifs des classes d'âge sont corrigés en fonction de la classe qui s'étend sur la plus courte durée.

Âge (ans)	*Coefficient de correction*	*P2*	*M2*
	× 1		
2.5		55.6	
3			80.4
	× 1		
5		41.7	60.3
	× 0.5		
9		27.8	40.2
	× 0.5		
13		13.9	20.1
	× 0.3		

Table 29.3. Répartition des classes d'âge en fonction de la hauteur de la P3 et de la M2 inférieures des équidés de JEM002. Les intervalles de hauteur de la couronne définissent les classes d'âge. La hauteur maximale correspond à la classe d'âge de 3–5 ans. Les trois quarts de la hauteur maximale correspondent à la classe d'âge de 5–9 ans. La moitié de la hauteur maximale correspond à la classe d'âge de 9–13 ans et le quart correspond à la classe d'âge de 13 ans et plus. Les effectifs des classes d'âge sont corrigés en fonction de la classe qui s'étend sur la plus courte durée.

Âge (ans)	*Coefficient de correction*	*P3*	*M2*
	× 1		
3		74.4	80.25
	× 1.5		
5		55.8	60.19
	× 0.75		
9		37.2	40.13
	× 0.75		
13		18.6	20.06
	× 0.25		

jugales. Le début du processus de formation des racines des dents jugales correspond à l'âge auquel l'émail est minéralisé et au moment où la dent atteint sa hauteur maximale (Hoppe *et al.* 2004). En se basant sur l'âge relatif aux stades de minéralisation des dents permanentes avancé par Hoppe *et al.* (2004), nous avons estimé l'âge des séries dentaires qui contiennent des dents dont les racines ne sont pas encore formées. La P2 et la M2 supérieures, la P3 et la M2 inférieures qui présentent la plus grande hauteur dans notre échantillon pour ces catégories de dents, se trouvent dans des rangées dentaires incluant des dents dont les racines ne sont pas encore formées et dont l'âge a été bien défini. Ainsi nous avons pu déterminer la hauteur maximale de ces dents au sein de notre échantillon. Nous avons ensuite divisé, à l'instar de Vila (1998[7]), la hauteur maximale de chaque catégorie de dent par quatre pour définir des intervalles de hauteur de la couronne auxquels correspond un âge réel (Table 29.2 et 29.3).

Le profil de mortalité, ainsi construit, montre que la majorité des individus appartient aux classes d'âge entre deux ans et demi et treize ans (Fig. 29.13). Un individu a moins de deux ans et un autre a plus de treize ans. Ces résultats obtenus à partir d'un référentiel basé sur la minéralisation des dents des chevaux sont à prendre avec précaution dans la mesure où notre assemblage est dominé par l'âne et que le développement et l'attrition des dents se déroulent à des rythmes différents entre l'âne et le cheval (Hoppe *et al.* 2004; Misk et Seilem 1997; Muylle *et al.* 1999). D'après Misk et Seilem (1997) la formation des racines des dents jugales supérieures de l'âne se fait dans l'ordre suivant: M1, M2, M3, P2, P4, P3 et celle des dents inférieures dans cet ordre: M1, M2, M3, P2 et P3, P4 (Misk et Seilem 1997, tab. 2, 32). Ces données contredisent nos observations sur les ânes de JEM002 chez qui les racines des dents jugales supérieures et inférieures se développent en premier au niveau de la M1, puis la P2, P3, M2, P4 et enfin la M3. La séquence de formation des racines établie par Misk et Seilem (1997) ne convient donc pas à notre assemblage. Nous avons retenu tout de même l'âge de cinq ans marquant la fin du processus de formation des racines ainsi que la corrélation entre les séquences d'usure et les âges correspondants proposée par ces auteurs (figs 1–16, 28–31) afin de vérifier la répartition par âge établie pour notre échantillon. A l'issue de l'adaptation de cette méthode à notre assemblage, il s'avère que le profil de mortalité ne change pas de manière significative, hormis l'élargissement de la première classe d'âge lié à l'impossibilité d'utiliser les séquences de formation des racines[8].

Sex-ratio

Nous complétons l'approche démographique par l'estimation du sex-ratio. Sur les 23 individus représentés par des dents jugales inférieures, seulement neuf présentent des diastèmes intacts permettant l'évaluation de la présence et du développement des canines. Dans cet échantillon, huit spécimens sont caractérisés par un développement important des canines, ce qui permet leur attribution à d'individus mâles. Le dernier spécimen dépourvu de canine est une femelle. Par ailleurs, sept canines isolées sont issues de mâles, ce qui donne un sex-ratio de 15 mâles pour une femelle. Si nous avons compté 15 canines bien développées sur un total de 23 individus, la forte fragmentation des diastèmes ne permet pas d'exclure la présence d'autres

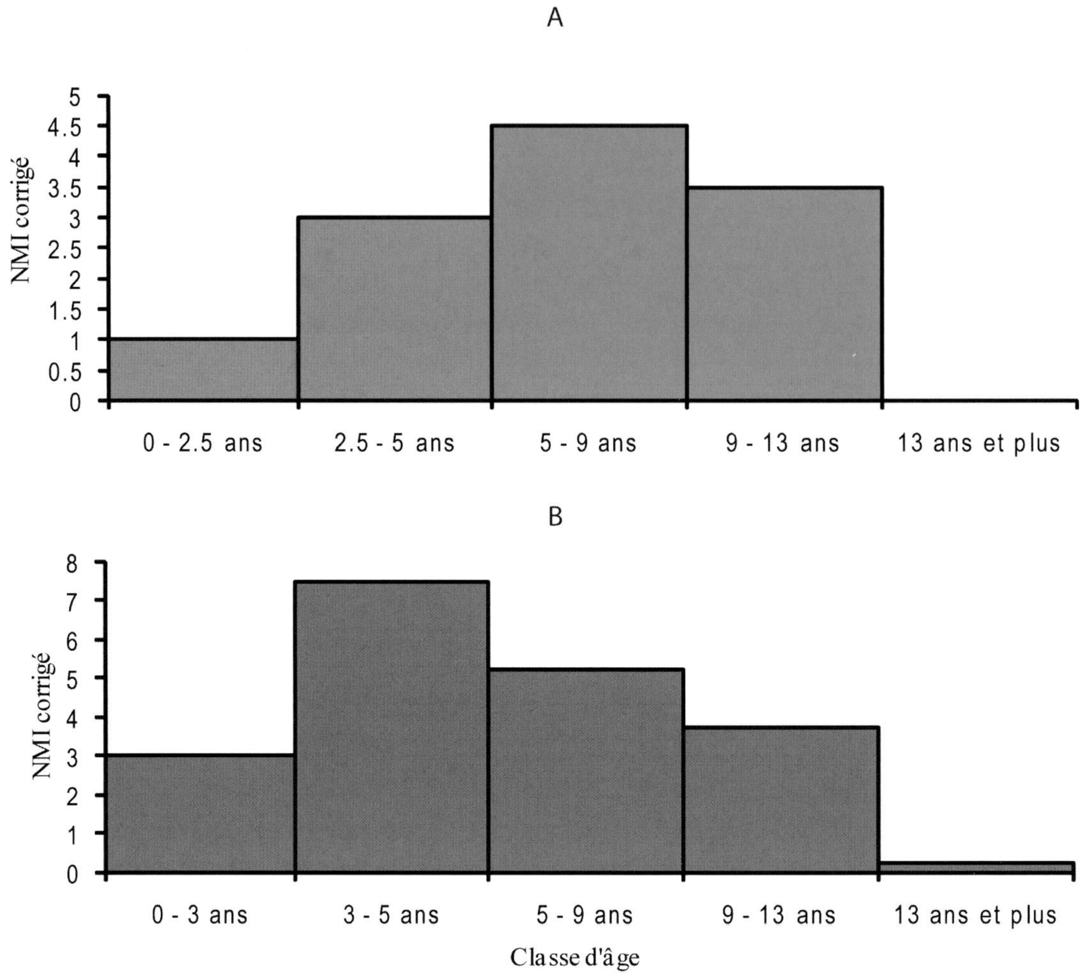

Fig. 29.13. A. Le profil de mortalité en NMI corrigé des équidés de JEM002 construit à partir des âges estimés sur les dents jugales supérieures (NMI=20); B. Le profil de mortalité en NMI corrigé des équidés de JEM002 construit à partir des âges estimés sur les dents jugales inférieures (NMI=21).

femelles. Néanmoins les mâles demeurent dominants avec au moins 15 individus sur 23.

Par ailleurs, la combinaison des données relatives à l'âge et au sexe souligne que le seul individu sénile est une femelle.

Stature et gracilité

Comparés à des données issues de quelques sites du Proche-Orient, les ânes de JEM002 se placent parmi les grands individus. La Figure 29.14 illustre la largeur du tibia distal des ânes de JEM002, ceux de Sagalassos (Ier–VIème siècle ap. J.-C.; De Cupere 2001), de Kamid el-Loz (Âge du Bronze récent et Âge du Fer; Bökönyi 1990) et de Tell Jennin (Âge du Bronze Ancien; Al-Zawahra et Ezzughayyar 1998). Les ânes de JEM002 sont de grande taille, mais ils ne diffèrent de manière significative que des spécimens de l'Âge du Bronze récent de Kamid el-Loz (test t, P=0.01).

La hauteur au garrot de l'âne de JEM002, estimée sur la phalange I (221), est de 1.19m.

Le cheval de JEM002 aurait une hauteur au garrot de 1.27m, estimée sur le métacarpe, selon le coefficient appliqué au cheval de Przewalskii et au cheval arabe. Il aurait 1.43m au garrot selon le coefficient appliqué au cheval de trait (Willoughby 1974; Eisenmann http://vera-eisenmann.com[9]). L'indice de gracilité[10] calculé pour le métacarpe est de 15.3. Comparé aux chevaux de la Gaule romaine (Lepetz 1996), le cheval de JEM002 se situe parmi les individus robustes.

Pathologies osseuses

Les équidés de JEM002 sont atteints de pathologies osseuses notamment au niveau des vertèbres et des côtes. Les membres sont peu affectés ; une seule pathologie a été relevée au niveau d'un métacarpe. Des pathologies orales ont été aussi observées.

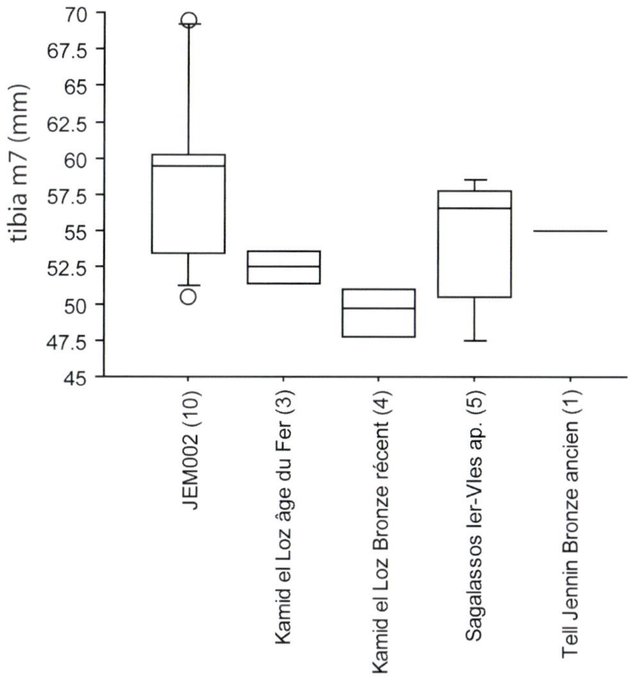

Fig. 29.14. La largeur de l'articulation distale du tibia (mesure 7) des ânes de JEM002, comparés à ceux de Sagalassos (Ier–VIème siècle après J.-C.; De Cupere 2001), de Kamid el-Loz (Âge du Bronze récent et Âge du Fer; Bökönyi 1990) et de Tell Jennin (Âge du Bronze ancien; Al-Zawahra et Ezzughayar 1998). La différence de taille est significative entre JEM002 et l'Âge du Bronze de Kamid el-Loz (test t, P = 0.01).

LA CAGE THORACIQUE

Les trois régions de la colonne vertébrale sont affectées de pathologies. Cependant peu de vertèbres, par rapport au nombre total, présentent des anomalies (16 sur un total de 288). La fragmentation a limité la possibilité de déterminer le nombre d'individus affectés.

Deux vertèbres cervicales isolées témoignent d'anomalies. L'une présente une malformation ou une cicatrisation suite à une fracture au niveau du processus articulaire. L'autre présente un développement d'un nouveau tissu osseux au niveau du tubercule ventral du processus transverse.

Sept vertèbres thoraciques isolées sont pathologiques. De l'exostose s'est développée sur la face ventrale des corps vertébraux (NR=3) et des ostéophytes ont proliféré sur la face ventrale de la tête articulaire et du disque vertébral caudal (NR=2). De l'exostose s'est formée au niveau d'un processus épineux. Enfin, une vertèbre présente une fissure horizontale au niveau du disque vertébral caudal.

Un ensemble en connexion anatomique composé de quatre vertèbres thoraciques et une côte présente des pathologies. Des ostéophytes se sont formées sur la face ventrale du disque vertébral caudal et autour de la tête articulaire. Le bord caudal du processus épineux présente du *lipping*. Une vertèbre montre une résorption de l'os sous forme de cupules (*pitting*) au niveau du corps vertébral et du processus épineux. Une tête articulaire présente une zone de dépression (Fig. 29.15; Bendrey *et al.* 2008). La côte présente un développement d'un nouveau tissu osseux sur la face viscérale et externe.

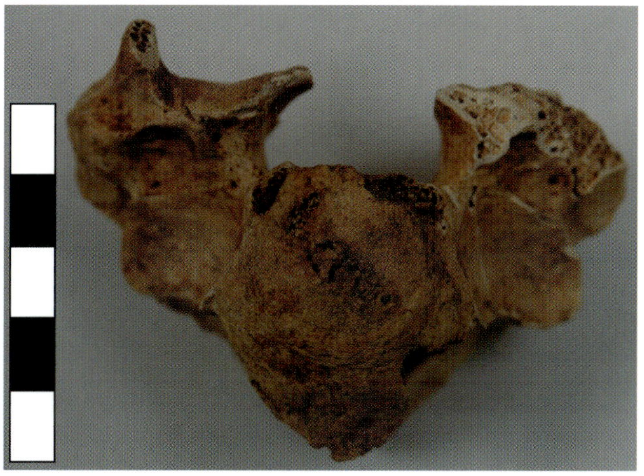

Fig. 29.15. Vue crâniale d'une vertèbre thoracique d'équidés montrant une zone de dépression au niveau de la tête articulaire.

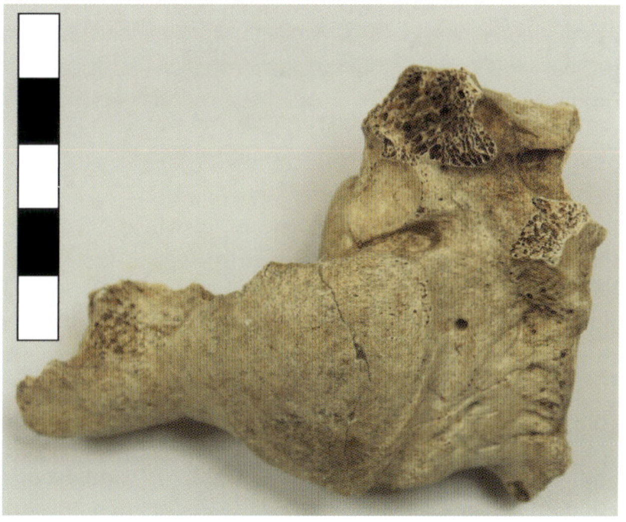

Fig. 29.16. Vue latérale d'une vertèbre d'équidés montrant une formation d'ostéophytes sur le corps vertébral.

Deux vertèbres lombaires présentent des exostoses. Deux autres vertèbres sont soudées, résultant d'une spondylose déformante (Fig. 29.16). Des anomalies sont observées au niveau de trois côtes isolées. Deux présentent une formation d'un nouveau tissu osseux avec des zones de destruction de l'os en forme de cupules. Enfin, une côte est soudée à la dernière vertèbre cervicale.

Les pathologies observées pourraient provenir de causes naturelles ou de l'exploitation des animaux par l'homme. Pour élucider l'étiologie de ces pathologies, nous tenterons de nous prévaloir de la structure de l'âge de l'échantillon affecté. Les ostéophytes qui ont proliféré

Fig. 29.17. Vue linguale de la P3 droite du crâne 36 montrant un renflement de la couronne.

sur les surfaces périarticulaires et l'exostose au niveau du processus épineux peuvent être liés à l'âge, à une anomalie congénitale ou à une mauvaise répartition du poids sur le dos de l'animal (Levine *et al.* 2000; Arbogast *et al.* 2002). Ces anomalies sont observées chez de vieux individus d'équidés sauvages âgés de 27 ans mais elles sont absentes chez les plus jeunes âgés de 12 ans (Bendrey 2007). Les équidés de JEM002 étant relativement jeunes, ces pathologies seraient plutôt causées par un poids agissant sur le dos des animaux. Aussi, les lésions sur le disque vertébral caudal des vertèbres thoraciques sont dues au poids exercé sur le dos de l'animal qui agit directement sur le processus épineux (comme le poids du cavalier) ou à la conduite d'un char dans le cas du cheval (Levine *et al.* 2000; Levine 2004). Enfin, la dégénérescence des disques intervertébraux résultant en une spondylose déformante a lieu suite à des microfractures des corps vertébraux (Miklikovà 2008; Rossel *et al.* 2008).

En outre, les destructions de l'os sous forme de petites cupules (*pitting*) au niveau des vertèbres et des côtes et le développement d'un nouveau tissu osseux pourraient résulter d'une maladie infectieuse (Bendrey 2008).

LE MÉTACARPE

Un métacarpe présente de l'exostose au niveau de l'extrémité proximale et une ossification des ligaments inter – osseux entre le canon et le deuxième métacarpe vestigial. Cette fusion des métacarpes survient généralement avec l'âge (Bendrey 2007). Cependant, les animaux de JEM002 étant pour la majorité jeunes, cette pathologie pourrait être due à une répétition de mouvements qui endommage les ligaments et le périoste des os (Bertone 2002). Elle résulterait donc de l'exploitation de l'animal.

LES PATHOLOGIES ORALES

En ce qui concerne les pathologies orales, une P2 inférieure présente une usure irrégulière où le bord mésial est peu usé par rapport à la partie distale de la dent. L'os sous-jacent présente des perforations d'origine infectieuse. Une autre anomalie orale se présente sous la forme d'un renflement de la couronne d'une P3 supérieure (Fig. 29.17).

Par ailleurs, des pathologies sont observées au niveau du corps mandibulaire. Certaines mandibules présentent des exostoses et du *pitting*. Des plages de destruction de l'os chez un jeune individu de deux ans et demi résulteraient d'infections plutôt que de l'âge.

Caractérisation taphonomique de l'assemblage

L'examen des traces d'origine anthropique sur les os et la fréquence des éléments du squelette fait ressortir des différences dans le traitement des carcasses des équidés, d'une part, et des espèces de la triade, d'autre part.

Les traces observées sur les os des équidés se présentent sous forme d'incisions fines et de cupules. Ce sont des stigmates laissés par le dépouillement, la désarticulation et le prélèvement de tendons et de la viande. Les traces de dépouillement sont localisées sur la phalange I (un spécimen sur trois). Les traces de désarticulation se trouvent sur le talus (quatre spécimens sur huit) et le calcanéum (sept spécimens sur onze). Le prélèvement de tendons a laissé des traces sur le métacarpe (un spécimen sur trois). Trois os coxaux (sur un total de quatorze),[11] dix côtes (sur un total de 214), quatre vertèbres (sur un total de 288) et une mandibule témoignent du prélèvement de la viande.[12]

Cette découpe est faite au couteau comme le révèlent les types de traces. L'utilisation du couperet est attestée sur une côte. Ceci témoigne du dépeçage de la cage thoracique. Toutefois les traces laissées sur les os sont rares; seulement 3% des os d'équidés portent des traces de découpe.[13] Sur des sites médiévaux en France, où l'hippophagie est attestée, 7.1 à 8.6% des os d'équidés portent des traces de découpe, contre 10 à 20% pour le bœuf (Yvinec 1999). Dans notre échantillon, 63% des os de bœuf portent des impacts de découpe. Ainsi, par rapport à la fréquence des traces de découpe sur les os de bœuf, les stigmates relevés sur les os des équidés de JEM002 s'avèrent être occasionnels.

A l'opposé, les traces de découpe observées sur les os du bœuf soulignent la multitude d'impacts d'outils, tout particulièrement du couperet. Les gestes du boucher sont stéréotypés suggérant une normalisation de la découpe. Par exemple, l'humérus porte des traces de décarnisation sur les quatre faces avec une récurrence des traces de raclage et d'incisions à l'emplacement des muscles grand rond et

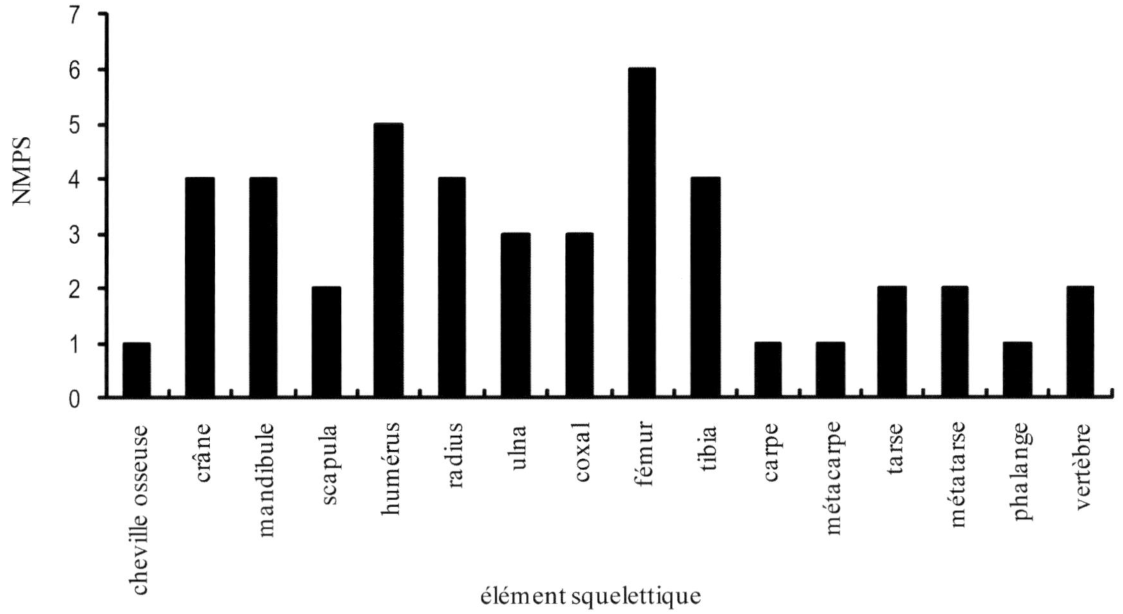

Fig. 29.18. Fréquence des éléments du squelette des bovinés en Nombre Minimal de Parties Squelettiques (NMI=6).

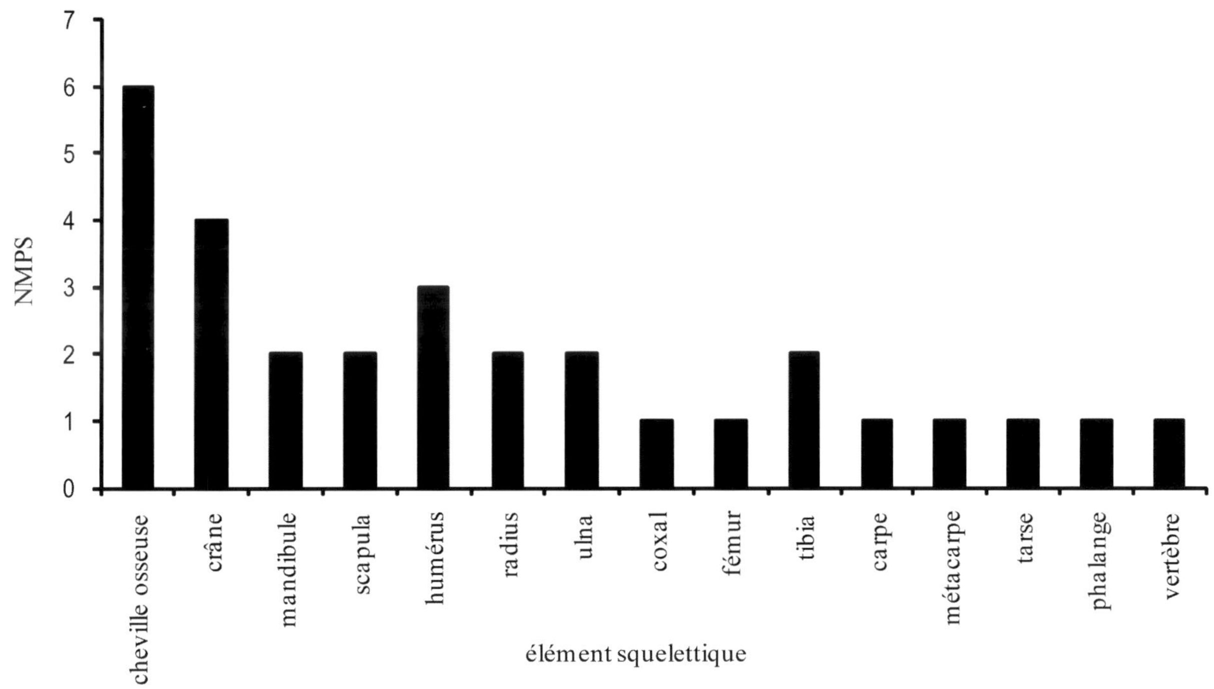

Fig. 29.19. Fréquence des éléments du squelette des caprinés en Nombre Minimal de Parties Squelettiques (NMI=6).

grand dorsal sur la face médiale (cinq humérus sur huit). Des récurrences de percussions au couperet et des traces de raclage sont localisées aussi sur la face crânio-médiale de l'os à l'emplacement du muscle brachio-céphalique (six humérus sur huit; Barone 1976).

Le petit bétail fait également l'objet d'un dépeçage au couperet et d'un désossage au couteau. Nous sommes en présence de techniques de transformation très différentes de celles qui sont pratiquées sur les carcasses d'équidés. Il est remarquable de constater que des animaux de même format comme les caprinés et le porc fassent l'objet d'une découpe similaire, tandis que les équidés et le bœuf, de gabarit proche, se singularisent par une technique de découpe différente.

Concernant la répartition des éléments du squelette, les animaux de la triade sont représentés principalement par la tête osseuse et les membres (à part la scapula de bœuf). Les extrémités, les côtes et les vertèbres sont moins fréquentes (Figs. 29.18–29.20). Les équidés sont essentiellement représentés par la tête osseuse, le membre

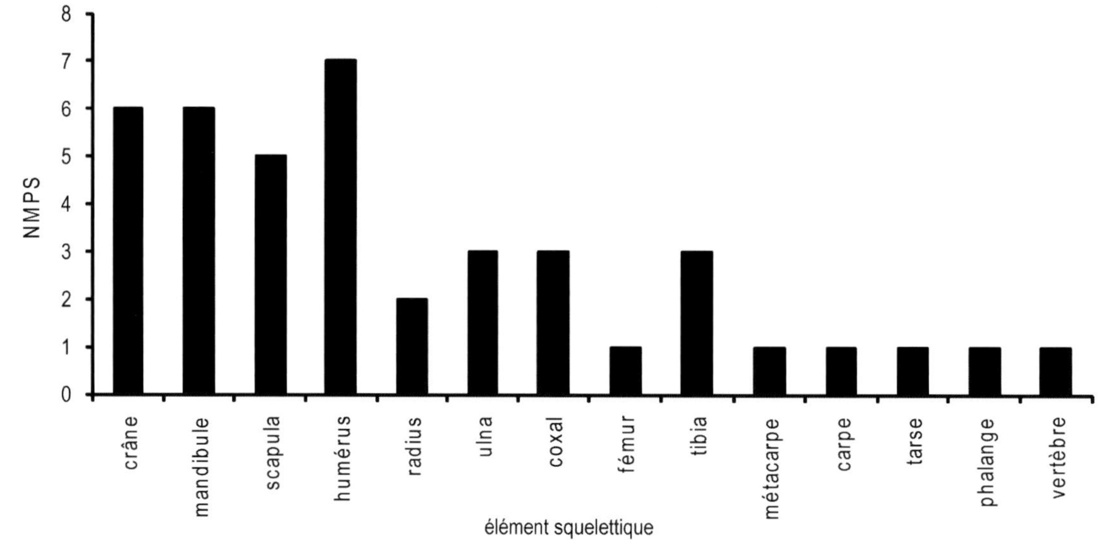

Fig. 29.20. Fréquence des éléments du squelette des suinés en Nombre Minimal de Parties Squelettiques (NMI=7).

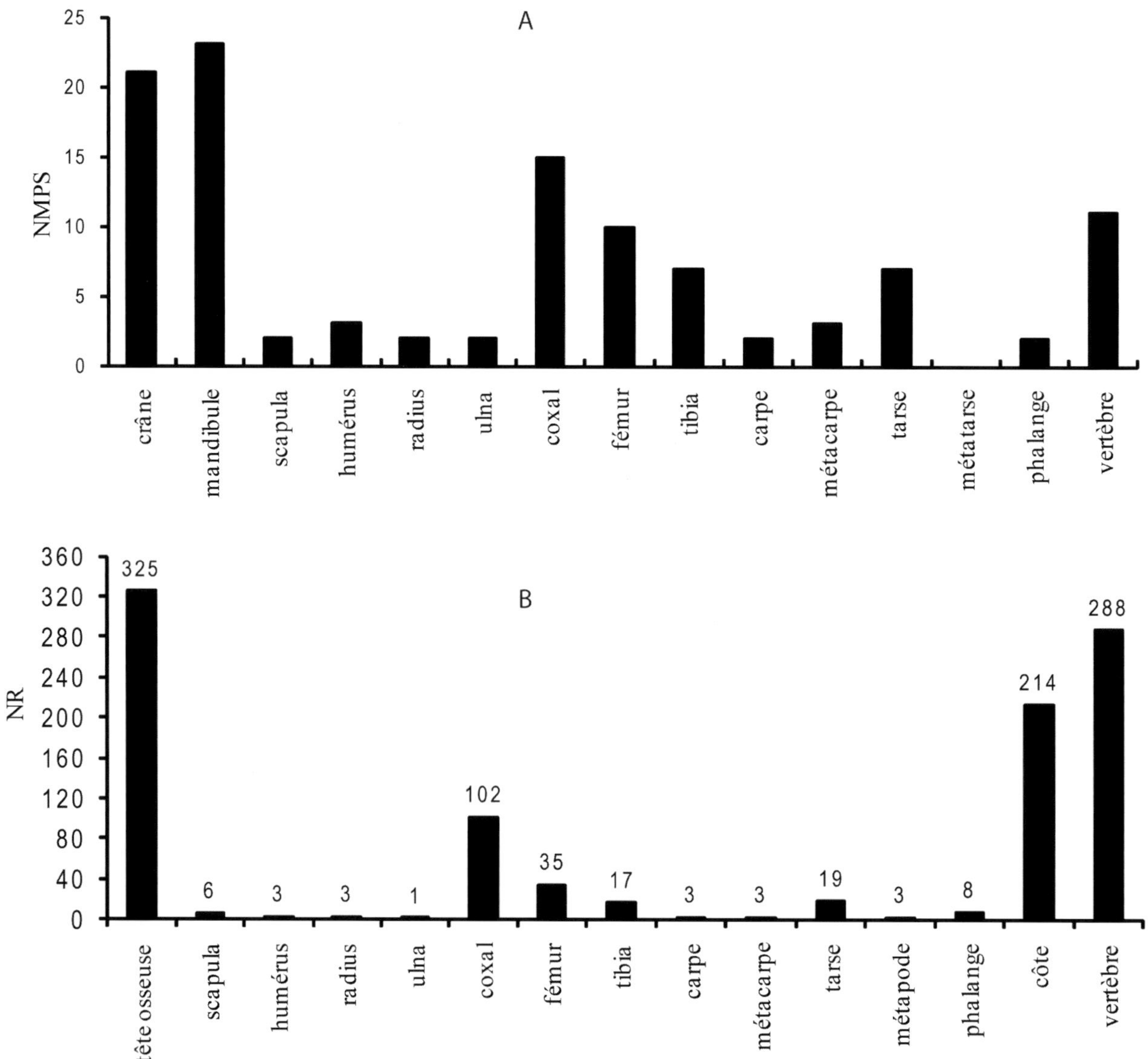

Fig. 29.21. Fréquence des éléments du squelette des équidés (NMI=23). A: en Nombre Minimal de Parties Squelettiques; B: en Nombre de Restes.

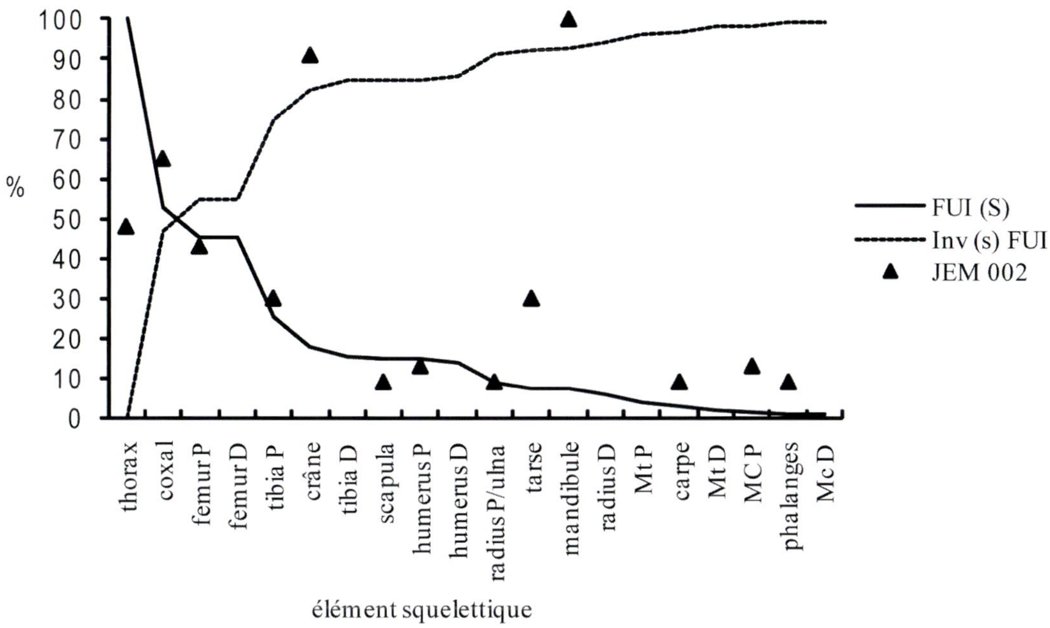

Fig. 29.22. Comparaison de la fréquence en NMPS des éléments du squelette des équidés exprimée en pourcentage de la plus haute valeur avec le (S) FUI (données d'après Outram et Rowley-Conwy 1998) et le FUI inverse. Les éléments sont placés sur l'axe des abscisses en ordre décroissant de FUI (Outram 2006). Les éléments squelettiques de JEM002 ne sont pas répartis en partie proximale et distale pour les os longs, nous les avons placés sur le graphique au niveau des parties proximales.

postérieur, les côtes et les vertèbres. Le membre antérieur et les extrémités sont peu représentés. Le métatarse et les os tarsiens distaux sont absents (Fig. 29.21).

Il ressort ainsi des différences dans le rejet des éléments de la carcasse entre les équidés et les autres taxons. Ceci suggère qu'au moment de la formation de ce dépotoir, nous sommes à des stades différents de la transformation des matières premières animales; les déchets issus des espèces de la triade d'une part et des équidés d'autre part relèvent d'activités différentes.

La technique de découpe et la répartition des éléments du squelette des animaux de la triade suggèrent que nous sommes en présence de déchets de boucherie. Les membres désossés, comme en témoignent les traces de découpe, et la tête sont rejetés à l'issue de cette opération. Le déficit au niveau du rachis suggère qu'il serait vendu au détail en portions de viande à l'os (Nin et Leguilloux 2003).

La répartition des éléments squelettiques des équidés ne relève pas de rejets de boucherie. Elle n'est pas non plus caractéristique des déchets de consommation (Fig. 29.22). La fréquence des éléments squelettiques suggère que tout le squelette est présent, à l'exception du membre antérieur et des bas de pattes qui ont été prélevés.

A la différence observée au niveau des traces de découpe et de la répartition des éléments du squelette, vient s'ajouter la présence de connexions anatomiques exclusivement dans le cas des équidés (Figs 29.23 et 29.24). Plusieurs trains de vertèbres et de côtes sont restés soudés (NR=10).

Du point de vue taphonomique, les os n'ont pas subi l'intempérisation en surface. Aussi, les traces d'intervention d'animaux charognards et commensaux

Fig. 29.23. Détail du dépotoir de JEM002. Vertèbres thoraciques et côtes d'équidés en connexion anatomique.

Fig. 29.24. Détail du dépotoir de JEM002. Vertèbres cervicales d'équidés en connexion anatomique.

comme le mâchouillement, les traces de griffes et de charriage sont très rares. Les traces de racines sont tout aussi rarement observées. Seulement 0.9% des os portent des traces d'origine climatique et biologique. Les altérations post-enfouissement résultent surtout de la nature argileuse du sol. La variation saisonnière de l'humidité a dû entraîner la dilatation et la rétraction de l'argile, ce qui aurait fragilisé les os. Ainsi les crânes, les côtes et les vertèbres, de morphologie complexe, sont les éléments les plus fragmentés.

Discussion

Ce dépotoir a été constitué par l'élimination de déchets osseux encombrants et leurs rejets à l'extérieur des murailles, dans une zone accueillant divers rebuts générés par la ville. La prédominance des restes équins est caractéristique de cet ensemble. Elle suggère une spécialisation des activités qui ont eu lieu à proximité de cette zone de décharge.

Les équidés sont représentés surtout par des ânes. Du point de vue démographique, il s'agit essentiellement de mâles adultes, soit des individus adaptés au portage et à la traction. En effet, historiquement, comme aujourd'hui, ces animaux sont initiés aux différentes tâches à partir de l'âge de trois ans (Varron IV, 2, 6).

Cette sélection des individus suggère que ces animaux sont fortement impliqués dans l'économie de la ville. Les sources antiques nous apprennent que l'âne est utilisé pour tourner la meule, pour le transport dans les travaux agricoles et aux labours des terres légères (Varron, *De Rustica* II, 6, 5). D'après Varron (*De Rustica*, II, 8, 5), il n'est pas utilisé pour le transport des passagers, rôle qui est assuré par la mule. Les sources nous renseignent aussi qu'un plus grand nombre d'ânes est requis pour le transport des denrées que pour le moulin, alors que le labourage est l'activité qui nécessite le moins d'animaux. Les troupeaux d'ânes ne sont constitués que pour les transports commerciaux (Caton, *De Agricultura*, XI–XII). Dans ce témoignage, l'effectif des ânes nécessité pour chaque tâche n'est pas précisé. Dans l'assemblage de JEM002, le nombre d'individus peut être considéré relativement élevé. En outre, les pathologies osseuses sont associées à la surcharge des animaux dans le cadre de leur utilisation. Ainsi, il est vraisemblable que ces animaux, ou un certain nombre d'entre eux, aient été utilisés comme bêtes de somme.

Le cheval de JEM002, vu sa robustesse, a probablement été réservé à des tâches de portage ou de traction dans le cadre des activités économiques urbaines ou dans le cadre des échanges commerciaux.

Certains sites archéologiques attestent de la sélection des équidés en fonction de leur âge, leur sexe ainsi que leur gabarit. Dans le fort militaire romain de Carnuntum (Kunst 2000), les chevaux et les mulets sont tous des mâles, âgés pour la plupart entre cinq et neuf ans[14] et de taille correspondant à la partie supérieure de la variabilité connue dans la région. Dans un autre contexte, l'écurie de la maison de Casti Amanti à Pompéi (Genovese et Cocca 2000), quatre ânes et un mulet mâles, utilisés comme bêtes de somme, sont âgés de quatre à neuf ans. La sélection des animaux semble être en rapport avec leur utilisation ; les équidés mâles appartenant à la tranche d'âge de quatre à neuf ans sont régulièrement exploités. A JEM002, les mâles sont bien attestés mais la fourchette d'âge est plus étalée. L'exploitation à plus long terme expliquerait le fait que des sujets soient atteints de pathologies, contrairement aux ânes sains de Casti Amanti.

D'autre part, l'exploitation de la viande des équidés est probable. Toutefois les carcasses équines ne relèvent pas pour autant de la même chaîne opératoire que le bœuf, les caprinés ou le porc. Par ailleurs, les données issues du site Bey002[15] révèlent que la viande des équidés est consommée de l'époque hellénistique jusqu'au début de l'époque romaine impériale (Oueslati 2004). Les traces de découpe observées sur ce site combinent les incisions fines laissées par le couteau et les impacts de couperet réservés au dépeçage de la carcasse. Les traces observées sur les os des équidés de JEM002 sont plus discrètes. Nous pouvons envisager dans ce cas une consommation opportuniste de la viande équine, prélevée directement sur les carcasses. D'autres contextes archéologiques témoignent de cette dualité des techniques de boucherie avec une prédominance du désossage qui se pratique au couteau et une fragmentation limitée des os (Yvinec 1999; Oueslati 2006).

En outre, les carcasses des équidés ont fait l'objet d'un prélèvement de matière première comme la peau, les tendons et certains os. Le membre antérieur et les extrémités, en déficit par rapport aux autres éléments du squelette, seraient transportés pour être transformés ailleurs, soit pour la viande soit pour la tabletterie. L'utilité de la scapula, du radius et des métapodes est connue pour l'artisanat de l'os (Arbogast *et al.* 2002). A cet effet, la désarticulation au niveau du talus et du calcanéus expliquerait l'absence des os tarsiens distaux et du métatarse.

Si notre assemblage n'a pas livré de rebuts issus d'os d'équidés, une accumulation contemporaine vient nous renseigner sur le tronçonnage de matière première. Il s'agit de rebuts liés au prélèvement de la diaphyse des métatarses de bœuf et d'équidés sous forme d'extrémités articulaires sciées, ainsi que des os de la rangée distale du tarse d'équidés (os naviculaire, cuboïde et grand cunéiforme). Cette accumulation, mise au jour sur le même secteur de la fouille, se caractérise également par une carbonisation totale de ces déchets qui ont été probablement utilisés comme combustible[16]. A ce témoignage du travail des os d'équidés et de bovins attesté sur notre site, nous pouvons associer d'autres accumulations découvertes *intra muros* notamment sur le site BEY002 où les rebuts de façonnage d'objets en os ont été mis au jour (Oueslati 2004). Ainsi la mise en évidence d'étapes situées plus en amont de la chaîne opératoire de transformation de l'os à l'extérieur de la ville et sur des carcasses équines partiellement exploitées n'est pas sans évoquer l'activité contemporaine de l'équarrissage. Une activité équivalente pourrait être à l'origine des stigmates laissés sur les os, notamment dans le cadre du dépouillage, et du déficit d'autres parties du squelette recherchées par les tabletiers. Ces derniers recherchent également des os de camélidés comme l'atteste une articulation radio-ulnaire proximale sciée dans notre assemblage.

D'un point de vue taphonomique, l'enfouissement des os s'est fait avant la décomposition totale des ligaments puisque des segments de vertèbres et de côtes sont restés en connexion anatomique. Cependant le dépôt est formé majoritairement d'éléments isolés. En l'absence de traces de désarticulation au niveau des membres et de la quasi-absence de traces d'intervention d'animaux commensaux, la dislocation des liens anatomiques pourrait être expliquée par deux faits distincts. Soit la carcasse a été désarticulée sans que ce processus n'ait laissé de stigmates sur les os. Soit le dépotoir a été généré par la remobilisation de déchets initialement accumulés dans un autre locus avec des carcasses partiellement décomposées. La cage thoracique, dont certaines connexions ont été préservées, est la partie du squelette dont les ligaments mettent le plus de temps à se décomposer (Hill et Behrensmeyer 1984).

Les cas de remobilisation de carcasses où certains ensembles anatomiques, notamment la colonne vertébrale, sont restés en connexion sont connus dans la littérature. Méniel (2008) interprète les amas osseux mis au jour dans les enclos de certains sanctuaires comme des dépôts secondaires issus de sacrifices d'animaux dont les carcasses ont été exposées et partiellement décomposées avant qu'elles ne soient transportées dans un ossuaire, en l'occurrence, un fossé d'enclos ou une fosse. Une intervention anthropique de ce type pourrait expliquer les caractéristiques de l'accumulation du site JEM002. L'absence de traces de désarticulation pourrait être expliquée par le fait qu'à l'issue du prélèvement des matières premières, certains membres sont restés en connexion. Puis, après leur décomposition partielle, les os seraient évacués dans cette zone de décharge. Dans ce cas, le temps de décomposition au niveau du dépôt primaire a été suffisamment bref pour que la colonne vertébrale ne soit complètement disloquée.

Du point de vue archéologique, la remobilisation de déchets peut être en relation avec l'assainissement d'une zone en vue de l'aménager pour de nouvelles fins. Le développement des constructions, comme en témoigne l'extension ultérieure de la ville (A. Seif et F. Béaino, comm. pers. 2009), pourrait être en relation avec la nouvelle gestion spatiale des déchets.

Ainsi, ce dépotoir nous renseigne sur la transformation des ressources animales et la gestion des déchets à Berytus. A l'échelle de la ville, la composition de cette accumulation est différente de celle des dépotoirs *intra muros* où les déchets sont surtout issus de la consommation et où abondent les caprinés, le bœuf, le porc et les restes de poissons (Rackham 1996; Oueslati 2008). Cette différence vient du fait que nous sommes loin de la zone d'habitations qui génèrent des déchets culinaires. La concentration de carcasses d'équidés en un locus près des remparts est produite par l'équarrissage qui génère un gros volume de déchets dont l'évacuation à l'extérieur de la ville paraît inévitable. Ce dépotoir a accueilli aussi des déchets de boucherie issus de caprinés, du porc et notamment du bœuf. Ce dernier fait l'objet d'une boucherie artisanale de production de masse. Ce traitement de la carcasse privilégie la rapidité de la découpe et génère une grande quantité de déchets (Oueslati 2006). Cette technique est connue à l'époque romaine et est diffusée dans de nombreuses provinces de l'empire (Maltby 1985; Wilkens 1997; Yvinec 1993; Oueslati 2004). Les dépotoirs typiques qui en résultent sont issus d'un traitement d'un grand nombre de bovins parfois atteignant plusieurs milliers d'individus (Oueslati 2005). Il en découle ainsi que le caractère spécialisé de l'accumulation de JEM002 est en rapport avec une exploitation non domestique des ressources animales telle que la boucherie, l'équarrissage et la tabletterie.

Conclusion

La prédominance de l'âne dans le dépotoir de JEM002

montre que cet animal faisait partie du paysage quotidien de Berytus où il devait jouer un rôle important dans l'approvisionnement de la ville et tout particulièrement dans le transport des denrées. Les mortalités survenues parmi ces bêtes de somme ont nécessité leur évacuation à l'extérieur de la ville. Notre dépotoir situé *extra muros* illustre l'accumulation de ces carcasses. Leur analyse taphonomique a révélé une exploitation partielle de la peau, de la viande et le prélèvement de matière première pour la tabletterie. Contrairement au bœuf, les traces relevées sur les os d'équidés se sont avérées rares soulignant des traitements distincts des deux animaux. Il est tentant d'associer les restes d'équidés mis au jour dans notre dépotoir avec des déchets issus de l'équarrissage, activité bien documentée à des périodes plus récentes. Mêlés à cette accumulation spécialisée, se trouvent différents témoignages de déchets de boucherie d'autres espèces moins abondantes qui ont permis de confirmer l'introduction des nouvelles techniques de découpe romaines à Berytus.

Du point de vue archéologique, cette zone périphérique de Berytus a été utilisée au début du Ier siècle pour disposer des déchets issus des diverses activités économiques de la ville. L'extension de Berytus à la fin de ce siècle modifiera la fonction de l'emprise du site qui se trouve phagocyté au sein de la trame urbaine.

Remerciements

Nous remercions les directeurs de la fouille du site JEM002, M. Assaad Seif et M. Fadi Béaino, pour nous avoir confié l'étude du matériel ainsi que les responsables à la Direction Générale des Antiquités pour avoir facilité l'accès à ce matériel. Nous remercions particulièrement Mme Véra Eisenmann pour avoir vérifié la détermination des équidés. Nous tenons à remercier Mme Marjan Mashkour pour ses conseils avisés. Enfin, nous remercions M. Robin Bendrey pour ses commentaires constructifs qui ont permis d'améliorer ce papier.

Notes

1 Le site a été fouillé dans le cadre d'une opération de sauvetage dirigée par M. Assaad Seif (Direction Générale des Antiquités, Liban) entre mai 2005 et août 2008.

2 La densité et l'enchevêtrement de ces vestiges ont justifié un carroyage de la zone de fouille et les déblais ont été intégralement tamisés sur des mailles de 5mm et 0.5mm. Ce matériel a été étudié par Y. Hourani en 2009 au cours de son master au Muséum national d'Histoire naturelle de Paris en collaboration avec le laboratoire d'Archéozoologie de Lille 3. Les tests d'invertébrés ne sont pas inclus dans cette étude.

3 NMI: Nombre minimal d'individus (Poplin 1976; Grayson 1984: MNI)

4 La somme de la longueur et de la largeur occlusales divisée par deux.

5 Les différentes opinions sur le caractère discriminant du cornet des incisives sont présentées dans Eisenmann (1979).

6 NR: Nombre de Restes (Poplin 1976; Grayson 1984: NISP)

7 Vila utilise les données de Levine (1982) pour définir l'âge auquel la couronne de la dent est usée au quart, à la moitié et aux trois-quarts.

8 Pour comparer la structure de l'âge de notre échantillon d'équidés aux données relatives à l'attrition des dents jugales chez l'âne, nous avons comparé nos spécimens aux radiographies des dents et à l'âge correspondant présentés par Misk et Seilem (1997). Les classes d'âge définies sont ainsi très relatives. La première classe d'âge s'étale de 0 à 5 ans et regroupe tous les individus présentant au moins une dent n'ayant pas les racines formées. Entre 5 et 8 ans toutes les dents ont des racines et la couronne est moyennement réduite. Entre 8 et 15 ans, sont regroupées les dents dont la couronne est bien réduite et dont les racines sont longues par rapport à la hauteur totale de la dent. Enfin, âgées de 15 ans et plus sont les dents dont la partie enchâssée dans l'os est complètement sortie et dont la couronne est très réduite, seules les racines sont encastrées.

9 <http://vera-eisenmann.com/withers-height-estimations-article001268.html

10 L'indice de gracilité est le rapport du diamètre transverse à la longueur de la diaphyse: diamètre transverse minimal de la diaphyse × 100/longueur totale

11 Les traces de découpe sur le bassin sont localisées sur la crête de la tubérosité ischiatique donc nous n'avons pris en considération que les fragments d'os coxaux où cette partie de l'ischium est présente pour déterminer la fréquence des traces de découpe. Les traces sont observées sur le côté droit dans les trois cas.

12 Le fémur, le tibia et l'humérus portent des traces anecdotiques, de fines et courtes stries. Nous n'avons pas pu conclure qu'il s'agissait de traces de découpe, même après leur observation au stéréomicroscope.

13 Nous n'avons pas pu investiguer d'éventuelles traces d'abattage en raison du mauvais état de conservation des crânes. La fragmentation de ces éléments a nécessité une consolidation au moment de la fouille. Le sédiment mêlé à la colle a rendu impossible l'observation d'impacts d'outils sur les crânes.

14 La fréquence de cette classe d'âge est connue dans d'autres contextes militaires romains. Pour plus de détails se référer à Kunst (2000).

15 Site fouillé en 1994 par l'Institut Français du Proche-Orient et situé au centre-ville actuel de Beyrouth et au cœur du centre antique de la ville.

16 Cette accumulation fera l'objet d'une étude ultérieure.

Références

Al-Zawahra, M. & Ezzughayyar, A. 1998, 'Equid remains from the Bronze Age Periods at site 4 of Tell Jennin (Palestine)', in A. M. Choyke, L. Bartosiewicz & H. Buitenhuis (eds), *Archaeozoology of the Near East III: proceedings of the third international symposium on the archaeozoology of southwestern Asia and adjacent areas*, 157–172. Centre for Archaeological Research and Consultancy, Groningen.

Arbogast, R.-M., Clavel, B., Lepetz, S., Méniel, P. & Yvinec, J.-H. 2002, *Archéologie du cheval.* Errance, Paris.

Armitage, P. L. & Chapman, H. 1979, 'Roman mules'. *London Archaeologist* 13(3), 339–346.

Barone, R. 1976, *Anatomie comparée des mammifères domestiques, Tome 1: Ostéologie atlas.* Vigot, Paris.

Bendrey, R. 2007, 'Ossification of the interosseous ligaments between the metapodials in horses: a new recording methodology and preliminary study'. *Journal of Osteoarchaeology* 17, 207–213.

Bendrey, R. 2008, 'A possible case of tuberculosis or brucellosis in an Iron Age horse skeleton from Viables Farm, Basingstoke, England', in Z Miklikovà & R Thomas (eds), *Current Research in Animal Palaeopathology: proceedings of the second ICAZ animal palaeopathology working group conference*, 19–26. British Archaeological Report International S1844, Archaeopress, Oxford.

Bendrey, R., Taylor, G. M., Bouwman, A. S. & Cassidy, J. P. 2008, 'Suspected bacterial disease in two archaeological horse skeletons from southern England: palaeopathological and biomolecular studies'. *Journal of Archaeological Science* 35, 1581–1590.

Bertone, A. L. 2002, 'The metacarpus and metatarsus', in T. S. Stashak (ed.), *Adam's Lameness in Horses* (5th edn), 800–830. Lippincott Williams & Williams, Philadelphia.

Bökönyi, S. 1990, *Kamid el-Loz 12. Tierhaltung und Jagd.* Saarbrücker Beiträge zur Altertumskunde 42, Rudolf Habelt GMBH, Bonn.

Boessneck, J. 1987, 'Tierknochenfunde vom Uch Tepe'. *Acta praehistorica e archaeological* 19, 131–163.

Burleigh, R. 1986, 'Chronology of some domestic equids in Egypt and western Asia', in R. H. Meadow & H.-P. Uerpmann (eds), *Equids in the Ancient World*, 230–236. Beihefte Zum Tübinger Atlas Des Vorderen Orients 1.

Churcher, C. S. & Richardson, M. L. 1978, 'Equids', in V. J. Maglio & H. B. S. Cooke (eds), *Evolution of African mammals*, 379–422. Harvard University Press, London.

Davies, R. W. 1969, 'The training grounds of the Roman cavalry'. *Archaeological Journal* 125, 73–100.

De Cupere, B. 2001, *Animals at Ancient Sagalassos. Evidence of the Faunal Remains.* Studies in Eastern Mediterranean Archaeology 4, Brepols, Turnhout.

Dive, J. & Eisenmann, V. 1991, 'Identification and discrimination of first phalanges from Pleistocene and modern *Equus*, wild and domestic', in R. H. Meadow & H.-P. Uerpmann (eds), *Equids in the Ancient World*, 278–333. Beihefte Zum Tübinger Atlas Des Vorderen Orients 2.

Ducos, P. 1968, *L'origine des animaux domestiques de Palestine.* Publication de l'Institut de Préhistoire de l'Université de Bordeaux 6, Delmas, Bordeaux.

Eisenmann, V. 1979, *Etude des cornets des dents incisives inférieures des Equus (Mammalia, Perissodactyla) actuels et fossils.* Paleontographia Italica 71. Industrie Grafiche V, Lischi & Figli, Pisa.

Eisenmann, V. 1980, *Les Chevaux (*Equus sensu lato*) fossiles et actuels: crânes et dents jugales supérieures.* Cahiers de Paléontologie, Editions du CNRS, Paris.

Eisenmann, V. 1981, *Etude des dents jugales inférieures des Equus (Mammalia, Perissodactyla) actuels et fossiles.* Palaeovertebrata 10, Fasc. 3–4, Montpellier.

Eisenmann, V. 1986, 'Comparative osteology of modern and fossil horses, half-asses, and asses', in R. H. Meadow & H.-P. Uerpmann (eds), *Equids in the Ancient World*, 67–115. Beihefte Zum Tübinger Atlas Des Vorderen Orients 1.

Eisenmann, V. & Beckouche, S. 1986, 'Identification and dicrimination of metapodials from Pleistocene and modern equus, wild and domestic', in R. H. Meadow & H.-P. Uerpmann (eds), *Equids in the Ancient World*, 116–154. Beihefte Zum Tübinger Atlas Des Vorderen Orients 1.

Eisenmann, V., Alberdi, M. T., De Giuli, C. & Staesche, U. 1988, *Studying Fossil Horses, volume 1: Methodology.* Mike Woodburne and Paul Sondaar, Leiden.

Eisenmann, V. & Mashkour, M. 2000, *Data Base for Teeth and Limb Bones of Modern Hemiones.* Fiches d'ostéologie animale pour l'archéologie, Série B: Mammifères 9.

Genovese, A. & Cocca, T. 2000, 'Internal organization of an equine stable at Pompéi'. *Anthropozoologica* 31, 119–123.

Grayson, D. K. 1984, *Quantitative Zooarchaeology.* Academic Press, Orlando.

Groves, C. P. 1986, 'The taxonomy, distribution, and adaptations of recent equids', in R. H. Meadow & H.-P. Uerpmann (eds), *Equids in the Ancient World*, 11–47. Beihefte Zum Tübinger Atlas Des Vorderen Orients 1.

Hill, A. & Behrensmeyer A. K. 1984, 'Disarticulation patterns of some modern east African mammals'. *Paleobiology* 10(3), 366–376.

Hoppe, K. A., Stover, S. M., Pascoe, J. R. & Amundson, R. 2004, 'Tooth enamel biomineralization in extant horses: implications for isotopic microsampling'. *Palaeogeography, Palaeoclimatology, Palaeoecology* 206, 355–365.

Johnstone, C. J. 2004, 'A biometric study of equids in the Roman World', PhD Thesis, University of York.

Kunst, G. K. 2000, 'Archaeozoological evidence for equid use, sex structure and mortality in a roman auxiliary fort (Carnuntum-Petronell, Lower Austria)'. *Anthropozoologica* 31, 109–118.

Lepetz, S. 1996, *L'animal dans la société gallo-romaine de la France du Nord.* Revue archéologique de Picardie Numéro spécial 12.

Levine, M. A. 1982, 'The use of crown height measurements and eruption-wear sequences to age horse teeth', in B. Wilson, C. Grigson & S. Payne (eds), *Ageing and Sexing Animal Bones from Archaeological Sites*, 223–250. British Archaeological Report S109, British Archaeological Reports, Oxford.

Levine, M. A. 2004, 'Exploring the criteria for early horse domestication', in M. Jones (ed.), *Traces of Ancestry: studies in honour of Colin Renfrew*, 115–126. McDonald Institute for Archaeological Research, Cambridge.

Levine, M. A., Bailey, G. N., Whitwell, K. E. & Jeffcott, L. B. 2000, 'Palaeopathology and horse domestication: the case of some Iron Age horses from the Altai Mountains, Siberia', in G. Bailey, R. Charles & N. Winder (eds), *Human Ecodynamics and Environmental Archaeology*, 123–33. Oxbow Books, Oxford.

Maltby, M. 1985, 'Assessing variations in Iron Age and Roman Butchery practices: The need for quantification', in N. R. J. Fieller, D. D. Gilbertson & N. G. A. Ralph (eds), *Palaeobiological Investigations. Research Design, Methods and Data Analysis*, 19–31. Biritsh Archaeological Report 266, British Archaeological Reports, Oxford,.

Mashkour, M. 2001, 'Chasse et élevage du Néolithique à l'Âge du Fer dans la plaine de Qazvin (IRAN). Etude archéozoologique des sites Zagheh, Qabrestan et Sagzabad', Thèse de doctorat, Université de Paris I Panthéon-Sorbonne.

Méniel, P. 2008, *Manuel d'Archéozoologie funéraire et sacrificielle. Âge du Fer.* Editions Infolio, France.

Miklikovà, Z. 2008, 'Skeletal alterations of animal remains from the early medieval settlements of Bajc, south-west Slovakia', in Z. Miklikovà & R. Thomas (eds), *Current Research in Animal Palaeopathology: Proceedings of the second ICAZ animal palaeopathology working group conference*, 63–73. British Archaeological Report S1844, Archaeopress, Oxford.

Misk, N. A. & Seilem, S. M. 1997, 'Radiographic studies in the development of cheek teeth in donkeys'. *Equine Practice* 19(2), 27–38.

Muylle, S., Simoens, P., Lauwers, H. & Van Loon, G. 1999, 'Age determination in mini-Shetland Ponies and donkeys'. *Journal of Veterinary Medicine* A46, 421–429.

Nin, N. & Leguilloux, M. 2003, 'La gestion des déchets à Aix-en-Provence dans l'Antiquité', in *La ville et ses déchets dans le monde romain: rebuts et recyclages*, 133–163. Actes du Colloque de Poitiers (19–21 Septembre 2002), éditions monique mergoil, Montagnac.

Oueslati, T. 2004, 'Acquisition, transformation and consumption of animal resources in Berytus, Place des Martyrs (BEY002)', unpublished paper given at the sixth international Symposium on the archaeozoology of Southwestern Asia and adjacent areas, Ankara.

Oueslati, T. 2005, *Les ossements animaux, l'archéozoologie et les professions de l'alimentation dans le nord de la Gaule romaine*. Revue du Nord – Archéologie de la Picardie et du Nord de la France 87, 175–183.

Oueslati, T. 2006, *Approche archéozoologique des modes d'acquisition, de transformation et de consommation des ressources animales dans le contexte urbain gallo-romain de Lutèce (Paris, France)*. British Archaeological Report 1479, Archaeopress, Oxford.

Oueslati, T. 2008, 'Spatial fluctuation of food habits in byzantine Beirut (BEY002, BEY028, BEY115)', in E. Vila, L. Gourichon, A. M. Choyke & H. Buitenhuis (eds), *Archaeozoology of the Near East VIII: proceedings of the eighth international symposium on the archaeozoology of southwestern Asia and adjacent areas*, 629–640. Travaux de la Maison de l'Orient et de la Méditerranée 49, Lyon.

Outram, A. K. & Rowley-Conwy, P. 1998, 'Meat and marrow utility indices for horse (*Equus*)'. *Journal of Archaeological Science* 25, 839–849.

Outram, A. K. 2006, 'Juggling with indices: A review of the evidence and interpretations regarding Upper Palaeolithic horse skeletal part abundance', in S. Oslen (ed.), *Horses and Humans: the evolution of human-equine relationships*, 49–59. British Archaeological Report S1560, Archaeopress, Oxford.

Peters, J. 1998, *Römische Tierhaltung und Tierzucht. Eine Synthese aus archäozoologischer Untersuchung und schriftlich-bildlicher Überlieferung*, Passauer Universitätsschriften zur Archäologie 5, Verlag Marie Leidorf GmbH, Rahden.

Poplin, F. 1976, 'A propos du nombre de restes et du nombre d'individus dans les échantillons d'ossements'. *Cahiers du Centre de recherches Préhistoriques* 5, 61–74.

Rackham, J. 1996, 'Bey 006 Animal bones and Site 1 Microsample Preliminary Report'. *Bulletin d'Archéologie et d'Architecture Libanaise* 1, 224–227. Publications du Musée National, Beyrouth.

Rossel, S., Marshall, F., Peters, J., Pilgram, T., Adams, M. D. & O'Connor, D. 2008, 'Domestication of the donkey: timing, processes and indicators'. *Proceedings of the National Academy of Sciences of the United States of America* 105(10), 3715–3720.

Simpson, G. G. 1941, 'Large Pleistocene felines of North America'. *American Museums Novitiates* 1136, 1–27.

Uerpmann, H.-P. 1986, 'Halafian equids remains from Shams ed-Din Tannira in northern Syria', in R. H. Meadow & H.-P. Uerpmann (eds), *Equids in the Ancient World*, Beihefte zum Tübinger Atlas des Vorderen Orients. Reihe A, 19(1), Wiesbaden: Dr. Ludwig Reichert, 246–265.

Uerpmann, H.-P. 1987, *The ancient distribution of ungulate mammals in the Middle East*. Beihefte zum Tübinger Atlas des Vorderen Orients, Reihe A, NR. 27. Wiesbaden: Dr Ludwig Reichert.

Uerpmann, H.-P. 1991, 'Equus africanus in Arabia', in R. H. Meadow & H.-P. Uerpmann (eds), *Equids in the Ancient World*, 12–33. Beihefte zum Tübinger Atlas des Vorderen Orients 2.

Uerpmann, M. & Uerpmann, H.-P. 1994, 'Maultiere in der römischen Armee zur Zeit der Eroberungdfeldzüge in Germanien', in M. Kokabi & J. Wahl (eds), *Beiträge zur Archäozoologie und Prähistorischen Anthropologie*, 353–357. Forshungen und Berichte zur Vor- und Frühgeschichte in Baden-Württemberg 53.

Vigne, J.-D. 1998, 'Faciès culturels et sous-système technique de l'acquisition des ressources animales. Application au Néolithique ancien méditerranéen', in A. D'ANNA & D. BINDER (eds.). *Production et identité culturelle. Actualité de la recherche. Actes 2e rencontres méridionales de Préhistoire Récente (Arles, 8-9 novembre, 1996)*, APDCA. Antibes, 24–45.

Vila, E. 1998, *L'exploitation des animaux en Mésopotamie aux IVe et IIIe millénaires av. J.-C.* Editions du CNRS, Paris.

Wilkens, B. 1997, 'La faune du site romain de Calvatone, Cremona (Italie)'. *Anthropozoologica – Actes du colloque international d'archéozoologie 7, Constance, Allemagne (26/09/1994)* 25–26, 611–616.

Willoughby, D. P. 1974, *The Empire of Equus*. Barnes, New York.

Yvinec, J. H. 1993, 'Le dépotoir de boucherie des thermes de Champlieu (Oise)'. RAP 1/2, 201–208.

Yvinec, J. H. 1999, 'Etude archéozoologique du site de Dury Le Moulin'. *Revue archéologique de Picardie* 1–2, 247–256.

Sources anciennes

Caton, *De l'agriculture* (éd. R. Goujard), Coll. Des Universités de France, Paris, 1975.

Varron, *Economie rurale*, livre II (éd. C. Guiraud), Coll. Des Universités de France, Paris, 1985.

Site électronique

http://www.vera-eisenmann.com/, dernière consultation le 28 mars 2010

30. The animal bone remains from Mar Nicola, a Byzantine–Islamic site at Beit Jala, Palestine

Mohammad Al-Zawahra

This report deals with the faunal remains that have been collected at the site of Mar Nicola, located in the centre of Beit Jala, 2km northwest of Bethlehem (Palestine). The majority of the animal bone assemblage represents two taphonomic groups, i.e. consumption refuse and bone-working refuse. Meat was mainly provided by domestic animals (ovicaprines, cattle and pig), while wild animals, more specifically gazelles, did not represent a significant contribution to the diet of the inhabitants. Bone-working was almost exclusively restricted to camel bone. In addition, specific skeletal bones were selected. No finished bone objects were found at the site, but the bone waste points towards the presence of a specialised workshop during the Islamic period at Mar Nicola.

Keywords Mar Nicola, Palestine, Byzantine, Islamic, bone-working

Introduction

The site of Mar Nicola is located at the centre of Beit Jala, 2km to the northwest of Bethlehem (West Bank – Palestine), at a height of 825m above sea level (Fig. 30.1).

Mar Nicola is a church of Greek Orthodox denomination and was founded in the place of the old cave that was once inhabited by St Nicholas. A mosaic floor was discovered in 2002 during works that were carried out by the Beit Jala Municipality, in order to put a pipe canal in the street. Thereupon, a salvage excavation was conducted by the Department of Antiquities and Cultural Heritage under the supervision of Dr Hamdan Taha (General Director) and Mr Wael Hamamreh. Two excavation areas were opened, i.e. area A, including the mosaic floor, and area B, to the south. The latter area was excavated in order to test the extent of the mosaic. Two phases were detected in area A, namely the Byzantine period and the Middle and Late Islamic period. Area B yielded remains that were possibly dated to the Roman period. This report deals with the faunal remains that were collected during the salvage excavation of Mar Nicola.

The faunal remains

The faunal remains were mainly collected by hand-picking in the excavation trench. Soil samples were taken from the floor surfaces and sieved using a 5mm mesh. The bulk of the animal remains were retrieved from area A, mostly from Islamic contexts. The other periods are rarely represented (Table 30.1).

The total assemblage is composed of 1195 specimens. The taxonomic composition of the animal remains is given in Table 30.2. A total of 710 specimens (59.4%

Table 30.1. The total number of bone fragments (N) from Mar-Nicola, arranged by period. Bones from the same individual are counted as one.

	Area B	*Area A*		*Total*
Period	*Roman ?*	*Byzantine*	*Mid–Late Islamic*	
N	12	27	1156	1195

Figure 30.1. Location of Mar Nicola (Basemap source: National Geographic/Esri).

of the faunal assemblage), including the human bone, could be identified to species level. From Table 30.2, it is clear that the majority of the animal remains are from mammals (both domestic and wild). Birds are represented by four bones only. In addition, there is one fragment of mother-of-pearl and one human bone. Most of the identifiable mammal remains are from domestic animals (99.6%); wild mammals are represented by three gazelle bones only. The unidentified animal remains (40.6% of the total assemblage) were sorted out according to the size of the animal they originated from, i.e. large and medium mammals.

Domestic mammals

A total of 704 specimens are from domestic animals. Ovicaprids are the best represented (58.4%), followed by cattle (24.7%) and dromedary (8.8%) (Table 30.2).

*Sheep (*Ovis ammon f. aries*) and goat (*Capra aegagrus f. hircus*)*

More than half of the domestic identifiable remains originate from sheep and goats (58.4%). Due to their fragmentary state and the difficulty of separating the bones of sheep and goat (Boessneck 1969; Prummel & Frisch 1986), part of their remains had to be classified as ovicaprid (O/C). Not all the skeletal elements of the ovicaprids are evenly represented. Mandible fragments, humerus, radius and tibia are the most abundant, while the other elements are present but to a much lesser degree. The fragmentation rate of the sheep and goat bones was very high as a result of various activities, from both natural and anthropogenic origin. Indeed, many ovicaprid bones show traces of butchery activities, including both cut and chop marks (see Fig. 30.2 and Appendix). In addition, one proximal ulna fragment of ovicaprid has a saw mark.

Table 30.2. The number of animal remains from the site of Mar-Nicola, Beit Jala, throughout the various periods.

	Roman ?		*Byzantine*		*Islamic*		*Total*	
	N	*%*	*N*	*%*	*N*	*%*	*N*	*%*
Molluscs								
Mother of pearl	0	0	0	0	1	0.1	1	0.1
Wild mammals								
Gazelle (*Gazella gazella*)	0	0	1	0.1	2	0.1	3	0.3
Domestic bird								
Chicken (*Gallus gallus* f. *domestica*)	0	0	0	0	3	0.3	3	0.3
Domestic mammals								
Dromedary (*Camelus ferus* f. *dromedarius*)	0	0	0	0	62	5.2	62	5.2
Donkey (*Equus africanus* f. *asinus*)	8	0.7	1	0.1	11	0.9	20	1.7
Horse (*Equus ferus* f. *caballus*)	0	0	0	0	2	0.1	2	0.1
Pig (*Sus scrofa* f. *domestica*)	0	0	2	0.1	30	2.5	32	2.7
Goat (*Capra aegagrus* f. *hircus*)	0	0	3	0.3	51	4.3	54	4.5
Sheep (*Ovis ammon* f. *aries*)	0	0	1	0.1	30	2.5	31	2.6
Sheep/Goat	5	0.4	5	0.4	316	26.4	326	27.3
Cattle (*Bos primigenius* f. *taurus*)	0	0	5	0.4	168	14.4	173	14.5
Zebu (*Bos namadicus* f. *indicus*)	0	0	1	0.1	0	0	1	0.1
Total domestic	13	1.1	18	1.5	673	56.3	704	58.9
Other								
Human	0	0	0	0	1	0.1	1	0.1
Unidentified								
Bird	0	0	0	0	1	0.1	1	0.1
Large-sized mammals	0	0	6	0.5	211	17.7	217	18.2
Medium-sized mammals	5	0.4	0	0	216	18.1	221	18.5
Indeterminates	1	0.1	2	0.1	44	6.7	47	3.9
Total unidentified	6	0.5	8	0.7	471	39.4	485	40.6
Total	19	1.6	27	2.3	1149	96.2	1195	100

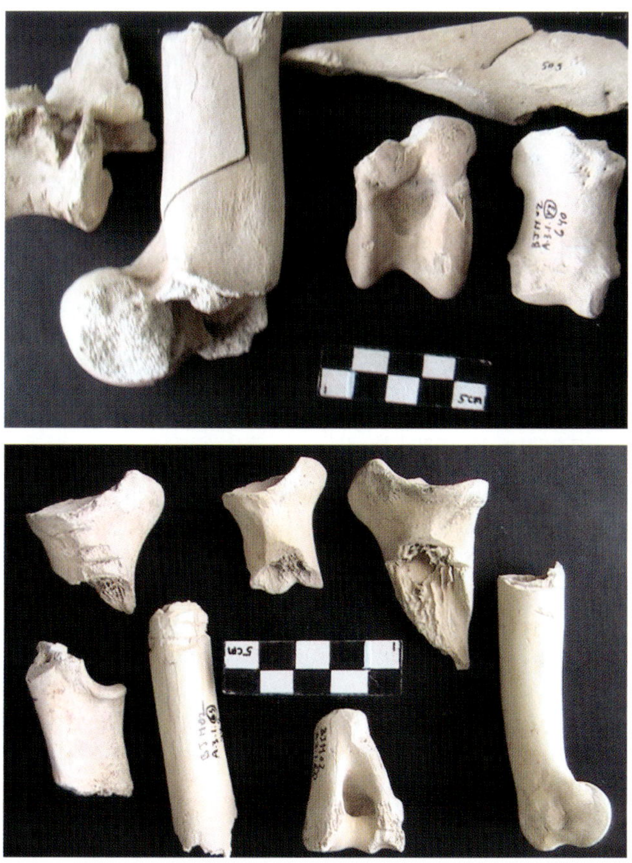

Fig. 30.2. Examples of cattle, sheep and goat bones with butchery marks.

Table 30.3. The age distribution of sheep and goats throughout the various phases of Mar Nicola; F: Fused; U: Unfused, N: number of specimens.

Age	*Elements*	*Byzantine*			*Islamic*			*Total*
		% U	*% F*	*N*	*% U*	*% F*	*N*	*N*
<1 y	scapula, acetabulum, dist. hum., prox. rad.	–	–	–	3.8	96.2	53	53
1–2 y	prox. ph1,2, dist. mp, dist tib.	0.0	100.0	1	26.5	73.5	49	50
2–3 y	prox. ulna, prox. fem, calc., dist. rad.	–	–	–	57.9	42.1	19	19
3–4 y	prox. hum., dist. fem., prox. tib.	–	–	–	0	100	5	5

Age at slaughter of the ovicaprids has not been established using the eruption and wear of the mandibular teeth (Grant 1982) since no sufficient data were available. Instead the fusion rates of the different skeletal elements (Silver 1969) have been recorded; the results are presented in Table 30.3.

Almost all of the data are from Islamic contexts. Although difficult to interpret, it can be concluded from these data that a small part of the sheep and goats died at a very young age. Considering the age and the absence of any kind of butchering mark on the juvenile remains, these individuals may have died a natural death (Payne 1973). A large part of the ovicaprid remains seem to have been slaughtered at a rather young age. Indeed, more than half of the articulations that close at an age of 2–3 years are still unfused. But, at the same time, about 75% of the articulations that close at an age of 1–2 years are already fused. Another large group reached full maturity, i.e. older than 3 years. These data show that most probably during the Islamic period, after a minor culling of young animals at an age of about 1 year, a selection was made around the age of 2 years and some of the animals were slaughtered for meat. The remaining animals were kept to maintain the herd, and for the production of milk, wool, hair, etc.

*CATTLE (*BOS PRIMIGENIUS F. TAURUS*)*

A total of 174 fragments could be identified as cattle; one specimen, a thoracic vertebra, has (tentatively?) been identified as belonging to humped cattle or zebu (Table 30.2). Most of the cattle bones (n=168) were recovered from the Islamic contexts from area A, while only six fragments, including the zebu specimen, were dated to the Byzantine period. Cranial fragments (skull, mandible and teeth) dominate the sample, followed by the limb bones. Similar to the ovicaprid remains, the cattle remains were highly fragmented and many specimens had butchery marks. A total of 24 bones had cut marks and 37 had chop marks (Fig. 30.2). No complete cattle long-bones have been recovered, to enable the estimation of their shoulder heights. Again, since no sufficient data on mandibular teeth were available, ageing data have been obtained using the age of epiphyseal closure of the different skeletal elements. Almost all of the long-bones were fused and indicate that most of the cattle were killed at an adult age. This indicates the herding of cattle for their 'secondary' products, e.g. milk and traction. When they could no longer be used, they were slaughtered for consumption.

One bifid spinous process of the thoracic vertebra was found from area A in a Byzantine context. Bifid spines are one of the osteological criteria that has been used to identify zebu, but this character can sometimes also be found in cattle (Stallibrass 1983). In Palestine, zebu or humped cattle remains were identified in Gerar, dated to the 6th century BC (Clason 1978). Zeuner (1963) assumes that the introduction of humped cattle into Palestine was due to Persian influence. Remains of zebu cattle were found from many other sites, e.g. Tell Jenin/Late Bronze Age (Al-Zawahra, 1999), Kherbet Belameh (Al-Zawahra, 2002), Tell Jemmeh/Hellenistic granary, 300–200 BC (Wapnish 1982) and Tell Dan/mid-9th century BC (Wapnish & Hesse 1991). All this archaeological evidence points towards the exploitation of zebu, which has adapted to very warm temperatures in Palestine since the late 2nd millennium BC onwards.

*PIG (*SUS SCROFA F. DOMESTICA*)*

Pig remains are quite uncommon in the faunal assemblage of Mar Nicola. This animal is represented by 32 bone fragments, mostly found in Area A (Table 30.2). The majority of the pig remains were recovered from Islamic contexts, only two fragments were dated to the Byzantine period. Five bones show chop marks, two fragments have cut marks, while a gnawing mark has been observed on one bone fragment. Except for two distal metacarpals, all of the long bones are fused. This indicates that these animals were mostly slaughtered when mature. In addition, there is a complete absence of foetal bones and the remains of immature and young individuals. The preponderance of mature pigs could well have to do with commerce strategies. Considering the fact that young pigs produce meat of a better quality, it is possible that these were sold and that only the meat of the older animals, e.g. sows that were no longer suitable for breeding, and thus also of lesser quality, was consumed at the site.

Fig. 30.3. General view of worked bone specimens from Mar Nicola.

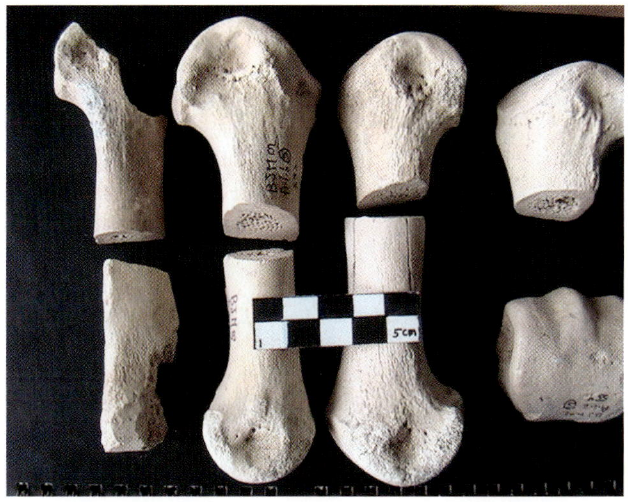

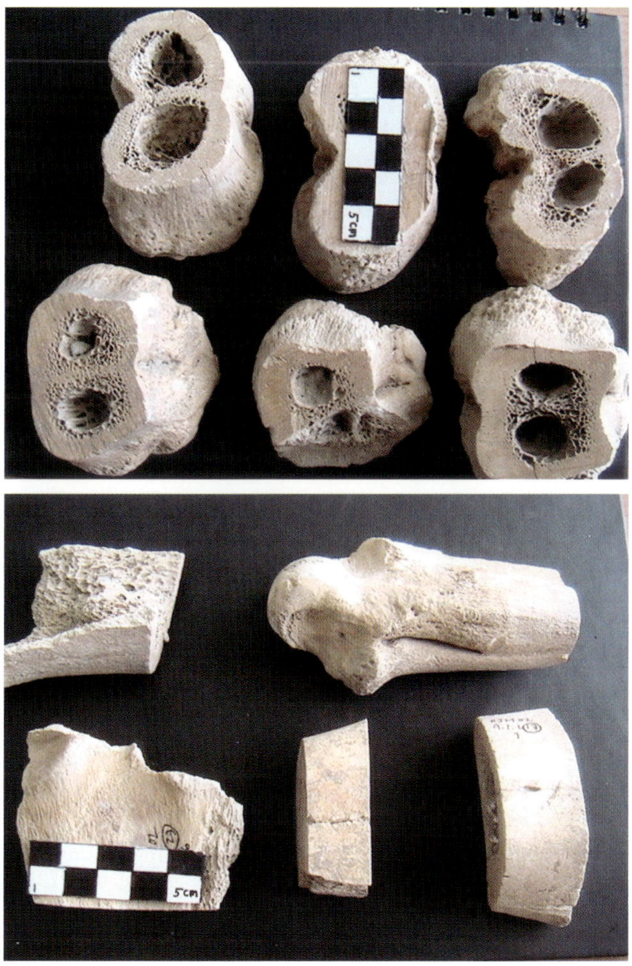

Fig. 30.4. Camel metapodial waste remains from Mar Nicola with manufacturing traces.

Fig. 30.5. Different type of camel bones, with manufacturing marks.

DROMEDARY (CAMELUS FERUS F. DROMEDARIUS)

In total, 62 camel bones were found at the site of Mar Nicola (Table 30.2). All of them were collected in area A and dated to the Islamic period. Considering the fact that the site falls within the geographical distribution of *Camelus dromedarius*, all bones were assigned to this species. The camel bones consist almost exclusively of the long bones of the lower part of the fore and hind limbs, i.e. radius, tibia, metacarpal and metatarsal. In addition, only the articular ends of these long bones are represented. With the exception of one mandible fragment, no other cranial or post-cranial elements of camel have been identified. Further, most of the camel remains (n=48) showed sawing traces and traces of working (Figs. 30.3–30.5 and Appendix). The worked

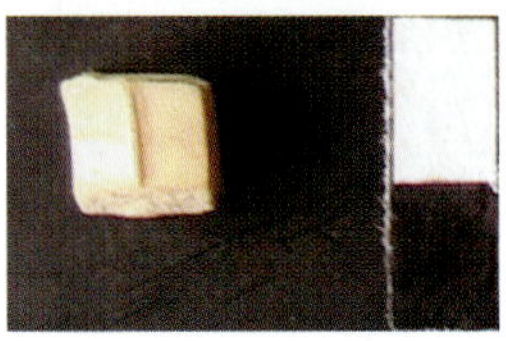

Fig. 30.6. Bones sawed into slices and into small round cylindrical objects for the preparation of rosary beads.

Fig. 30.7. The manufacturing of rosary beads (photos are kindly provided by Prof. Alice Choyke).

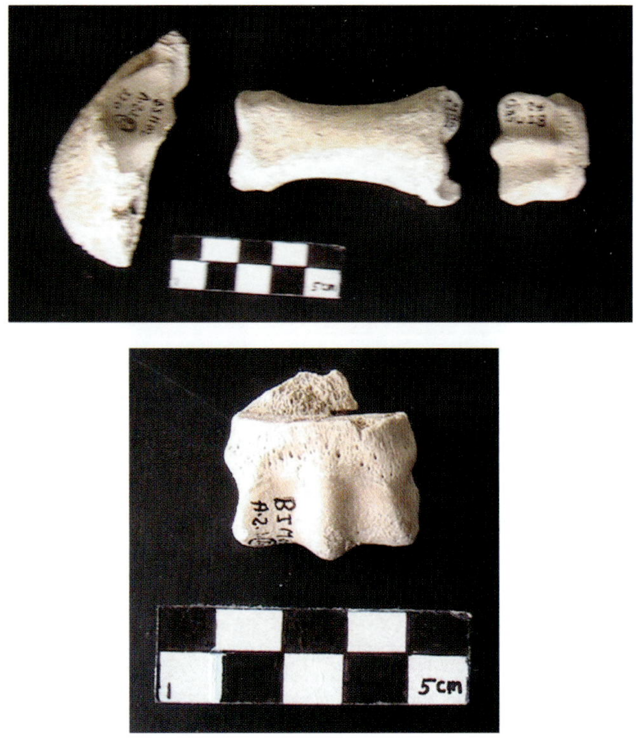

Fig. 30.8. a. Articulated donkey foot indicating the presence of carcasses at Mar Nicola; b. donkey distal metapodial with saw mark.

bone fragments were included among others bone slices that had been bilaterally engraved in order to produce round cylindrical objects, with a diameter of about 5–7mm. This may correspond to the manufacturing of rosary beads or can be connected with the use of bone *tesserae* in mosaics. Polished bone has indeed a unique colour and shine that is not found in *tesserae* made of stone. However, the use of bone in mosaics still has to be investigated. It has been reported that bone is often used for the manufacturing of rosary beads (Alice Choyke, pers. comm. 2008) (Figs. 30.6 & 30.7).

So far, no butchery traces have been observed on the camel bones and, with the exception of one distal metapodial and one humerus, all bones are from adult individuals. In addition, it is clear from the skeletal distribution that only straight bones with a relatively thick shaft are ideal for bone-working.

This indicates that the camel bones may have been brought to the site of Mar Nicola exclusively for manufacturing activities.

EQUIDS

Among the equid remains from Mar Nicola, donkey (*Equus africanus* f. *asinus*) is much more common than horse (*Equus ferus* f. *caballus*). Both species are represented by respectively 20 and two bone specimens (Table 30.2). The cranial remains of donkey and horse were separated using the enamel patterns of the mandibular teeth (Clutton-Brock 1989).

The excavation at area B yielded in one context a metarsal with articulating phalanges of donkey. This context was probably dated to the Roman period. More bones of donkey (n=11) were collected from area A and dated to the Islamic period; only one specimen is dated to the Byzantine period. No butchery evidence could be detected from these bones, but one donkey distal metapodial bone had been sawn (Fig. 30.8)

Horse is represented by one carpal bone and one third phalanx. Both of them were dated to the Islamic period.

*CHICKEN (*GALLUS GALLUS F. DOMESTICA*)*

Four bird fragments have been recovered at the site of Mar Nicola, three of them – a sternum, a pelvis and a femur – originate from domestic fowl. All four bird bones are dated to the Islamic period.

The chicken bones were checked for the presence of medullary bone but it could not be detected. The presence of medullary bone is restricted to female individuals during a specific time of the year, more precisely during the egg laying season (Lentacker & Van Neer 1996). The absence of medullary bone in the material of Mar Nicola indicates that the sample is composed of either male chickens or females which were killed before or after the egg laying season.

*GAZELLE (*GAZELLA GAZELLA*)*

Three bones were identified as gazelle. All three were collected from area A and included two fragments – a metatarsal and a 1st phalanx – that were dated to the Islamic period; one metacarpal bone was found from a Byzantine context. The latter specimen has a cut mark. According to the zoogeographical distribution of gazelle (Tchernov *et al*. 1986/7), these gazelle bones could be attributed to the mountain gazelle, *Gazella gazella*.

SHELLS

Only one shell was found at the site of Mar Nicola. It is a fragment of mother-of-pearl and was retrieved from area A in an Islamic context. The shell is sawn at both sides. This points towards the use of mother of pearl as a raw material for the manufacturing of objects.

THE UNIDENTIFIED BONES

A total of 485 bone fragments, 40.6% of the whole assemblage, could not be identified to taxon. But, they were sorted according to animal size. About half of them could be classified as medium-sized mammals, while the other half originated from large-sized mammals (Table 30.2). The relatively high proportion of unidentified bones is a result of the high degree of fragmentation. Many of these unidentified fragments have also cut (n=18), chopping (n=15) and sawing (n=10) marks (cf. Appendix).

Taphonomy

The preservation state of the animal bones is very good. However, they are quite fragmented and show many anthropogenic modifications (cf. Appendix). As already mentioned above, many specimens show butchery and manufacturing traces: 18.6% of the bone fragments show evidence of butchery (Fig. 30.2), while 5.2% show working traces, mostly saw marks, resulting from manufacturing activities (Figs. 30.3–30.5, 30.8). Few specimens (1%) show traces of carnivore damage.

The majority of the animal remains can be considered as consumption refuse. All of the butchery marks that are represented on the sheep/goat bones are the result of dismembering the animals and chopping these parts into smaller pieces. Cattle bones, on the other hand, show also traces of skinning and defleshing. In addition, chop marks on vertebrae show that the vertebral column was divided in a bilateral direction. It can be deduced, from the width and the depth of the marks, that strong and large axes were used for butchery activities.

A second taphonomic group consists of the waste of artisan activities. The sawing marks on the camel bones and on a few other species, indicates that bone was used for the manufacturing of objects at the site of Mar Nicola. It is possible that a specialised workshop for the production of rosary beads was located here. However, no finished objects have been found, only unfinished objects and waste fragments related to bone working were collected. Mother-ofpearl was used as a raw material at the site.

The faunal assemblage of Mar Nicola also contained the remains of carcasses, or animals that were not eaten. The equid remains belong to this third taphonomic group, as for example indicated by the articulated bones of donkey in area B. It is possible that bones from donkey and camel carcasses were collected for manufacturing activities. Probably, the latter may have been deposited elsewhere, as only a limited number of skeletal elements were represented at the site.

Summary and conclusion

The site of Mar Nicola is located in the centre of Beit Jala, 2km northwest of Bethlehem, Palestine. The animal bone assemblage from this site consists of three taphonomic groups, i.e. 1) refuse from food consumption, 2) waste of artisanal activities, and 3) carcasses (Gautier 1987). Meat was primarily provided by domestic mammals (ovicaprids, pigs and cattle), while wild animals, more specifically gazelle, did not represent a significant contribution to the diet of the inhabitants. Sheep and goat remains are the most abundant among the domestic animals, followed by cattle. No signs of butchery could be found on the equid remains, suggesting that they were not consumed but that they were only used for transport. The waste of artisanal activities predominantly included camel bones. These comprised the proximal and distal ends of the long straight limb bones, and almost all of them exhibited saw traces. Other animals with traces of sawing included sheep/goat, donkey and mollusc. Due to the presence of a Byzantine church, bone and mother of pearl may have been brought to the site as a raw material for manufacturing religious objects such as rosary beads and necklaces, or decoration objects, such as *tesserae* for mosaics. No finished objects were found within the bone assemblage, only unfinished object fragments and related waste were discovered. This shows that the site of Mar Nicola was a specialised centre for the production of handicrafts and religious items during the Islamic period.

Acknowledgments

I would like to thank my colleagues at the Ministry of Tourism & Antiquities for allowing me to present the paper on the faunal material of Mar Nicola. I would also like to thank the organisers of the ASWA conference, Dr. Marjan Mashkour and Dr. Mark Beech, for their efforts. My thanks are also due to the Abu Dhabi Authority for Culture & Heritage for their support and hospitality. Moreover, my thanks go to everybody who made my participation possible at the ASWA conference in al-Ain.

References

Al-Zawahra, M. A. M. 1999, 'The faunal remains from Tell Jenin, northern West Bank-Palestine', MA thesis, Leuven.

Al-Zawahra, M. A. M. 2002, 'The faunal remains from Kherbet Belameh', Unpublished report (not accessible).

Boessneck, J. 1969, 'Osteological differences between Sheep (*Ovis aries*) and goat (*Capra hircus*)', in D. Brothwell & E. S. Higgs (eds), *Science in Archaeology*, 331–358. Thames & Hudson, Harlow.

Clason, A. T. 1978, 'Late Bronze Age-Iron Age zebu in Jordan?'. *Journal of Archaeological Science* 5, 91–93.

Clutton-Brock, J. 1989, 'A dog and a donkey excavated at Tell Brak'. *Iraq* 51, 217–224.

Free, J. P. & Howard, F. 1992, *Archaeology and Bible History.* Zondervan, Grand Rapids.

Gautier, A. 1987, 'Taphonomic groups: how and why?'. *Archaeozoologia* 1, 47–52.

Grant, A. 1982, 'The use of tooth wear as a guide to age of domestic ungulates', in B. Wilson, C. Grigson & S. Payne (eds), *Ageing and Sexing Animal Bones From Archaeological Sites*, 91–108. British Archaeological Report 109. British Archaeological Reports, Oxford.

Lentacker, A. & Van Neer, W. 1996, 'Bird remains from two sites on the Red Sea coast and some observations on medullary bone'. *International Journal of Osteoarchaeology* 6, 488–496.

Payne, S. 1973, 'Kill-off patterns in sheep and goats: the mandibles from Asvan Kale'. *Anatolian Studies,* 23, 281–303.

Prummel, W. & Frisch, H. J. 1986, 'A guide for the distinction of species, sex and body size of sheep and goat'. *Journal of Archaeological Science* 13, 567–577.

Schmid, E. 1972, *Atlas of Animal Bones*. Elsevier, Amsterdam.

Silver, I. A. 1969, 'The ageing of domestic animals', in D. Brothwell & E. S. Higgs (eds), *Science in Archaeology*, 283–289. Thames & Hudson, Harlow.

Stallibras, S. 1983, 'A bifid thoracic vertebral spine from a bovine in the Roman Fenland'. *Journal of Archaeological Science* 10, 265–266.

Tchernov, E., Dayan, T. & Yom-Tov, Y. 1986/7, 'The paleogeography of *Gazella gazella* and *Gazella dorcas* during the Holocene of the southern Levant'. *Israel Journal of Zoology* 34, 51–59.

Wapnish, P. 1982, 'Archeozoology at Tell Jemmeh: taphonomy and paleoeconomy in historic archaeology'. *Natural Geographic Society Research* 21, 57–80.

Wapnish, P. & Hesse, B. 1991, 'Faunal remains from Tell Dan: perspectives on animal production at a village, urban and ritual centre'. *Archaeozoologia* 4(2), 9–86.

Zeuner, F. E. 1963, *A History of Domesticated Animals*. Hutchinson, London.

Appendix: inventory of all traces observed on the faunal remains of Mar Nicola (Beit Jala)

(LM: large sized mammal; MM: medium-sized mammal; R: right; L: left; A: axial; PF: proximal fused; DF: distal fused; DFG: distal fusing; AF: acetabulum fused; C: cut mark; CH: chop; B: burnt; G: gnaw; P; pathology; PO: polish; S: saw marks)

Cont.	*Num*	*Elem*	*Species*	*Frag*	*S*	*%*	*Fus*	*C*	*Ch*	*B*	*G*	*P*	*Po*	*S*	*Remarks*
A.1.1.58	468	Astr	*Bos*	Comp	R	100		C							On art. facet/ head
A.2.1.27	338	Astr	*Bos*	Incomp	R	75		C							Vent. mid.
A.3.1.43	637	Astr	*Bos*	Incomp	L	75		C							Cs at med. mid.
A.1.1.10	76	Axis	*Bos*	Ant.	A	25		C							Long. Cut on wing
A.1.1.17	50	Axis	*Bos*	Frag	A	50		C							Mid. lat.
A.3.1.43	659	Carp	*Bos*	Comp		100		C							Many parallel, ant. art. facet
A.1.1.56	291	Cran	*Bos*	Temp	L	<25		C							Condyle joint
A.2.1.27	336	Cran	*Bos*	Pmax	L	<25		C							C. many dorsal
A.1.1.5	587	Hum	*Bos*	Dist	L	25	DF	C	CH						C med./ch. med.
A.1.1.58	477	Hum	*Bos*	Dia		<25		C							Many on lat.
A.3.1.63	462	Hum	*Bos*	Dist	R	<25	DF		CH						Med.obl.
A.1.1.10	69	Mand	*Bos*	Ramus	R	<25		C							Lat. wide at caronoid
A.1.1.13	164	Mand	*Bos*	Ramus	L	<25			CH						Post. mid. ramus
A.1.1.13	171	Mand	*Bos*	Ramus	L	<25			CH						Post. ramus 3 wide marks
A.1.1.56	303	Mand	*Bos*	Condyl	R	<25			CH						Ext. close art. facet
A.1.1.58	478	Mand	*Bos*	Ramus	R	<25			CH						Lat. & between processes
A.3.1.63	444	Mand	*Bos*	Ramus	L	<25			CH						Lat. med.
A.1.1.10	125	Mc	*Bos*	Prox	L	<25	PF	C							Many dors.
A.1.1.3	703	Mc	*Bos*	Prox	R	<25	PF		CH						Mid. med. & lat.
A.2.1.27	332	Mc	*Bos*	Prox	R	75	PF		CH						
A.3.1.43	639	Mc	*Bos*	Prox	R	25	PF	C							Med./ant. close art. facet
A.1.1.13	163	Mp	*Bos*	Dist	R	<25	DF	C							Ant. on articular facet
A.3.1.43	674	Mp	*Bos*	Dia		<25		C							
A.1.1.10	68	Mt	*Bos*	Prox	R	25	PF	C							Close to art. facet
A.1.1.56	310	Mt	*Bos*	Prox	L	50	PF	C							Close art. facet
A.1.1.56	309	Mt	*Bos*	Prox	R	50	PF		CH						Mid. shaft
A.1.1.58	474	Mt	*Bos*	Prox	L	25	PF		CH						Close to art. facet, med.
A.3.1.43	633	Mt	*Bos*	Prox	L	25	PF	C	CH						CH. at med. C. close art.facet
A.3.1.43	634	Mt	*Bos*	Prox	R	50	PF		CH						Mid. lat.
A.1.1.13	168	Pelv	*Bos*	Ilium		<25		C							
A.1.1.13	165	Pelv	*Bos*	Ilium	R	<25		C	G						C base ilium
A.1.1.3	704	Pelv	*Bos*	Pubis		<25	AF		CH						On art. facet & body
A.1.1.5	600	Pelv	*Bos*	Pubis		<25	AF	C							
A.1.1.10	107	Ph1	*Bos*	Comp	R	100	PFDF	C							Lat.
A.1.1.13	177	Ph1	*Bos*	Incomp	L	75	PFDF	C	CH						C. ant./CH. lat.
A.1.1.13	178	Ph1	*Bos*	Comp	L	100	PFDF	C							C.lat. on art. facet
A.1.1.17	9	Ph1	*Bos*	Comp	R	100	PFDF	C							On artic. facet
A.3.1.43	636	Ph1	*Bos*	Comp	R	100	PFDF	C							Ant. mid.
A.3.1.43	640	Ph1	*Bos*	Comp	R	100	PFDF	C							Ant. close on art. facet
A.3.1.63	441	Ph1	*Bos*	Comp	R	100	PFDF	C							Dors. prox. facet
A.1.1.58	509	Pmax	*Bos*	Incomp	L	<25		C							C. on pmax. caud.
A.1.1.13	158	Rad	*Bos*	Prox	L	<25	PF	C							Ant. med. close to art. facet
A.1.1.17	3	Rad	*Bos*	Prox	L	<25	PF	C							Post.med.
A.1.1.56	311	Rad	*Bos*	Dia	L	<25			CH						
A.1.1.58	504	Rad	*Bos*	Dia	L	<25			CH						Lat. prox
A.2.1.27	339	Rad	*Bos*	Prox	L	25	PF	C	CH						C. med., CH. Ant.
A.3.1.63	464	Rad	*Bos*	Prox	R	<25	PF		CH						Ant. 1cm from art. facet
A.1.1.13	185	Scap	*Bos*	Dist	R	<25	DF		CH						Ch. at neck
A.1.1.17	55	Scap	*Bos*	Dist		<25		C							Close to artic. facet
A.1.1.13	159	Tarsal	*Bos*	Incomp	R	75			CH						Many lat.
A.2.1.27	337	Tarsal	*Bos*	Comp	L	100		C							At both sides/navic.
A.1.1.58	470	Tib	*Bos*	Dist	L	<25	DF		CH						Lat. 8cm of dist. end
A.1.5.32	263	Tib	*Bos*	Dist	L	<25	DF	C							Lat.
A.3.1.63	443	Tib	*Bos*	Dist		<25	DF	C	CH						Med. art. facet
A.1.1.10	75	Ulna	*Bos*	Prox	R	25	PF	C							
A.1.1.17	52	Ulna	*Bos*	Prox	L	<25	PN		CH						Dors.
A.2.1.27	321	Hum	Camel	Prox		<25	PN	C							On art. facet
A.1.1.13	190	Mc	Camel	Prox	L	<25	PF							S	
A.1.1.16	752	Mc	Camel	Prox		<25	PF							S	
A.1.1.5	589	Mc	Camel	Prox	R	<25	PF							S	
A.1.1.51	773	Mc	Camel	Prox	L	<25	PF							S	Working piece
A.2.1.27	318	Mc	Camel	Prox	L	<25	PF	C						S	C. ant. art. facet

Cont.	*Num*	*Elem*	*Species*	*Frag*	*S*	*%*	*Fus*	*C*	*Ch*	*B*	*G*	*P*	*Po*	*S*	*Remarks*
A.2.1.27	319	Mc	Camel	Prox	L	<25	PF							S	
A.2.1.35	618	Mc	Camel	Prox	L	<25	PF	C						S	C. on art. facet
A.3.1.43	635	Mc	Camel	Prox	R	<25	PF		CH					S	Lat. close art. facet
A.1.1.10	65	Mp	Camel	Dist	R	<25	DF							S	3cm an art. facet
A.1.1.10	66	Mp	Camel	Dist	R	<25	DF			G				S	4cm an art. facet
A.1.1.12	748	Mp	Camel	Dia		<25								S	M: 35
A.1.1.13	191	Mp	Camel	Dist		<25	DF							S	
A.1.1.13	192	Mp	Camel	Dia		<25								S	Sawed bilaterally
A.1.1.13	193	Mp	Camel	Dia		<25								S	
A.1.1.17	48	Mp	Camel	Dist	L	<25	DF							S	
A.1.1.17	47	Mp	Camel	Dist	R	<25	DF							S	
A.1.1.5	553	Mp	Camel	Dist	L	<25	DF							S	
A.1.1.5	555	Mp	Camel	Dist	L	<25	DF							S	
A.1.1.5	552	Mp	Camel	Dist	R	<25	DF	C						S	C. At articular facet, lat.
A.1.1.5	556	Mp	Camel	Dist	R	<25	DF							S	
A.2.1.27	313	Mp	Camel	Dia		<25								S	Sawed piece
A.2.1.27	320	Mp	Camel	Dia		<25								S	Working piece
A.2.1.68	736	Mp	Camel	Dia		<25								S	Working piece
A.3.1.43	642	Mp	Camel	Dist		<25	DF							S	
A.3.1.63	445	Mp	Camel	Dist	R	<25	DF							S	Swaed
A.1.1.10	64	Mt	Camel	Prox	R	<25	PF	C						S	2cm an art.facet, c.med.
A.1.1.10	63	Mt	Camel	Prox	R	<25	PF							S	Sawed 1cm an art. facet
A.1.1.13	188	Mt	Camel	Prox	R	<25	PF	C						S	
A.1.1.13	189	Mt	Camel	Prox	R	<25	PF							S	
A.1.1.5	590	Mt	Camel	Prox	R	<25	PF							S	
A.1.1.51	772	Mt	Camel	Prox	R	<25	PF							S	Working piece
A.2.1.27	316	Mt	Camel	Prox	R	<25	PF	C				P		S	C. ant. med.
A.2.1.27	317	Mt	Camel	Prox	R	<25	PF							S	
A.2.1.68	738	Mt	Camel	Prox	L	<25	PF							S	
A.2.1.27	335	Ph1	Camel	Incomp		75	PFDF		CH						Long. chopping
A.1.1.17	2	Rad	Camel	Dia	L	<25		C	CH						Cs lat/ant. CH. Lat
A.1.1.17	1	Rad	Camel	Dia	R	<25								S	Working piece sawed bi.
A.1.1.4	745	Rad	Camel	Dia		<25								S	
A.1.1.5	609	Rad	Camel	Dia										S	Working piece
A.1.1.5	610	Rad	Camel	Dia										S	
A.1.1.51	771	Rad	Camel	Dia		<25								S	Working piccc
A.2.1.27	315	Rad	Camel	Prox	L	<25	PF							S	
A.2.1.27	326	Rad	Camel	Dia	L	<25								S	
A.2.1.27	314	Rad	Camel	Dist	R	<25	DF		CH					S	Med. caud. On art. facet
A.2.1.27	325	Rad	Camel	Dist	R	<25	DF							S	
A.2.1.35	619	Rad	Camel	Prox	L	<25	PF							S	
A.2.1.68	737	Rad	Camel	Dia		<25								S	Working piece
A.3.1.63	457	Rad	Camel			<25							PO	S	
A.2.1.68	740	Tarsal	Camel	Comp		100		C							
A.1.1.5	551	Tib	Camel	Dist	R	<25	DFG							S	
A.1.1.58	467	Tib	Camel	Dist	R	<25	DF							S	Sawed
A.1.1.10	132	Calc	*Capra*	Comp	L	100	PN	C							Below dist. art. facet
A.1.1.13	235	Cornu	*Capra*	Frag	L	50			CH						Mid.
A.2.1.27	394	Cornu	*Capra*	Incomp	L	50			CH						At base
A.1.1.13	236	Cran	*Capra*	Occ.	A	25			CH						Right below horn base
A.1.1.10	89	Hum	*Capra*	Dist	R	<25	DF	C							Med. lat. close trochlea
A.3.1.43	651	Hum	*Capra*	Dist	L	50	DF		CH						On art. facet/mid. med.
A.3.1.43	653	Hum	*Capra*	Dist	R	<25	DF	C	CH						On art. facet & neck
A.1.5.32	261	Mt	*Capra*	Prox	R	<25	PF	C							Med. close art. facet
A.1.1.13	218	Ph2	*Capra*	Comp	L	100	PFDF	C							C. ant. close art. facet
A.1.1.5	595	Rad	*Capra*	Prox	R	50	PF		CH						Lat. on art. facet
A.2.1.27	350	Rad	*Capra*	Prox	L	25	PF	C							Post. close art. facet
A.2.1.27	347	Rad	*Capra*	Prox	R	25	PF	C							Ant. med. close art. facet
A.1.1.17	17	Scap	*Capra*	Dist	L	25	DF	C	CH						C. med. on artic. facet
A.1.5.32	275	Scap	*Capra*	Dist	L	<25	DF	C	CH						C. art. facet & CH. lat. At neck
A.3.1.43	672	Scap	*Capra*	Dist		<25	DF		CH						Med. lat.
A.2.1.27	331	Mp	Donkey	Dist	L	<25	DF							S	Sawed with technique
A.1.5.32	268	Mc	Gazel	Prox	R	75	PF	C							Med.
A.1.1.58	481	Fem	Indet	Dist		<25	DF		CH		G				On condyle
A.3.1.63	453	Indet	Indet										PO	S	Swaed
A.3.1.63	454	Indet	Indet										PO	S	
A.3.1.63	456	Indet	Indet										PO	S	
A.3.1.63	459	Indet	Indet										PO	S	Bead core
A.3.1.63	460	Indet	Indet										PO	S	
A.3.1.63	461	Indet	Indet										PO	S	
A.3.1.63	463	Indet	Indet										PO	S	

Cont.	Num	Elem	Species	Frag	S	%	Fus	C	Ch	B	G	P	Po	S	Remarks
A.1.1.10		Lm	Indet	7					CH						1 with chopping mark
A.1.1.5	615	Tool	Indet											S	
A.1.1.51	760	Tool	Indet											S	Working piece
A.1.1.58		Costa	Lm	3					CH						1 with chopping mark
A.3.1.63		Costa	Lm	3					CH			P			2 CH./1 path.
A.1.1.17	5	Fem	Lm	Dist	R	<25			CH						Wide mark
A.3.1.43	641	Hum	Lm	Dia		<25		C	CH						
A.1.1.13		Indet	Lm	18				C							Two with Cs
A.1.1.3		Indet	Lm	4										S	All sawed
A.1.1.5	573	Indet	Lm						CH						Many parallel marks
A.1.1.5	611	Indet	Lm	Dia										S	
A.1.1.58		Indet	Lm	8				C							2 with cut
A.2.1.68	742	Indet	Lm											S	Working piece
A.3.1.43	649	Indet	Lm	Dia		<25							PO	S	
A.1.1.13	247	Mand	Lm	Ramus		<25			CH						
A.1.1.17	11	Mp	Lm	Dia		<25			CH						Uniside sawed
A.1.1.10	131	Tool	Lm	Dia										S	Sawed
A.1.1.5	612	Tool	Lm	Dia										S	
A.1.1.5	613	Tool	Lm	Dia										S	
A.1.1.5	614	Tool	Lm	Dia										S	
A.1.1.58	549	Tool	Lm			<25			CH						A chesil frag
A.1.1.58	550	Tool	Lm						CH						
A.1.1.13		Vert	Lm	6				C							One with C
A.1.1.17		Vert	Lm	6				C							Three with C marks
A.3.1.63		Vert	Lm	4					CH						1 CH.
A.1.1.58		Costa	Mm	4				C							1 with cut
A.1.1.13		Indet	Mm	21				C							Three with cut
A.1.1.58		Scap	Mm			<25			CH						
A.1.1.13		Vert	Mm	7				C							Two with C
A.1.1.17		Vert	Mm	5				C							One with C
A.1.1.4		Vert	Mm	Frag	A	<25		C							C. on arch
A.3.1.63		Vert	Mm	3				C	CH						1 c./ 1 ch.
A.1.1.56	292	Atlas	O/c	Frag	A	50		C							Ant. at wings
A.1.1.17	51	Axis	O/c		A	50		C							Long. C
A.2.1.27	388	Axis	O/c	Frag	A	50		C							Dors.
A.3.1.63	451	Cran	O/c	Frontal	L	<25			CH						Horn base, to remove it
A.1.1.10	133	Fem	O/c	Dist	R	<25	DF	C							
A.1.1.10	81	Fem	O/c	Dist	R	<25	DF		CH						Dors. on art. facet
A.3.1.63	412	Fem	O/c	Dia		<25	DF		CH					S	
A.1.1.10	90	Hum	O/c	Dia	L	<25			CH						Mid.
A.1.1.13	224	Hum	O/c	Dist	L	<25	DF	C							Lat. on art. facet
A.1.1.13	234	Hum	O/c	Dist	L	<25	DF	C							Ant. close to troch.
A.1.1.13	227	Hum	O/c	Dist	L	25	DF		CH						Mid. med.
A.1.1.13	233	Hum	O/c	Dist	R	<25	DF	C							On trochlea
A.1.1.13	225	Hum	O/c	Dia	R	25			CH						Mid.
A.1.1.13	229	Hum	O/c	Dia	R	50			CH						Many parallel med.
A.1.1.17	35	Hum	O/c	Dia		<25		C							
A.1.1.17	30	Hum	O/c	Dia	R	<25		C							Cs deep dors.
A.1.1.5	569	Hum	O/c	Dist	L	25	DF	C	CH						C. ant./lat & CH. med./dist
A.1.1.5	568	Hum	O/c	Dist	L	<25	DF	C							Lat. dors.
A.1.1.5	597	Hum	O/c	Dist	L	<25	DF	C							Med.
A.1.1.58	486	Hum	O/c	Dia	R	25			CH						
A.2.1.27	384	Hum	O/c	Dist	L	25	DN		CH						Med. mid. of shaft
A.3.1.43	655	Hum	O/c	Dia		<25			CH						Mid.
A.3.1.43	656	Hum	O/c	Prox		<25			CH						On prox. area dors.
A.3.1.43	654	Hum	O/c	Dia	L	<25		C							Mid. lat.
A.3.1.43	657	Hum	O/c	Dist	R	<25	DF		CH						Med. neck
A.3.1.63	411	Hum	O/c	Dist	R	<25	DF		CH						Lat. dor.
A.1.1.56	282	Mand	O/c	Trans	L	75		C							C ext.
A.3.1.43	689	Mand	O/c	Ramus	L	<25		C							Ext. on the neck
A.1.1.10	97	Mc	O/c	Prox	L	25	PF	C							Lat. ant. close art. facet
A.1.1.10	115	Mc	O/c	Prox	R	75	PF	C							Ant. med.
A.1.1.51	764	Mp	O/c	Dist		<25	DN		CH						Close to dist.
A.1.1.58	514	Mt	O/c	Dist	R	50	DF		CH						At lat. dist.
A.2.1.27	377	Mt	O/c	Prox	L	25	PF	C							Ant. med.
A.1.1.10	139	Pelv	O/c	Ilium	L	<25	AF		CH						3 parallel lat.
A.1.1.10	136	Pelv	O/c	Ilium	R	<25	AF	C							Dors. pretuberosity
A.1.1.13	250	Pelv	O/c	Acet	R	<25			CH						Base of ilium
A.1.1.17	53	Pelv	O/c	Ilium	R	<25		C							Med. ant. to acet.
A.1.1.17	57	Pelv	O/c	Ischium	R	<25	AF	C							Post. to acet

Cont.	*Num*	*Elem*	*Species*	*Frag*	*S*	*%*	*Fus*	*C*	*Ch*	*B*	*G*	*P*	*Po*	*S*	*Remarks*
A.1.1.58	523	Pelv	O/c	Ischium	L	<25	AN	C							Before lip
A.2.1.35	625	Pelv	O/c	Ischium		<25	AF		CH						
A.1.1.10	124	Rad	O/c	Dist	L	50	DF	C	CH						CH. dors. mid./C. on art. facet
A.1.1.10	91	Rad	O/c	Dist	L	50	DF	C							Lat. mid. artic. facet
A.1.1.10	92	Rad	O/c	Dia	R	25		C							Lat.
A.1.1.10	121	Rad	O/c	Dia	R	25		C							Ant.
A.1.1.10	122	Rad	O/c	Dist	R	<25	DF	C							Ant. many
A.1.1.13	214	Rad	O/c	Prox	L	25	PF	C							Med.
A.1.1.5	577	Rad	O/c	Dia	L	75		C							Ant./med.
A.1.1.58	497	Rad	O/c	Dia	R	50		C							Med.
A.1.1.58	496	Rad	O/c	Dia	R	50			CH						Mid./dist, med.
A.1.5.32	265	Rad	O/c	Prox	R	25	PF		CH						Med. mid.
A.2.1.27	380	Rad	O/c	Prox	R	25	PF	C							Dors. on art. facet
A.2.1.27	382	Rad	O/c	Dist	R	25	DN	C							Med.
A.2.1.27	348	Rad	O/c	Prox	R	25	PF		CH						Head, lat.
A.2.1.27	381	Rad	O/c	Dia	R	25			CH						
A.3.1.43	670	Rad	O/c	Prox	L	<25	PF		CH						Ant.
A.3.1.43	666	Rad	O/c	Dia	R	<25			CH						Med.
A.3.1.43	667	Rad	O/c	Dist	R	25	DF		CH						Mid. both sides
A.3.1.63	448	Rad	O/c	Dia	R	25			CH						Mid all around
A.1.1.13	237	Scap	O/c	Blade	R	<25			CH						Neck
A.1.1.58	490	Scap	O/c	Dia	R	<25			CH						Base of spine
A.2.1.27	359	Scap	O/c	Dist	R	25		C							Close art. facet
A.2.1.27	362	Scap	O/c	Dist	R	<25	DF	C							Caud. on necke
A.3.1.63	414	Scap	O/c	Dist	L	<25	DN	C							Caud. neck
A.3.1.63	417	Scap	O/c	Dist	R	25		C							Caud. neck
A.3.1.63	415	Scap	O/c	Dist	R	25	DF		CH						Dors. neck
A.1.1.10	85	Tib	O/c	Dia	R	50		C							Cuts ant. post. distally
A.1.1.10	112	Tib	O/c	Dia	R	25		C							
A.1.1.10	113	Tib	O/c	Dia	R	25			CH						Many parallel, med.
A.1.1.13	253	Tib	O/c	Dia	R	50			CH						Med. prox.
A.1.1.13	255	Tib	O/c	Dia	R	<25			CH						Med.
A.1.1.13	256	Tib	O/c	Dist	R	25	DF		CH						Med. close to dist.
A.1.1.17	40	Tib	O/c	Dia	L	<25		C							Cuts med. mid.
A.1.1.17	37	Tib	O/c	Dist	L	50	DF		CH						Mid.
A.1.1.17	39	Tib	O/c	Dia	R	25		C							Cuts mid. dors.
A.1.1.5	579	Tib	O/c	Dia	L	75			CH						Many parallel/ dors.
A.1.1.5	564	Tib	O/c	Dia	R	25			CH						Med. mid
A.1.1.58	501	Tib	O/c	Dist	L	25	DF		CH						Lat. med.
A.1.1.58	503	Tib	O/c	Dia	L	<25			CH						Med. mid
A.1.1.58	506	Tib	O/c	Dia	L	50			CH						Mid. dors/med.
A.2.1.27	352	Tib	O/c	Dist	L	25	DN	C							Med.
A.2.1.27	399	Tib	O/c	Dist	L	<25	DF		CH						Ch. med.
A.2.1.27	392	Tib	O/c	Dia	R	25			CH						Lat.
A.2.1.35	623	Tib	O/c	Dist	R	25	DF		CH						Med. mid. of shaft
A.3.1.43	681	Tib	O/c	Dist	L	<25	DF		CH						
A.3.1.43	665	Tib	O/c	Dia	R	25			CH						Med.
A.3.1.63	458	Tib	O/c	Dia		<25			CH						Mid. dors.
A.1.1.58	499	Ulna	O/c	Semilu	R	<25		C							Dorsal
A.3.1.53	728	Astr	Ovis	Incomp	L	75		C							On head dist
A.1.1.10	88	Hum	Ovis	Dist	L	<25	DF		CH						Lat. at artic. facet
A.1.1.13	226	Hum	Ovis	Dist	L	<25		C							Med.
A.1.1.13	228	Hum	Ovis	Dist	L	<25		C							Med.
A.1.1.17	31	Hum	Ovis	Dist	L	<25	DF	C							Lat. dor.
A.3.1.43	652	Hum	Ovis	Dist	R	25	DF		CH						On art. facet
A.1.1.17	25	Mc	Ovis	Prox	R	25	PF	C							Two parallel Cs
A.1.5.32	262	Mp	Ovis	Dist	L	25	DF		CH						Lat. many
A.1.1.10	119	Rad	Ovis	Prox	L	25	PF	C							Med. on art. facet
A.1.1.13	211	Rad	Ovis	Prox	R	25	PF	C							On articular facet
A.1.1.13	212	Rad	Ovis	Prox	R	25	PF		CH						Dors. ulna join area
A.2.1.27	349	Rad	Ovis	Prox	L	<25	PF		CH						Post. med.
A.1.1.10	79	Scap	Ovis	Dist	L	<25	DF		CH						4 paralllel at neck
A.1.1.13	222	Scap	Ovis	Dist	L	25	DN	C							Caud. close art. facet
A.2.1.27	360	Scap	Ovis	Dist	L	<25	DF		CH						Dors. half
A.1.1.17	18	Ulna	Ovis	Prox	R	<25	PN		CH						Dors. una arch
A.1.1.50	712	Shell	Shell	1										S	Sawn at both sides
A.1.1.17	44	Hum	Sus	Dist	R	<25	DF	C							Med. lat.
A.1.1.5	582	Hum	Sus	Dia	L	<25			CH						Ant.
A.1.5.32	266	Hum	Sus	Dia	L	<25			CH						Dors.
A.1.1.10	154	Pelv	Sus	Acet	L	<25	AF		CH						Dors. base of ischium
A.1.1.58	526	Pelv	Sus	Acet	L	25	AF	C	CH						Both at base ilium
A.1.1.5	560	Tib	Sus	Dist	R	<25	DF		CH		G				

31. Faunal analysis of the Castle of Aqaba (Jordan): preliminary results

Bea De Cupere, Anton Ervynck, Mircea Udrescu, Wim Van Neer & Wim Wouters

Excavations at the Castle of Aqaba have yielded a large amount of animal remains mainly dating to the Mamluk, Ottoman and Modern periods. The archaeozoological analysis of this assemblage provided an insight into the food provisioning of the site's inhabitants. The bulk of the animal food products were provided by sheep and goat, complemented by camel and with very few cattle. Hunting played only a minor role in the food economy, while fishing and harvesting shellfish must have been more important. The faunal data suggest diachronic changes in the herd management and fishing strategies.

Keywords archaeozoology, Jordan, Aqaba Castle, food provisioning, Mamluk period, Ottoman period, Modern period

Introduction

After environmental archaeology had slowly gained a place within the archaeological fieldwork in the Near East, and Jordan in particular, more efforts were spent on prehistoric sites. Only recently, later periods also came into the focus of attention, e.g., with the archaeozoological work at Petra (Nabatean to Late Roman, medieval; Studer 1994; 2002a; 2007; 2008; Schmidt & Studer 2003), Jebel an-Nabi Harûn (Byzantine; Studer 2001; 2002b), Deir 'Ain 'Abata (Byzantine; Beech 1992) and central Jordan (Roman; Toplyn 1994; 2006). The present study of the animal remains from the castle at Aqaba follows this trend and perhaps presents a new scope, by including for the first time Jordanian material from the post-medieval period, even the early 20th century. The aim of the analysis introduced in the present report, is to understand at least a part of the site's economy, and especially the organisation of the food supply as far as meat is concerned. In a second step, a diachronic comparison between the consumption patterns of the different occupation phases of the site is undertaken.

The Castle of Aqaba

Aqaba is situated in southern Jordan on the coast of the Red Sea, more precisely at the northern end of the Gulf of Aqaba (Fig. 31.1). The region lies within the Syro-African Rift Valley that goes through the Gulf of Aqaba and the Wadi Araba up to the Dead Sea and the Jordan valley. The contemporary climate is arid, with a mean annual rainfall of 40mm; precipitation falls only in the winter months. Summers are characterised by high temperatures, which often exceed 40°C. At its continental side, Aqaba is surrounded by deserts, i.e. the Sinai (W), the Negev (N) and the Hisma (NE-E-S). The coastal plain is encircled by mountains but has access to the north by the Wadi Arabah, which is accessible through the Wadi al-Yutum to the northeast of Aqaba. Vegetation in the Wadi Arabah is controlled by the soil moisture, which is conditioned by topographical factors and the ground-water level, located just below the surface in the southern part of the valley. The extreme aridity of the region results in the lack of a permanent watercourse in the wadi (Zohary 1945; Niemi, cited in Parker 1996).

Excavations have been carried out at several locations within the present city of Aqaba. The three historical sites

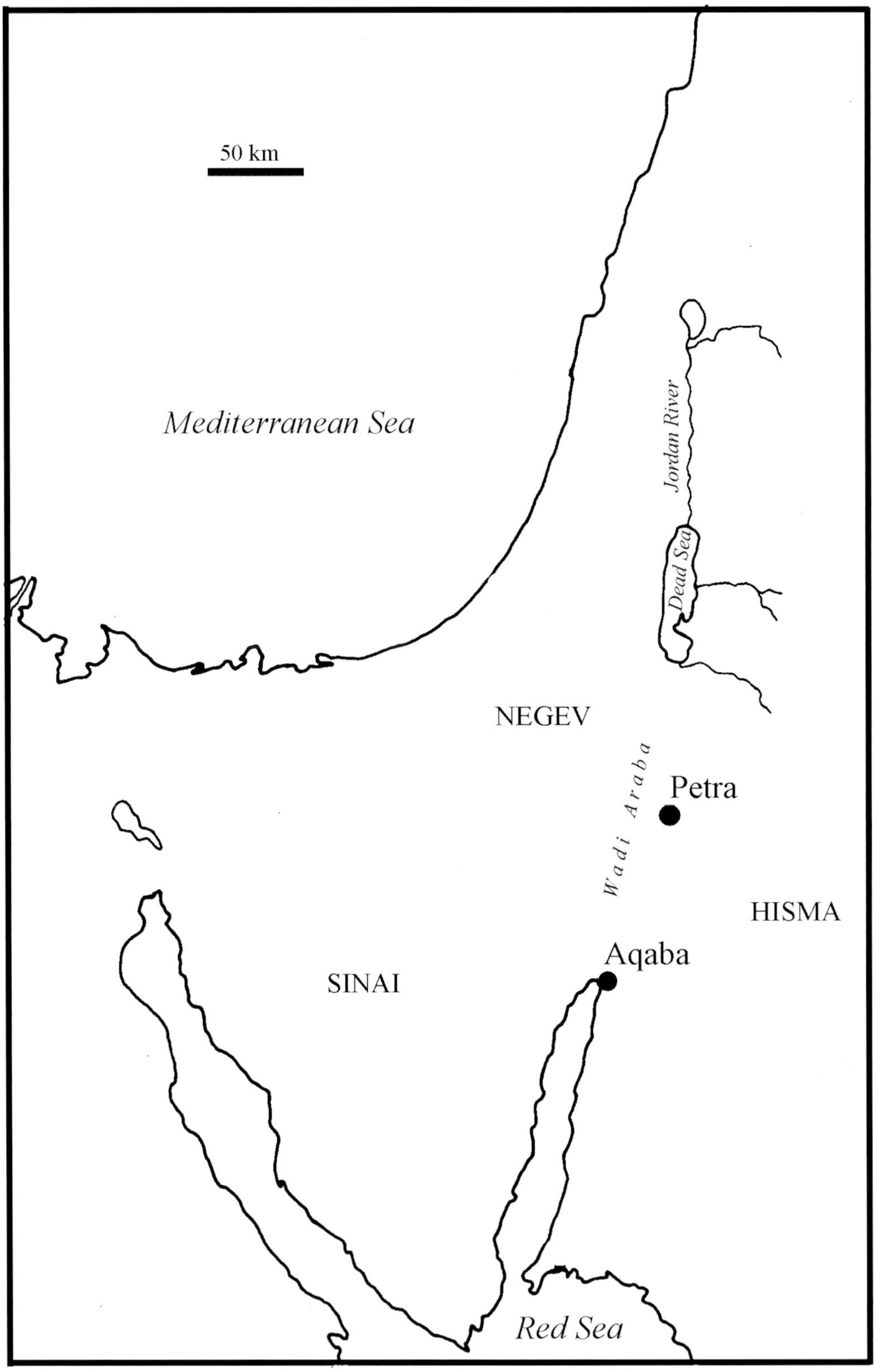

Fig. 31.1. Location of Aqaba.

include the Nabatean/Roman/Byzantine Aila (under the direction of S. T. Parker, North Carolina State University), the Early Islamic/medieval Ayla (under the direction of D. Whitcomb, Oriental Institute Chicago) and the medieval to recent times castle (Fig. 31.2). The excavations of the castle have been carried out within the framework of the 'Aqaba Castle Project' (2000–2003 under the direction of J. Demeulemeester, Ministère de la Région wallonne (Belgium), and D. Pringle, Cardiff University (UK); 2005–2008 under the direction of J. Demeulemeester

Fig. 31.2. Aerial view of the Castle of Aqaba (from Kennedy & Brewley 2004).

and R. al-Shqour, both Ghent University (Belgium)). Additionally, the 'Islamic Aqaba Project' was set up, which constitutes a research programme that aims to link the former American excavations of the medieval town of Ayla (1986–1993) and the excavations of the castle. So far, only preliminary reports are available for the castle excavations (De Meulemeester & Pringle 2001; 2002; 2004; 2005; De Meulemeester & al-Shqour 2006). Moreover, the most recent data on the chronology and development of the site provide new, albeit again preliminary, insights (al-Shqour *et al.* 2008). The chronological framework for the site used in the present report must thus be regarded as a first approach, without doubt to be refined by future research.

The castle is commonly believed to have been built in the 12th century AD by the crusaders. It would have been captured by Saladin in 1170, and subsequently rebuilt. However, renewed analysis of the historical documents now suggests that this crusader castle was located on the island of Jazirat Fara'un, along the Egyptian coast. The historical origin of Aqaba Castle thus remains unknown; its standing structures mainly show a Mamluk lay-out, representing building activities from the beginning of the 16th century, and more recent alterations. However, older structures are buried under the standing building complex. Material has been excavated dating back to the Early Islamic (Ummayid, Abbasid, Fatimid) periods (7th–12th century AD) and the Ayyubid period (end of the 12th–13th century), but the nature of this material (and its relation to the buildings) is not yet well understood. Some Roman artefacts have even been recovered; unfortunately their stratigraphic and taphonomic nature cannot yet be interpreted properly. As far as known at present, possible Roman activity at the site bears no relationship with the later medieval occupation. In any case, the site has a long history, possibly starting in the Early Islamic period and ranging into recent times (see Table 31.1).

The aim of the fieldwork was to undertake a complete survey and structural interpretation of the building remains and the underlying structures, and to document the chronological sequence of occupation at the site. The reconstruction of aspects of daily life did not take prime importance, but from the large amount of animal remains the chance was taken to study at least part of the food economy, as it was evidenced for different occupation phases. The main interpretation problem when doing so, remains, however, that the nature of the many different uses of the Aqaba site are poorly understood. The excavations and the available historical information show that the building must have been used during the medieval period as a military stronghold, a *khan*, a fortified caravanserai, and a trading post. It was indeed strategically well located, not only along the main Islamic pilgrimage routes, but also along trade routes leading from Egypt to Arabia and from Syria to the Red Sea. During the Ottoman period, the site must also have served many purposes before it became a military fort around 1840. During the First World War, it was captured by the Arabian Revolt, an event best known in the West by the contribution of T. E. Lawrence, also known as 'Lawrence of Arabia'.

The varying nature of the occupants of the site makes it difficult to go into details when reconstructing dietary patterns. Their status remains unclear (local elite, tradesmen, soldiers, pilgrims) just like any social differentiation within the site. Moreover, most animal remains were derived from secondary contexts of reworked consumption refuse and are thus hard to relate to functional areas within the building. The following analysis must thus be seen only as a broad diachronic approach to the archaeozoology of the site and its region.

Material and methods

A first analysis of the animal material was performed on site during a study session for two weeks in 2005 (by AE). Due to time constraints, it was decided to focus only on the material from the 2003 and 2005 campaigns. The contexts excavated during these field seasons have yielded the largest finds assemblages and are the best understood stratigraphically. The animal remains of the subsequent 2006 and 2007 campaigns (with exception of the shells) were exported to Belgium and are now stored at the RBINS (Royal Belgian Institute of Natural Sciences). The material from these campaigns has consequently been studied there (by MU & BDC). A second session of field work was then carried out in 2008 (by BDC). The animal remains that were collected during that campaign were studied on that occasion, as well as the shells from the 2006 and the 2007 campaigns. The entire assemblage of fish remains was examined at the RBINS (by WVN & WW).

The animal remains have been almost exclusively hand-collected. Although manual recovery of animal material inevitably implies the loss of small remains, it should be noted that the hand collecting during the excavations has been performed with considerable care. This is at least suggested by the often small dimensions of the finds and by the fragility of some of the material, e.g., the fish remains. Additionally, during the study session of 2005, soil samples (ca. 7 litres) were taken in order to check the extent of the loss of small material. After drying in the sun, these samples were sieved using a 0.5mm mesh width. The residue was then subdivided into a '>2mm' fraction and a '<2 mm' fraction. The former part of the residue included some small but sometimes identifiable animal remains, while in the latter fraction only bone and shell fragments were found that could not be identified. In both fractions, the finds density was rather low. Given the time constraints of the excavations, more extensive sieving efforts thus seemed hardly justifiable. However, an exception was made when a ceramic container was found that must have served as a sinkhole in one of the rooms' floors. The contents of this object have been recovered completely and have been dry sieved on a 0.5mm mesh. The same was done when a small water reservoir was found.

Table 31.1. Chronological units retained during analysis (as material from the Ayyubid and Mamluk periods could not always be differentiated easily, these periods have been grouped in the discussion of the finds).

Period	*Date (centuries AD)*	*No.finds*
Early Islamic (Ummayid, Abassid, Fatimed)	7th–12th	655
Ayyubid period	end 12th–13th	50
Mamluk period	end 13th–15th	10692
Ottoman period	16th–19th	14523
Modern period	beginning 20th	9727
Undated (also inc. (sub)recent & mixed material)		1155

For this study the fauna was lumped into five chronological units, dating from the Roman period up to the early 20th century (Table 31.1). Most of the finds (almost 69 %) belong to the Mamluk and Ottoman periods, and there is also a substantial portion of Modern material (26.4 %). The early phases (*i.e.* Early Islamic and Ayyubid/Mamluk) are very poorly represented (1.9 %). The scarce Roman material has not been retained in the tables and is only occasionally mentioned, as this material is not related to the castle. Some (sub) recent material was collected from the top, and this material has been lumped together with the undated or mixed material. Since the pottery study is not yet finished, dates in this report are preliminary.

Description of the faunal material

In the following paragraphs the different animal taxa identified will be briefly commented upon. Although small pieces and large blocks of coral have been found in many contexts, no attempt was made to identify these specimens to species level. These corals are fossils, and are not part of the archaeozoological assemblage. They are therefore omitted from this report.

Molluscs

The identification of the mollusc shells has produced a long list of taxa (Table 31.2). The shells have been identified using Sharabati (1984). The continuous changes

Table 31.2. Taxonomic representation of the molluscs and Echinodermata and the number of their remains (: almost all collected from the same context).*

	Early Isl	*Ayy/Mam*	*Maml*	*Ott*	*Mod*	*undated*
Cephalopoda						
Sepiidae	–	–	7	1	–	–
Polyplacophora						
Chitonidae						
(*Acantopleura haddoni*)	3	–	18	2	46*	–
Gastropoda						
Vermetidae						
Dendropoma maxima	–	–	–	–	1	–
Vermetus sp.	–	–	5	4	7	–
Fissurellidae						
Emarginula rugosa	–	–	–	1	–	–
Patellidae						
Patella sp.	–	–	4	7	2	–
Cellana eucosmia	–	–	1	–	1	–
Planaxiidae						
Planaxus sulcatus	–	–	–	–	1	–
Neritidae						
Nerita polita	–	–	2	–	15*	–
Nerita albicilla	2	–	27	16	30	1
Trochidae						
Trochus maculatus	–	–	–	2	–	–
Tectus dentatus	6	1	63	39	15	1
Tectus virgatus	–	–	–	2	5	1
Clanculus pharaonius	–	–	–	6	1	–
Turbinidae						
Turbo radiatus	8	–	42	43	11	1
operculum cf. *Turbo radiatus*	3	–	23	23	6	1
Cerithidae						
Cerithium sp.	–	–	1	–	1	–
Cerithium erythraeonense	2	–	11	7	4	–
Strombidae						
Strombus gibberulus albus	4	–	23	32	27	1
Strombus fasciatus	–	–	1	–	1	–
Strombus urceus	–	–	–	1	2	–
Strombus bulla	–	–	1	–	–	1
Lambis truncata sebae	3	1	28	31	11	1
Strombus tricornis	11*	–	24	54	–	11
Strombus decorus persicus	–	–	1	–	–	–
Strombus sp.	–	–	2	5	–	–
Cypraeidae						
Cypraea annulus	–	–	1	–	–	–
Cypraea pantherina	–	–	–	–	2	–
Cypraea caurica	–	–	7	6	1	–
Cypraea turdus	–	–	12*	10	–	–
Cypraea moneta	–	–	2	1	3	–
Cypraea nebrites	1	–	8	4	2	2
Cypraea sp.	1	–	18	11	9	2
Ovulidae						
Calpurnus verrucosus	–	–	–	–	1	–
Naticidae						
Polinices tumidus	1	–	4	3	–	–
Polinices melanostomus	1	–	3	–	3	–
Polinices sp.	–	–	1	–	–	–
Cassidae						
Casmaria ponderosa	–	–	–	2	2	–
Tonnidae						
Malea pomum	–	–	1	1	–	–
Bursidae						
Tutufa bubo	–	–	–	1	–	–
Cymatiidae						
Charonia tritonis	–	–	2	4	–	–
Distorsio anus	–	–	–	1	–	–
Cymatium aquatile	–	–	2	1	–	–
Cymatium rubeculum	–	–	–	–	1	–
Cymatium trilineatum	–	–	–	–	3	–
Muricidae						
Chicoreus virgineus	–	–	–	–	3	–
Chicoreus ramosus	–	–	1	–	–	–
Chicoreus sp.	–	–	5	–	–	–
Muricidae sp.	–	–	1	2	–	–

	Early Isl	*Ayy/Mam*	*Maml*	*Ott*	*Mod*	*undated*
Thaididae						
Morula granulata	–	–	–	–	1	–
Drupa morum morum	–	–	–	1	–	–
Thais savignyi	–	–	1	–	–	–
Nassa francolina	–	–	3	1	–	–
Rapana rapiformis	–	–	–	1	–	–
Buccinidae						
Engina mendicaria	–	–	–	2	–	–
Engina cf. *alveolata*	–	–	1	–	–	–
Nassariidae						
Nassarius cf. *arcularius*	–	–	1	–	–	–
Fasciolariidae						
Latirus polygonus	2	–	5	7	1	–
Pleuroplaca trapezium	1	–	6	5	–	–
Fusinus leptorhynchus	–	–	–	1	7	–
Fusus polygonoides	–	–	3	6	1	–
Colubrariidae						
Colubraria cf. *obscura*	–	–	–	1	–	–
Harpidae						
Harpa amouretta	1	–	1	1	2	–
Turbunellidae						
Vasum turbinellum	–	–	2	1	–	–
Costellariidae						
Vexillum amabilis	–	–	1	–	–	–
Vexillum sp.	–	–	–	1	–	–
Conidae						
Conus taeniatus	–	–	–	–	1	1
Conus cf. *fulgetrum*	–	–	–	3	–	–
Conus virgo	–	–	–	1	–	–
Conus arenatus	–	–	4	3	4	–
Conus tessulatus	–	–	1	2	1	–
Conus generalis maldivus	–	–	1	2	4	1
Conus vexillum	–	–	1	1	1	–
Conus namocanus	–	–	–	2	–	–
Conus striatellus	–	–	–	1	–	–
Conus 'pennaceus'	–	–	1	2	2	–
Conus textile	–	–	1	3	3	–
Conus sp.	5	–	25	37	31	–
Terebridae						
Terebra affinis	–	–	1	1	1	–
Terebra crenulata	–	–	–	3	1	–
Terebra dimidiata	–	–	–	1	–	–
Pyramidellamorpha						
Pyramidella 'sulcata'	–	–	3	1	–	–
Pyramidella sp.	–	–	1	–	–	–
Bivalvia						
Arcidae						
Barbatia tenella	1	–	–	–	–	–
Barbatia cf. *bistrigata*	–	–	–	2	–	–
Barbatia fusca	–	–	–	1	–	–
Barbatia helblingi	–	–	–	–	1	–
Barbatia sp.	–	–	–	1	–	–
Anadara antiquata	–	–	5	11	4	–
Anadara uropigimelana	–	–	4	11	2	–
Anadara sp.	–	–	2	1	–	–
Mytilidae						
Brachidontes variabilis	–	–	3	1	–	–
Modiolus auriculatus	–	–	–	–	5	–
Mytilidae sp.	1	–	1	2	–	–
Pteriidae						
Pinctada margaritifera	–	1	25	14	16	1
Pinctada radiata	–	–	1	4	–	–
Pinctada squamosa	–	–	1	–	–	–
Pinctada sp.	–	–	3	2	7	–
Ostreidae						
Dendrostea foilium	–	–	–	–	1	–
Gryphaeidae						
Hyotissa numisma	–	–	6	3	–	–
Limidae						
Lima paucicostata	–	–	–	1	–	–

	Early Isl	*Ayy/Mam*	*Maml*	*Ott*	*Mod*	*undated*
Chamidae						
Chama rupelli	1	–	7	5	17	1
Chama pacifica	–	–	–	5	–	–
Chama sp.	3	–	22	39	–	–
Spondylidae						
Spondylus aurantius	–	–	1	–	–	–
Spondylus cf. *marisrubri*	–	–	–	1	–	–
Spondylus hystrix	–	–	1	–	–	–
Spondylus sp.	–	–	1	5	–	–
Tridacnidae						
Tridacna maxima	2	6	75	121	11	1
Tridacna squamosa	–	1	39	67	2	–
Tridacna sp.	13	4	130	56	149	6
Cardiidae						
Trachycardium 'flavum'	–	–	3	1	–	–
Laevicardium papyraceum	–	–	1	2	–	–
Cardiidae sp.	–	–	2	1	–	–
Mesodesmatidae						
Atactodea glabrata	–	–	3	10	3	–
Glycymeridae						
Glycymeris petunculus	–	–	21	31	8	–
Lucinidae						
Ctena divirgens	–	–	1	–	–	–
Veneridae						
Circe calipyga	–	–	1	6	2	–
Circe corrugata	–	–	1	4	–	–
Venus verrucosa	–	–	1	1	–	–
Marcia hiantina	–	–	2	5	1	–
Veneridae sp.	–	–	–	2	–	–
Tellinidae						
cf. *Telinella staurella*	–	–	1	–	–	–
cf. *Angulus corbis*	–	–	–	1	–	–
Plicatulidae						
cf. *Plicatula plicata*	–	–	1	–	–	–
Ostreidae						
Saccostrea cuccullata	1	–	7*	–	41	–
Ostreidae sp.	–	–	–	3	8	1
mother of pearl fragments	–	–	15	7	1	–
Identified molluscs	77	15	802	841	557	35
Unidentified molluscs	36	1	225	143	118	20
Echinodermata						
Echinoidea	1	–	8	13	11	–
Clypeasteroida	–	–	–	1	–	–

in nomenclature, the determination of new species etc. (see e.g. Rusmore-Villaume 2008) implies that some of the scientific names used during the fieldwork might now be incorrect. Nevertheless, with the exception of the giant clams (see further), it has been decided to continue with the morphotypes and taxonomic names as given in Sharabati (1984).

The giant clams are the best represented among the mollusc shells. The two species that occur in the Gulf of Aqaba according to Sharabati (1984) are *Tridacna squamosa* and *T. crocea*. However, more recent research (e.g. Roa-Quiaoit 2005) reveals that the latter identification is incorrect. *T. crocea* (as defined by recent taxonomy) is not present in the Red Sea but *T. maxima* is. Both species are indeed present in the finds collections and are described as living on the surface of shallow coral reefs, although adult specimens of *T. maxima* are more common on the reef flat (<3m) than on the deeper fore-reef (9–15m) whereas adult *T. squamosa* are common on the deeper fore-reef. Over-exploitation of the giant clam as a food resource, together with habitat loss has depleted stocks (Roa-Quiaoit 2005). Many of the giant clam finds consist of heavily fragmented shells, and therefore, part of the remains has not been identified to species level, especially during the first field season when mainly Modern material was dealt with. Interestingly enough *T. maxima* and *T. squamosa* occur in about the same ratios during the Mamluk and the Ottoman period (approximately 65% and 35% respectively). This corresponds to their natural abundance in the Gulf of Aqaba (mean abundance of *T. maxima* being 0.5±0.3 ind/m^2 and of *T. squamosa* 0.3±0.2 ind/m^2; Roa-Quiaoit 2005, 29). Considering the fact that the species inhabit different areas within the reef – reef flat *versus* deeper fore-reef – this indicates that the harvest of giant clams was undertaken within the whole reef area.

Other frequently identified taxa belong to, among others, the top shells (*Tectus dentatus*), turbans (*Turbo radiatus*), conch shells (*Strombus tricornis*, *Strombus gibberulus albus*, *Lambis truncata sebae*), pearl oysters (*Pinctada* sp.), jewel boxes (*Chama* sp.) and bittersweet clams (*Glycymeris* sp.). In fact, the conch shells, the turbans and the top shells together constitute more than 50% of the gastropods. They can all be found on shallow reef flats or in intertidal water (Sharabati 1984) and are easily accessible. Therefore, the majority of the molluscs can most probably be considered as consumption refuse, as it was the case with the giant clams. This is also indicated by the fact that in several instances concentrations of shells have been found, e.g., *Nerita polita*, *Strombus tricornis* and *Saccostrea cucullata*. However, there were also shells with an eroded surface, probably due to transportation by the tides, showing that not all the molluscs were eaten. Still other molluscs cannot be considered as consumption refuse because they are poisonous, such as the cone shells. Their shells, just like the complete cowries that were found in the castle, may have been collected for decorative purposes, eventually as raw material for shell working. Mother of pearl – listed separately in Table 31.2 – was also used as raw material, as shown by some worked items, but the pearl oysters may also have been a source of food. The spectrum of the different shell taxa identified at the castle corresponds to that observed by AE in 2005 on the rocky shores of Jazirat Fara'un along the Egyptian Red Sea coast.

In a few contexts internal shells of squids (Sepiidae) have been found (n=8), and several assemblages contained plate fragments of mail shells (Chitonidae). In the latter case, there was even a concentration found – in a Modern context – which suggests that they were also eaten.

Echinodermata

Fragments of the body skeleton of a sea urchin (Echinoidea) were found in a few cases, possibly representing consumption refuse. The body skeletons of sea urchins are rather fragile, which might explain the fact that they have not been collected frequently. Many finds loci, however, contained solid, heavy sea urchin spines with a diameter of up to 7–8 mm. It is, however, believed that these spines may also form part of the geological record. No species identifications have been attempted. In one case the body of a sand dollar (Clypeasteroida) was collected.

Fish

More than 20 different fish taxa have been identified. With one exception, all taxa are marine and are mainly bony fish, but there are also some cartilaginous taxa such as sharks (Carcharinidae and Sphyrnidae) and rays (Rhynchobatidae). The taxonomic distribution of the fish remains for each period is given in Table 31.3 and the size reconstructions, based on direct comparison with modern fish of known body size, are given in Figure 31.3. The fish bones from the sieved sediment samples are listed separately (Table 31.4). The best represented groups of the hand-collected list are – in almost equal numbers – the groupers (Serranidae), the emperors (Lethrinidae) and the jacks (Carangidae), followed by the parrotfishes (Scaridae). These four groups constitute together more than 89 % of the identified fish remains which totally come up to 1513 specimens.

Most of the fish taxa listed in Table 31.3 can be classified as either coastal or reef-associated fish. Only a few taxa, i.e. the jacks, the thunnids (Scombridae) and the sharks and rays occur in the open sea. However, it is known that some jack and shark species can be found closer to the coast, especially in the case of juvenile animals (Desse & Desse-Berset 2003). Considering the main size of the jacks, i.e. 50cm and larger (Fig. 6), and the large size of the requiem sharks (Carcharinidae), i.e. more than 1.5m, it is assumed here that they were captured offshore.

From the data in Table 31.3 it is clear that the coastal fishes are much better represented than those from the open sea, respectively around 75% and 25% (Fig. 31.4). However, a steady increase in the proportion of the open sea fishes is visible from the Mamluk period into Modern times (Fig. 31.4). The proportion of the groupers (Serranidae) and parrotfishes (Scaridae) also varies through time (Fig. 31.5). While during the Mamluk period the parrotfishes are the most common, their numbers decrease in favour of the groupers during the Ottoman period. In Modern times both taxa become about equally important. The proportion of the emperors remains more or less constant through time.

When the standard lengths of the main four taxa are considered, diachronic shifts in size are noticed (Fig. 31.3). For example, smaller-sized groupers are represented during the Ottoman period compared to the Mamluk period; while the emperors attain larger sizes for the same period. In the case of the jacks, the size range is narrower during the Ottoman period with fishes occurring almost exclusively in the size intervals of 40–60cm and 60–80cm.

In an Ottoman context the presence of the catfish (*Clarias* sp.) is attested by a suboperculum of an animal that measured 70–80cm SL. This freshwater fish cannot be derived from local fishing, but must have been imported to the site. The closest possibility is the Jordan basin. On the other hand, considering the location of Aqaba along the trade route from Egypt to Arabia, it may well have been imported from the Nile. Long distance trade of catfish has been observed on numerous other archaeological sites (Arndt *et al.* 2003; Van Neer *et al.* 2004).

Reptiles

Fifteen specimens – almost all dated to the Mamluk period – could be identified as marine turtle (Cheloniidae). With the exception of a vertebra and a coracoid, these

Table 31.3. Taxonomic representation of the hand-collected fish and the number of their remains (: bones from the same individual, counted in total as one).*

	Early Isl	*Ayy/Maml*	*Maml*	*Ott*	*Mod*	*undated*
Chondrichtyes	–	–	1	–	–	–
Carcharinidae	–	–	–	2	1	–
Sphyrnidae						
Sphyrna sp.	–	–	–	–	2	–
Rhynchobatidae						
Rhynchobatos sp.	–	–	–	–	1	–
Scombridae						
Thunnus sp.	–	–	–	–	–	1
Chanidae						
Chanos chanos	12	–	–	–	–	–
Chanos sp	–	–	–	1	–	–
Clariidae						
Clarias sp.	–	–	–	1	–	–
Belonidae	–	–	1	1	–	–
Tylosaurus sp.	2	–	4*/1	–	–	–
Fistulariidae						
Fistularia sp.	–	–	1	3	2	–
Sphyraenidae						
Sphyraena sp.	–	–	1	1	–	1
Mugilidae	.	–	4	7	–	–
Serranidae	6	4	43	10*/246	6*/52	6
Rachycentridae						
Rachycentron canadum	–	–	–	2	–	–
Carangidae	9	–	58	179	32	9
Caranx sp.	–	–	7	3	20	12*
Carangoides sp.	–	–	8	–	27	1
Gnathanodon speciosus	–	–	1	–	–	–
Seriola sp.	–	–	2	–	–	–
Lutjanidae	–	–	3	28	5	–
Lethrinidae	3	1	122	152	74	6
Sparidae	–	–	4	18	2	–
Acanthopagrus sp.	–	–		2	–	–
Argyrops sp.	–	–	2	13	7	4*/2
Rhabdosargus sp.	–	–	1	–	–	–
Labridae	–	–	1	8	–	–
Scaridae	–	–	33	8	9	–
Cetoscarus sp.	–	–	8	5	1	–
Hipposcarus sp.	–	–	14	1	–	–
Scarus sp.	8	–	110	40	25	7
Coris aygula	–	–	3	–	–	–
Coris sp.	–	–	2	1	3*/2	–
Siganidae	–	–	1	–	–	–
Balistidae	–	–	1	5	6	1
Diodontidae						
Diodon sp.	–	–	–	1	–	–
Identified fish	40	5	433	729	270	36
Unidentified fish	10	2	120	338	84	6

Table 31.4 Taxonomic representation of the fish and the number of their remains, from the sieved samples.

	Early Islamic	*Mamluk*	*Modern*	*Mamluk sinkhole (AQ 2005 Y11)*
mesh width	2mm	2mm	2mm	5mm
Mugilidae	–	4	–	1
Serranidae	–	1	–	–
Carangidae	–	–	–	13
Lethrinidae	–	–	–	46
Sparidae	–	–	–	7
Scaridae			–	
Scarus sp.	2	–	–	4
Siganidae	–	–	–	8
Siganus sp.	–	5	–	15
Identified fish	2	10	0	94
Unident. fish	15	63	6	753

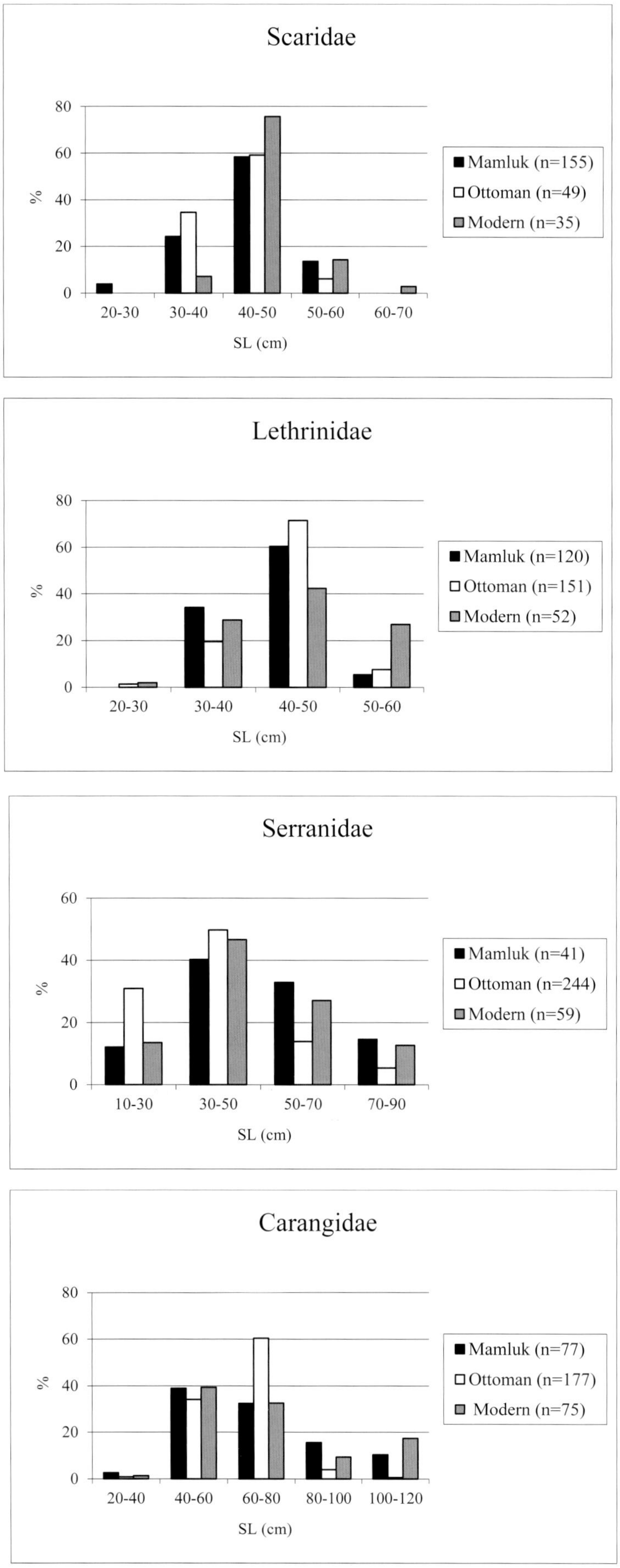

Fig. 31.3. Distribution of the standard length (SL) of the main fish taxa.

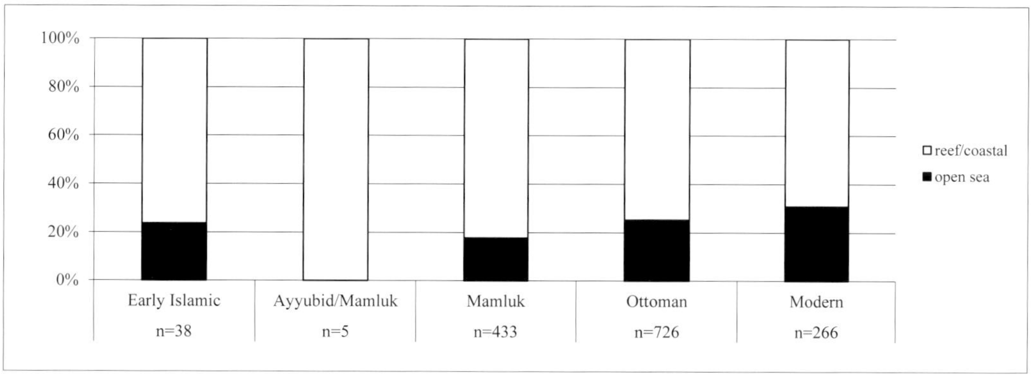

Fig. 31.4. Relative abundance of the fish from the open sea versus coastal fish and fish from the reefs; based on the hand collected material.

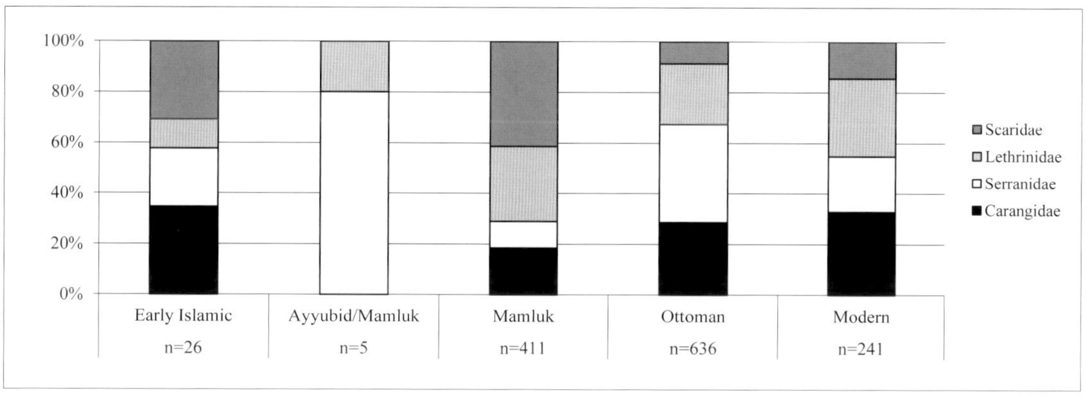

Fig. 31.5. Relative abundance of the four best represented fish families; based on the hand collected material.

are exclusively fragments from the carapace. Due to the fragmentary nature of these finds, species identification was impossible. Although there are four possible candidates (i.e. the green turtle (*Chelonia mydas*), the hawksbill turtle (*Eretmochelys imbricata*), the loggerhead turtle (*Caretta caretta*) and the olive ridley turtle (*Lepidochelys olivacea*; Marquez 1984), the green turtle and the hawksbill turtle are now the most common turtle species in the Red Sea (Frazier & Salas 1984)). Considering the low number of their remains, turtles probably only occasionally got stuck in fishing nets and were then brought to land, where they may have been consumed. Alternatively, the remains may come from animals that were stranded, or killed when they came ashore to lay eggs. In any case, the scarcity of the turtle remains does not point towards an organised fishing strategy for these species.

Birds

The relative abundance of bird bones compared to the total faunal assemblage – molluscs excluded – is low. More than half of the bird bones belong to chicken but, still, a rather large number of species could be identified (Table 31.5). This variety can be explained by the fact that Jordan lies on one of the world's major bird migration routes, from Africa to Eurasia. Also, the geographic location of the castle implies that both sea- or coastal birds and inland birds can be expected.

Egg shells of ostrich (*Struthio camelus*) have been collected three times, twice from a Mamluk and once from an Ottoman context. No bone fragments of this species have however been found. The great black cormorant (*Phalacrocorax carbo*) is represented by several specimens. Only two bones could be measured, an ulna (Bp 12.8mm; Dp 11.0mm) and a tarsometatarsus (GL 61.4mm; Bp 14.0mm; SC 7.0mm; Bd 15.5mm), both from (sub)recent contexts. The latter specimen had cut marks on its dorsal side. Another specimen with cut marks is a tibiotarsus from a pelican that was dated to the Modern period. Only the white pelican (*Pelecanus onocrotalus*) occurs in the region (Andrews 1995), although some field guides (e.g. Heinzel *et al.* 1987) suggest that the Gulf of Aqaba falls within the geographical distribution of both the white pelican and the Dalmatian pelican (*P. crispus*). Since there is size overlap between the bones of both species (Lorch 1992), the Aqaba finds could not be identified any further. The great bittern (*Botaurus stellaris*) is represented by an incomplete skeleton which was dated to the Mamluk period. Only the coracoid could be measured (GL 66.2mm; Lm 62.7mm; Bb 22.9mm; BF 18.8mm). Six bones were identified as stork (*Ciconia* sp.). Two species occur in Jordan, the white stork (*C. ciconia*) and the black stork (*C. nigra*). White stork is an abundant migrant through Jordan, especially in autumn, while the black stork is rather uncommon (Andrews 1995). Since there is size overlap between both species, morphological characteristics should be used to distinguish bone material from both species (Gruber 1990). One specimen, a coracoid dated to the Modern period, most closely resembles that of black stork (cf. Gruber 1990, Abb. 28:13). The other finds are too fragmentary to be identified to species level. Cut marks were observed on a proximal end of a coracoid dated to the Mamluk period.

Despite the large number of ducks and geese that occur in Jordan, the Anatidae family is poorly represented. Seven bones have been identified as greylag goose (*Anser anser*); five of them were collected from the same Early Islamic context and are probably from one individual. In terms of morphology and size, the duck remains fall within the variation of common teal (*Anas crecca*), mallard (*Anas platyrhynchos*) and pintail (*Anas acuta*), but were not identified with certainty. A humerus (GL 60.8mm; Bp 13.4mm; SC 4.7mm; Bd 9.4mm) was attributed to the common teal, a carpometacarpus (GL 56.1mm; Bp 12.9mm; Bd 7.3mm) to the pintail, and a skull and a sternum to the mallard.

Remains of a large vulture were tentatively identified as lammergeier (*Gypaetus barbatus*). Some elements, most probably from the same individual, could be measured, i.e. an ulna (Bp 25.7mm; Dip 27.1mm; SC 11.5mm; Bd 20.8mm), a radius (GL ±235mm; SC 11.3mm; Dd 15.9mm) and a carpometacarpus (GL 112.6mm; Bp 27.8mm; Bd 21.4mm). One of the two specimens that were dated to the Mamluk period, a carpometacarpus, showed cut marks.

The carcass of an Egyptian vulture (*Neophron pernocteropus*) was collected from a Mamluk context. Within the same trench, a skull was found that probably belongs to the same individual. Several elements could be measured, i.e. a coracoid (GL ±54mm; Lm 48.6mm), two femurs (GL 69.0 mm/-; Bp 19.3mm/18.6mm; Dp 11.6mm/11.7mm; SC 8.4mm/-; Bd 18.9mm/-; Dd 13.0mm/-), a tibiotarsus (SC 7.3mm/-; Bd 15.2mm/14.7mm; Dd 10.5mm/10.3mm) and a tarsometatarsus (Bp 15.7mm).

Several raptor bones comparable with *Buteo* sp. were found. Three species of buzzard occur in Jordan, i.e. common buzzard (*Buteo buteo*), rough-legged buzzard (*B. lagopus*) and long-legged buzzard (*B. rufinus*). The common buzzard could be identified with certainty in three instances. Measurements taken on these bones are from a coracoid (GL 41.0mm; Lm 36.7mm; Bb 19.6mm; BF 15.9mm) and a tibiotarsus (SC 5.4mm; Bd 11.2mm). Two tarsometatarsals (GL 71.6mm/72.9mm; Bp 12.7mm/12.4mm; SC 5.4mm/5.6mm; Bd 12.9mm/13.0mm) exclude identification as long-legged buzzard and were hence labelled *Buteo buteo/lagopus*. Measurements of *Buteo* sp. were taken on a humerus (GL 100.0mm; Bp ±20.5mm; SC 7.7mm; Bd 18.0mm) and a carpometacarpus (GL 59.0mm). One accipitrid bone could not be identified to genus level; this specimen was of a rather large size.

The chukar (*Alectoris chukar*) and the sand partridge (*Ammoperdix heyi*) – both resident and fairly common in Jordan (Andrews 1995) – were found in contexts dated to the Mamluk and Ottoman periods. Their remains are (almost) completely absent in the Modern period. Measurements of chukar and sand partridge are given in the appendix.

Unlike the moorhen (*Gallinula chloropus*) and the coot (*Fulica atra*), which are winter visitors, the corncrake (*Crex crex*) is a spring migrant (Andrews 1995). A skeleton of the latter was found in an Ottoman context, while several other isolated finds have been found in contexts of the Mamluk, Ottoman and Modern periods. Measurements were taken on the bones of the moorhen, i.e. on a humerus (GL 46.9mm; Bp 9.0mm; SC 3.0mm; Bd 6.0mm) and an ulna (GL 42.5mm; Bp 4.6mm; Dip 5.7mm; SC 2.1mm; Did 4.6mm), and in the case of coot, on a femur (GL 54.4mm; Bp 10.1mm; Bp 7.2mm; Dip 4.4mm; Bd 9.8mm; Dd 8.0mm). Measurements of the corncrake are given in the appendix.

More than 10 different gull species (*Larus* sp.) can be seen today at the Gulf of Aqaba. Considering the size overlap of some of these species and the lack of adequate reference material, no attempt was made to identify the gull remains from the Aqaba Castle to species level. In size, however, most were comparable with the lesser black-backed gull (*Larus fuscus*). Large flocks of this gull are frequently seen on spring passage – April/June – while fewer birds pass through in autumn, from late September to mid October (Andrews 1995). Measurements were taken on some post-cranial elements and included coracoid (Lm 48.7mm), scapula (Dic 11.5mm), humerus (SC 7.3mm; Bd 17.4mm), ulna (Bp 12.9mm/12.0mm) and carpometacarpus (GL 73mm; Bp 14.7mm). The skeletal remains of a carcass were found in an Ottoman context, while one specimen, a distal humerus, showed cut marks on the distal articulation surface. The latter was also dated to the Ottoman period.

In terms of size almost all pigeon remains could be attributed to rock or stock pigeon (*Columba livia/oenas*), while the presence of the larger-sized woodpigeon (*Columba palumbus*) could be excluded. Some bones were identified as dove (*Streptopelia* sp.). The skeletal remains dated to the Ottoman period, should most probably be classified as turtle dove (*S. turtur*). Indeed, other species from this genus, i.e. the collared dove (*S. decaocta*) and the palm dove (*S. senegalensis*) have only recently colonised Jordan; the same is true for the Namaqua dove (*Oena capensis*) (Andrews 1995). Measurements are given in the appendix.

The barn owl (*Tyto alba*) is represented by a tarsometatarsus (GL 61.8mm; Bp 9.3mm; SC 3.9mm; Bd 10.3mm); the specimen has been dated to the Ottoman period. Although breeding has been reported in the north-west of the country, birds seen elsewhere are more likely to be passage migrants (Andrews 1995). Four bones of eagle owl (*Bubo bubo*) have been identified. An ulna, dated to the Ottoman period, could be measured (Bp 14.1mm; Dip 16.2mm; SC 5.7mm) as well as a carpometacarpus

*Table 31.5. Taxonomic representation of the bird and the number of their remains (+: egg shell; *: bones from the same individual, counted in total as one).*

	Early Isl	*Ayy/Mam*	*Maml*	*Ott*	*Mod*	*undated*
Wild						
Struthionidae						
Struthio camelus	–	–	+	+	–	–
Phalacrocaricidae						
Phalocrocorax carbo	–	–	3	1	2	2
Pelecanidae						
Pelecanus sp.	–	–	–	–	1	–
Ardeidae						
Botaurus stellaris	–	–	11*	–	–	–
Ciconidae						
Ciconia sp.	–	–	3	2	1	–
Anatidae						
Anser anser	5	–	2	–	–	–
Anas cf. *crecca*	–	–	–	–	1	–
cf. *Anas plathyrhynchos*	–	–	–	1	1	–
Anas cf. *acuta*	–	–	–	–	1	–
Accipitridae						
cf. *Milvus migrans*	–	–	1	–	–	–
cf. *Gypaetus barbatus*	–	–	2	5*	–	–
Neophron pernocteropus	–	–	15*/1	–	–	–
Accipiter gentilis	–	–	v	1	–	–
Buteo buteo	–	–	1	–	1	1
Buteo buteo/lagopus	–	–	–	1	–	–
Buteo sp.	–	–	–	3	–	–
Accipitridae sp.	–	–	–	–	1	–
Galliformes						
Alectoris chukar	–	–	4	12	1	–
Ammoperdix heyi	–	–	1	3	–	–
Rallidae						
Crex crex	–	–	4	7*/5	3	–
Gallinula chloropus	–	–	–	2	–	–
Fulica atra	–	–	–	1	–	1
Laridae						
Larus sp.	–	–	1	7*/2	1	–
Columbidae						
Columba livia/oenas	1	–	9	7	2	–
Streptopelia sp.	–	–	–	4*	1	–
Strigidae						
Tyto alba	–	–	–	1	–	–
Bubo bubo	–	–	1	2	1	–
Corvidae						
Corvus corax	1	2	8	23	6	1
Domestic						
Gallus gallus f. domestica	1	–	33	99	68	12
Meleagris gallopavo f. domestica	–	–	–	–	–	1
Identified bird	8	2	76	169	92	18
Unidentified bird	1	–	21	19	16	5

from the Modern period (GL 70.8mm; Bp 15.5mm; Did 11.9mm). A tibiotarsus from the Mamluk period showed a chop mark at its distal end.

The raven (*Corvus corax*) is the best represented wild bird at the Castle of Aqaba. Although several species of *Corvus* are described for Jordan, ravens can easily be recognised by their size (Tomek & Bocheński 2000). When the complete long bones of Aqaba are considered, they all fall well above the size range of the rook (*C. frugilegus*) and the hooded crow (*C. corone*) (see appendix). Incomplete bones also seemed to belong to ravens. The measurements of the Aqaba specimens are, however, smaller than those of the European raven (for measurements see Tomek & Bocheński 2000). The smaller fan-tailed raven (*C. rhipidurus*) and the brown-necked raven (*C. ruficollis*), which are common today in the region of Aqaba, seem to be

Table 31.6. Taxonomic representation of the mammals and the number of their remains (: bones from the same individual; counted in total as one).*

	Early Isl	Ayy/Mam	Maml	Ott	Mod	undated
Wild						
Lepus capensis	–	–	8	10	9	–
Rattus rattus	–	–	1	–	–	–
Rodentia	–	–	–	1	–	–
Procavia capensis	–	–	5*	3*/1	–	–
Hyaena hyaena	–	–	4	11	2	–
Gazella dorcas	–	–	–	–	–	1
Gazella subgutturosa	–	–	–	1	–	–
Gazella sp.	1	–	33	96	18	17
Domestic						
Felis silvestris f. catus	–	–	5	4*/2*/2*/37	6*/19	2
Canis lupus f. familiaris	–	–	5	10	10	–
Equus africanus f. asinus	–	–	10	4*/83	4	2
Equus ferus f. caballus	1	–	1	8	2	–
Equidae	–	–	6	14	5	–
Camelidae	34	9	875	1276	368	82
Ovis ammon f aries	7	1	334	584	293	22
Capra aegagrus f. hircus	13	2	272	892	433	21
sheep/goat	153	13	3202	6668	4236	484
Bos primigenius f. taurus	4	–	46	29	10	1
Identified mammal	213	25	4802	9642	5410	632
Unidentified mammal	269	–	4205	2727	3169	403

absent among the bird remains of the castle. One coracoid, dated to the Ottoman period, has cut marks.

Only two species of domestic birds could be identified at the site, i.e. the chicken (*Gallus gallus* f. domestica) and the turkey (*Meleagris gallopavo* f. domestica). The latter was represented by one bone only, a distal tibiotarsus, which was found in a (sub)recent context. Chicken bones make up 51% of the identified bird bone assemblage. Measurements display a large variation in size (see appendix) and it is quite obvious, for example in the case of the ulna and the femur, that the specimens represent chickens of any size between very small to very large. Not only sexual dimorphism but also the existence of chickens belonging to different types or breeds might be responsible for this (De Cupere *et al.* 2005). Both adult and young individuals were consumed, although the latter group was less well represented.

Wild mammals

The number of wild mammal species is very low (Table 31.6). Remains of hare (*Lepus capensis*) have been collected from contexts dated to the Mamluk period into Modern times, albeit in small numbers only. The black rat (*Rattus rattus*) is represented by a complete skeleton in one single context, dated to the Mamluk period. The black rat is not a burrower and must therefore not be considered as intrusive (Ervynck 2002), but as contemporary to the Mamluk occupation. The mandible of this animal showed a fracture which may explain why the incisors were not able to wear off properly. As a consequence the upper incisors kept growing and formed a complete circle, with the left incisor penetrating the palatal bone (Fig. 31.6). The remains of rock hyrax (*Procavia capensis*) included an incomplete skeleton of a subadult animal, dated to the Mamluk period while for the Ottoman contexts an incomplete skeleton of a juvenile animal and an isolated humerus from a second subadult individual are evidenced. The latter specimen showed fine cut marks, just above the distal articulation, indicating that hyrax indeed was consumed. However, the presence of partial skeletons suggests that remains may also be included from animals that died naturally on the spot.

Most of the remains of the striped hyena (*Hyaena hyaena*) originate from Ottoman contexts; some measurements could be taken (see appendix). Interestingly enough, more than half of the specimens showed cut and/or chop marks. These elements included a scapula, a humerus and a radius from the Mamluk period, a mandible, an axis, two humeri and a pelvis from the Ottoman period as well as a humerus from the Modern period. They could be attributed to both the left and the right side of the body. Cut and chop traces on the

Fig. 31.6. Skull of black rat, with unworn upper incisors.

bone remains of the striped hyaena were also observed at other sites, e.g. Julfar (8th–17th century) in Ras al-Khaimah (UAE) (Desse & Desse-Berset 2000) and medieval Siraf (4th–16th century) in Iran (von den Driesch & Dockner 2002). In the case of Julfar, the hyaena skulls had fine cut marks on the condyles. The consumption of striped hyaena has been reported at the beginning of the 20th century among the Bedouins of Arabia, although more in terms of a medicinal treatment than being part of a regular meal (Smith 1907, 231). The use of the (striped) hyaena for medical purposes is also described by Ibn al-Baytar (13th century, translated by L. Leclerc 1987, 388–398) and has been listed as '*halal*' by the 14th century Egyptian writer ad-Damiri in his encyclopaedic work '*Hayat el-hayawaan*' or 'The life of animals' (Smith 1907). The consumption of hyaena represents an exception to the rule that predatory animals are not to be eaten. Indeed, hyaenas are considered to have weakly developed canines (!) and do not dominate other animals by mean of these teeth. In addition, hyaenas do not attack man, although they were known to be a danger for the domestic livestock. As such their meat – as well as that of fox, weasel, marten and hyrax – is allowed to be consumed (Benkheira 2000, 110–111). It is sometimes mentioned that only the right side of the animal was used for consumption (M. Alzawahra, pers. comm.) but the Aqaba finds do not corroborate this.

Gazelles are best represented among the wild mammals. On the basis of their geographical distribution, three species of *Gazella* sp. can be expected in Jordan, the dorcas gazelle (*G. dorcas*), the mountain gazelle (*G. gazella*) and the goitred gazelle (*G. subgutturosa*) (Harrison & Bates 1991). The mountain gazelle is a moderately large gazelle; it has a very slender and graceful stature with relatively long limbs. The dorcas gazelle is a small species, of less graceful build than the mountain gazelle and with shorter legs. The goitred gazelle is a large gazelle, heavier built than the mountain gazelle and the dorcas gazelle. At Aqaba, the presence of goitred gazelle is evidenced by a horn-core, while a small horn has tentatively been identified as belonging to *G. dorcas*. As E. Vila (1991) and others stated repeatedly, postcranial elements are impossible to separate. The majority of the gazelle remains are from similarly sized animals, but there are also indications for smaller individuals, e.g. in case of the femur (see appendix). Comparison of the Aqaba data with the measurements of *G. gazella* (Horwitz *et al.* 1990) shows that the finds mostly fall within the size range of the mountain gazelle. The smaller individuals are possibly from the smaller dorcas gazelle.

Fig. 31.7. The 'greyhound of Akkaba' (Smith 1840).

Domestic mammals

Remains of cat (*Felis silvestris* f. catus) have been collected more frequently than those of dog (*Canis lupus* f. familiaris). In the case of *Felis*, articulating elements could be found in several instances; one bone, a humerus dated to the Modern period, displayed a healed fracture. Measurements of both cat and dog elements are given in the appendix. The dog remains may have belonged to the so-called 'greyhound of Akkaba' (Fig. 31.7) or, more generally, the 'Bedouin greyhound', a large and fierce dog, resembling the Saluki (Richardson 1857, 57). The very few measurements that could be taken on the Aqaba dog remains are comparable to those of the Saluki Luman (see Clutton-Brock 1989).

Donkey (*Equus africanus* f. asinus) is much better represented than horse (*Equus ferus* f. caballus). It is of

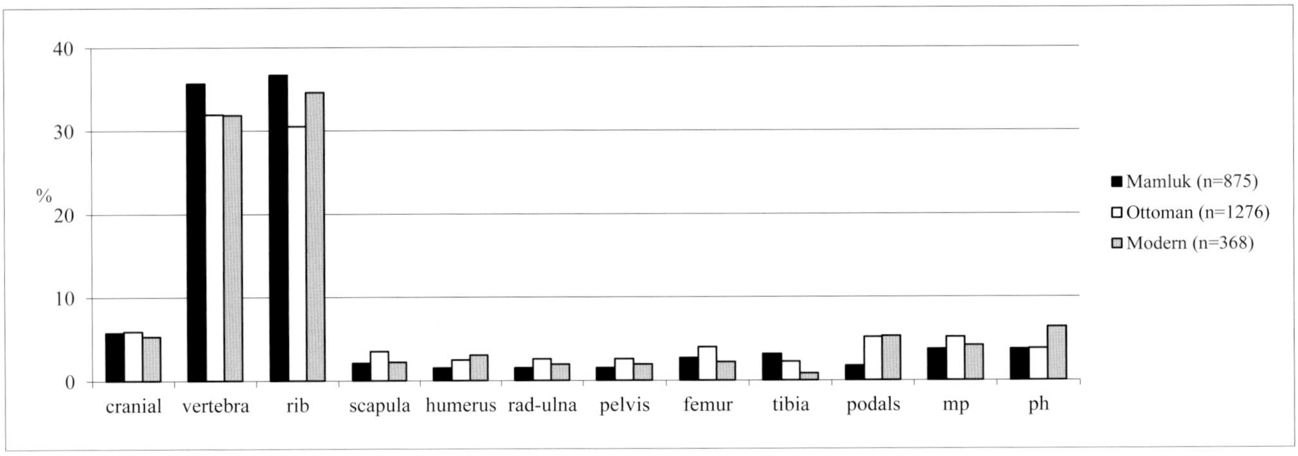

Fig. 31.8. Skeletal distribution of camelids through time.

course not excluded that remains of mule and/or hinny are present in the equid assemblage but supportive evidence was completely missing. Three complete metatarsals – all dated to the Ottoman period – allowed the calculation of the withers height for donkey, giving 1.07m, 1.15m and 1.16m respectively (using the multiplication factor of Kiesewalter, cited in von den Driesch & Boessneck 1974). No complete long-bones of horse were present. While several of these equid bones were burnt, traces of cutting or chopping were not observed, as in the bones of cat and dog.

Camelids are very well represented at the site of Aqaba (n=2644). Their remains were – unlike those of the other mammals – brittle and not well preserved. Complete bones or measurable parts of bones mainly included compact bones such as carpals, tarsals, proximal and distal ends of metapodials, and phalanges. Nevertheless, vertebrae and ribs were also very well represented. Indeed, when looking to the skeletal element distribution, the long-bones of the fore and hind limb were under-represented (Fig. 31.8). Chop and cut marks were regularly observed both on adult and subadult specimens. Articulating bones from the wrist and ankle joint, in combination with a chopped proximal metapodial, were found in several contexts. Some vertebrae and ribs probably also originated from the same individual. Traces of fire have been frequently noted; in these cases the bones were only slightly charred (showing a brown to black colour). Although suggested by the archaeologists, there was no true evidence for the deposition of camelid carcasses at the site. All camelid remains were interpreted as consumption refuse.

The identification of the camel remains to species – Bactrian camel (*Camelus ferus* f. bactrianus) or dromedary (*Camelus thomasi* f. dromedarius) – was carried out using the osteomorphological differences described in literature (Wapnish 1984; Steiger 1990; Studer & Schneider 2008). These largely corroborate an identification of this material as dromedary, as expected when the geographical distribution of the two species is considered (Mason 1984, 112). However, one specimen – a proximal metacarpus, dated to the Ottoman period – exhibited more closely the characteristics of Bactrian camel, while another find – a first phalanx – that showed intermediate characteristics of camel and dromedary was interpreted as a hybrid. Some skeletal elements, on the other hand, clearly showed all morphological characteristics of dromedary, but were very large, falling outside the 'normal' size range of dromedary. These included, among others, two proximal radii, a complete metacarpus, three 1st phalanges and three 2nd phalanges, all dated to the Ottoman period, as well as several carpals and tarsals. It is believed that these large camelid bones may derive from castrated dromedaries, rather than from hybrids. Indeed, males can have a bad temper and are easier to manage when castrated (Wilson 1998, 76–77). Such large animals, morphologically similar to dromedary, have also been identified at Nabatean Petra (Studer & Schneider 2008). The presence of the wild form, which had already become extinct some millennia before, can be excluded (Peters 2001). Considering the location of the Castle of Aqaba along major trade routes, it is possible that Bactrian camels may have come as far south as Aqaba with caravans. The possible presence of camel hybrids was already attested for Jordan at Early Islamic Pella (Köhler-Rollefson 1989); other archaeological sites include Troy (Turkey) and Mleiha (UAE) (Uerpmann 1999). Hybrids are much larger and stronger than either parent, and are therefore much appreciated as a draft animal (Potts 2004). A full discussion of the camelid finds, together with a more detailed description and an analysis of the measurements taken, is still in preparation (by BDC) and will be published elsewhere.

The mammal bone assemblages are without exception dominated by the remains of sheep (*Ovis ammon* f. aries) and goat (*Capra aegagrus* f. hircus), which form more than 86% of the identified mammal bone assemblage (see Fig. 31.7). The majority of the sheep must have been of a hornless breed; only a few isolated horn cores were encountered and, in addition, several skull fragments of sheep without horn-cores have been observed. In contrast,

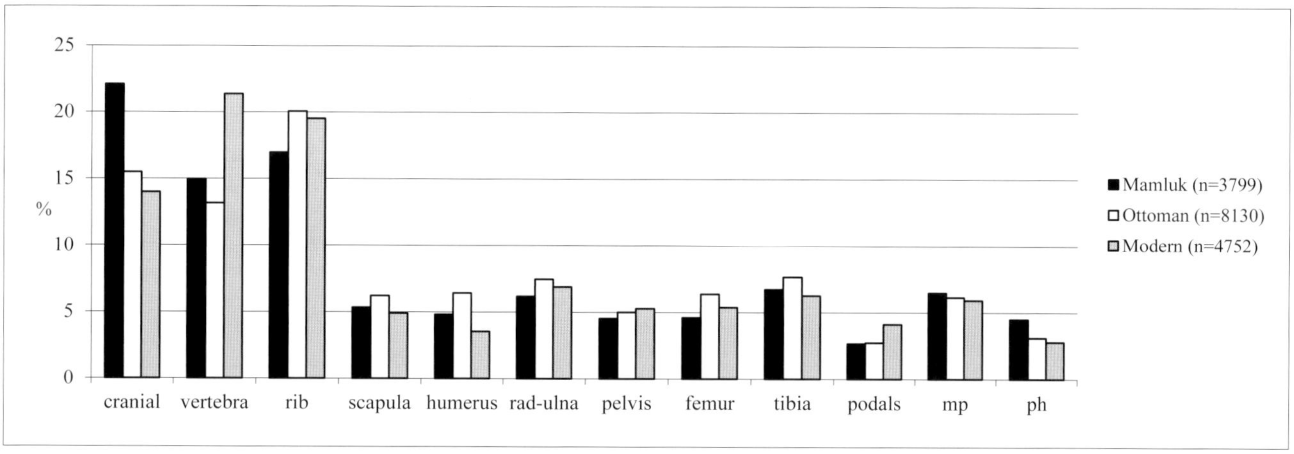

Fig. 31.9. Skeletal distribution of sheep/goat through time.

from goats, many well preserved horn-cores and cranial fragments with the base of a horn-core were found. All long-bones from the skeleton are present in about equal frequencies and their relative abundance remains constant through time (Fig. 31.9). The majority of the sheep/goat remains are heavily fragmented and show many cut and/or chop marks, indicating their status as consumption refuse. As is the case for the camelid remains, burnt bones are not uncommon, and most frequently they are only slightly charred and not cremated (which would result in a white colour). A large proportion of the articular ends of the long bones were unfused, indicating the young age of the animals consumed (Fig. 31.10). Age at slaughter of the sheep and goat has also been established using Grant (1982) and varies widely, from very young animals to very old individuals. Slaughter ages and survival curves for the different periods are given in Figure 31.11. Since there is only limited material from the Early Islamic period, this period has been omitted here. When sheep and goat are considered together, the Ottoman and the Modern periods show a similar distribution of the age classes, with a preponderance of very young and old animals. That intermediate age classes are less well represented is also shown in the survival curves (Fig. 31.12) by a plateau around mandibular wear classes (MWS) 15–30. The group of young animals (up to a MWS of 15) corresponds with the moment that the first molar and the second molar erupt, or an age of up to 1 year (Silver 1963, data for improved breeds). The group of older animals (from approximately MWS 35 onwards) corresponds with animals in which the third molar is moderately to heavily worn; or an age of over 4 years. Clearly, animals of intermediate age, between 1 and 4 years were thus less frequently slaughtered.

This model is characteristic for a herd from which some of the young (probably mostly males) are culled for their meat, and some are kept until older ages. Once mature, the latter portion (probably mostly females) of the herd might provide milk and wool, and sustain the size of the flock. Only when they are no longer suitable for one of these purposes, are they slaughtered. The different age classes are more evenly distributed during the Mamluk period. As a consequence the plateau in the survival curve is not visible, and the curve represents a more or less straight line (Fig. 31.12). This may be the result of differences in herd management compared to the Ottoman and Modern periods. Indeed, as sheep and goats may have been kept for different purposes, it is highly possible that slaughter ages differ according to the species considered. Consequently, all the results that present combined data for both sheep and goats have to be considered with caution.

Given the diachronic differences in the slaughtering patterns observed, an attempt was made to evaluate the possible impact of shifting species composition within the 'sheep/goat' group, by identifying the mandibles to species. For this purpose, the criteria described by Payne (1985) for the morphology of the Pd4 were used, together with those defined by Halstead *et al.* (2002), i.e. taking into account the position of the foramen with respect to the P2–4 and the presence or absence of a hollow immediately posterior of the M3. The absence of the foramen, which is described by Halstead *et al.* (2002) as a characteristic for goat, has not been applied here. Indeed, the material of Aqaba comprised many mandibles that, based on the Pd4, were identified as sheep but were not showing a foramen. Identification as sheep was maintained in these cases. Mandibular wear stages of the individual species were then again evaluated for the different periods (Fig. 31.13). Only a few specimens were available for the Modern period; they have therefore been omitted here. The mandibular wear stages in goat show a more or less similar distribution for both periods, i.e. a dominance of a group of young animals, up to MWS 15, and of a group of old animals. The sheep, on the contrary show a distribution that changes from the Mamluk to the Ottoman period. During the Mamluk period two large peaks can be distinguished around MWS 14 and 25. These correspond with the eruption of the M2 and M3, at an age of approximately 1 and 2 years respectively (cf. Silver 1963, data for improved breeds). Clearly, only a

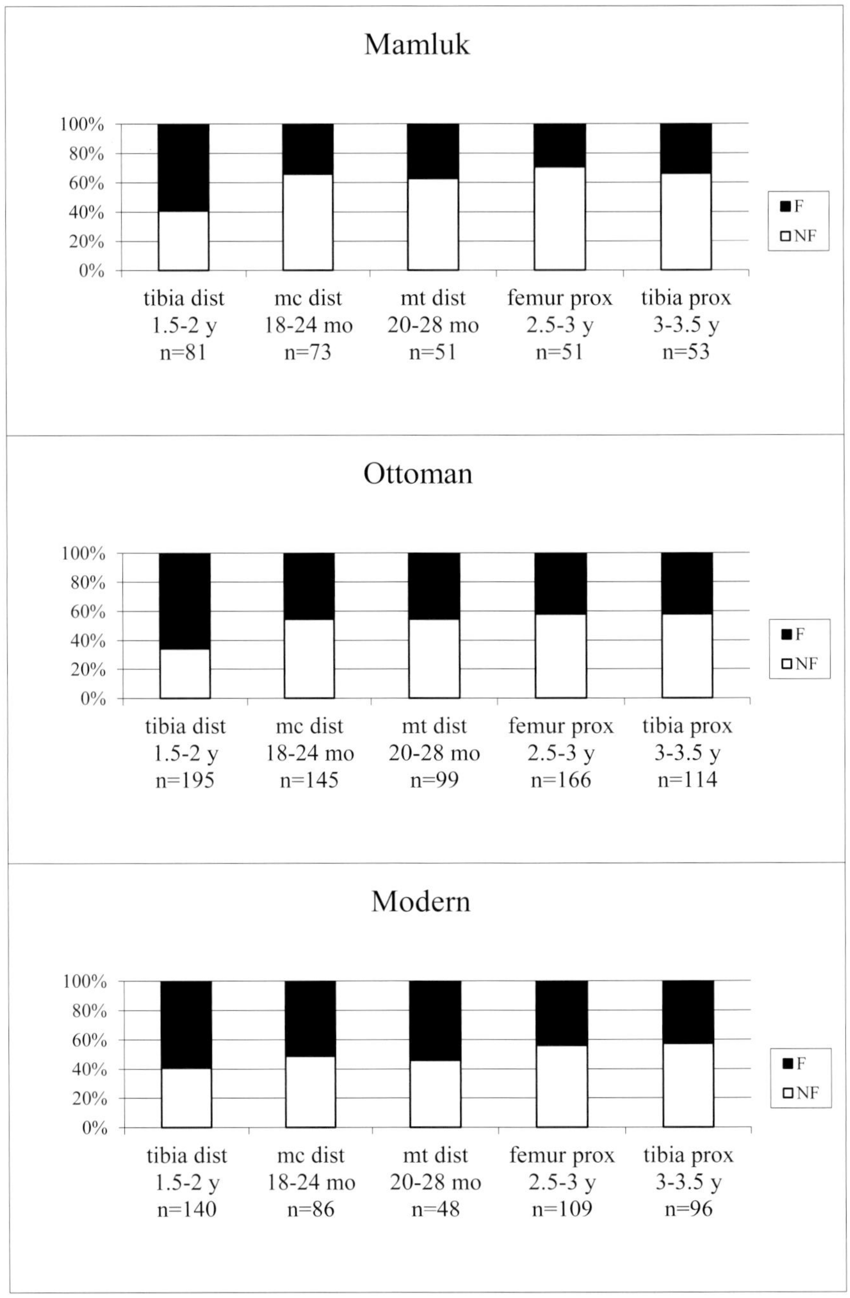

Fig. 31.10. Fusion stages of long bones of sheep/goat (F: fused; NF: unfused).

smaller portion of the sheep reached maturity. The Ottoman period, however, shows a large peak around MWS 3–8 (M1 erupting to slightly worn), corresponding to an age of around 6 months. The peaks of around 1 and 2 years, as seen for the Mamluk period, are lacking, and more animals were kept into old age. Considering the fact that the Pd4 was used for species identification, it is likely that the portion of the young animals is somewhat overestimated but the differences in slaughtering patterns observed must certainly be related to different management strategies for the sheep and the goat herds (Payne 1973). The pattern observed for the goats at Aqaba is typically what can be expected when milk production is aimed for. In the case of wool production, both sexes will be kept into adulthood (Payne 1973). The slaughtering pattern of sheep at the castle shows a clear preponderance of young animals, which corresponds better with the model of meat production. Indeed, in the latter case, young animals of both sexes will be slaughtered (Payne 1973). While subadult animals, which render more meat, were preferred in the Mamluk period, an emphasis on lambs is observed in the Ottoman period. It is possible that this pattern is a reflection of changing management strategies and culinary practices.

The ratio of sheep and goat also varies through time (Fig. 31.14). While sheep are better represented than goats during the Mamluk period, the opposite is true for the

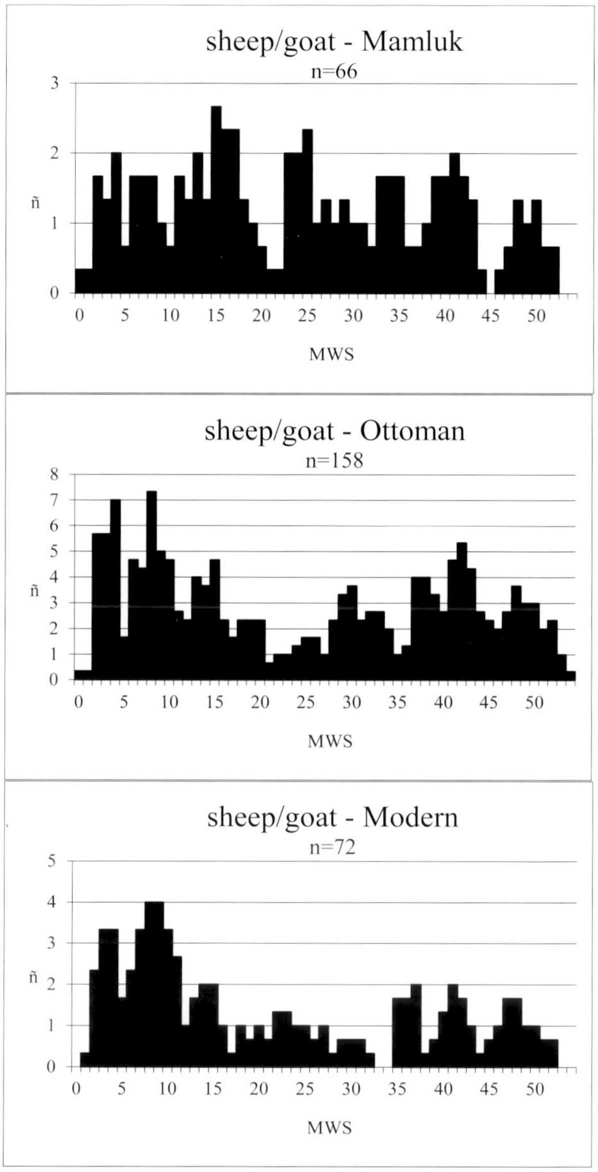

Fig. 31.11. Mandibular wear stages (MWS) of sheep/goat, according to Grant (1982); ñ: running mean.

Ottoman and the Modern period. The early periods also show a preponderance of goat, but the numbers involved are very low and the ratios are therefore not regarded as reliable. Considering the fact that the majority of the sheep must have been of a hornless breed, it seemed more correct for the calculation of the ratio not to include the horn cores. Mandibles, too, have been excluded, as the criteria to distinguish both species are not always unambiguous and mostly applicable on young specimens, i.e. with a Pd4 (see above). Therefore, only postcranial elements have been counted.

A summary of selected measurements is given in the appendix. Only the most abundant and complete bones are included, i.e. radius and metapodials. Height at the withers could be established from the total lengths of these long-bones, ranging between 44cm and 74cm in the case of the sheep, and between 51cm and 70cm in the case of the goats. One extremely large metacarpal of goat, dated to the Mamluk period and indicating a height at the withers of 85cm, may well be from ibex (*Capra nubiana*), which still occurred in the region at the beginning of the 20th century (Harrison & Bates 1991, 182). Species identification, however, remains uncertain. In general, a small size decrease could be observed for the small livestock from the Mamluk to Modern times (see appendix).

Cattle remains are very scarce; from a total number of 92 almost 40% were identified as metapodials and phalanges. They were relatively more abundant in the older levels. Indeed, while in the Early Islamic period cattle still account for about 2% of the domestic consumed mammals, their share decreases to about 1% in the Mamluk period, over 0.3% in the Ottoman period, to less than 0.2% in the Modern period. Almost all remains are from adult individuals.

Sieved samples

A sieved sample from a thin refuse layer, belonging to one of the older occupation periods, contained a number

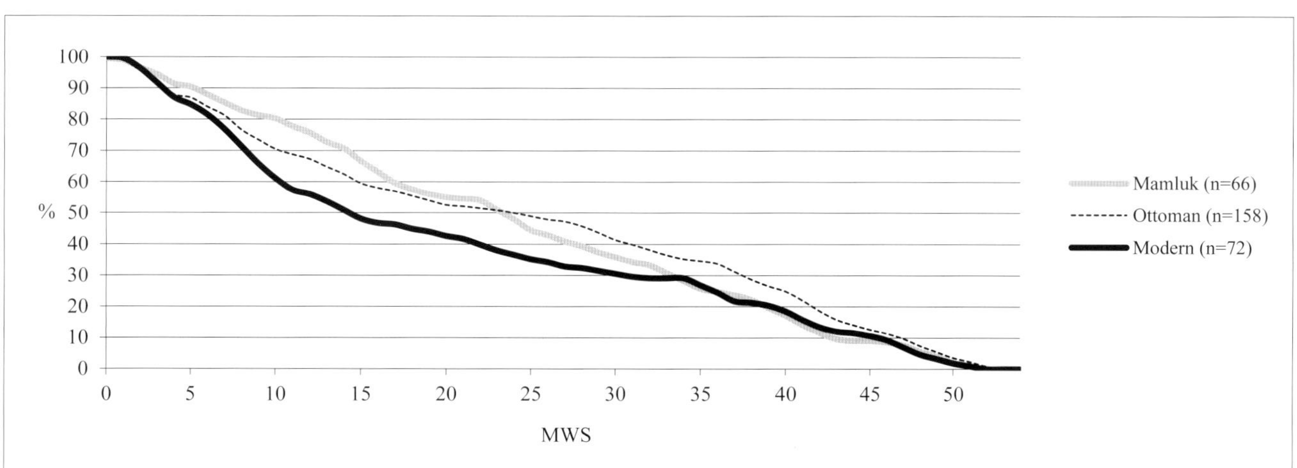

Fig. 31.12. Survival curves of the sheep/goat for the different periods.

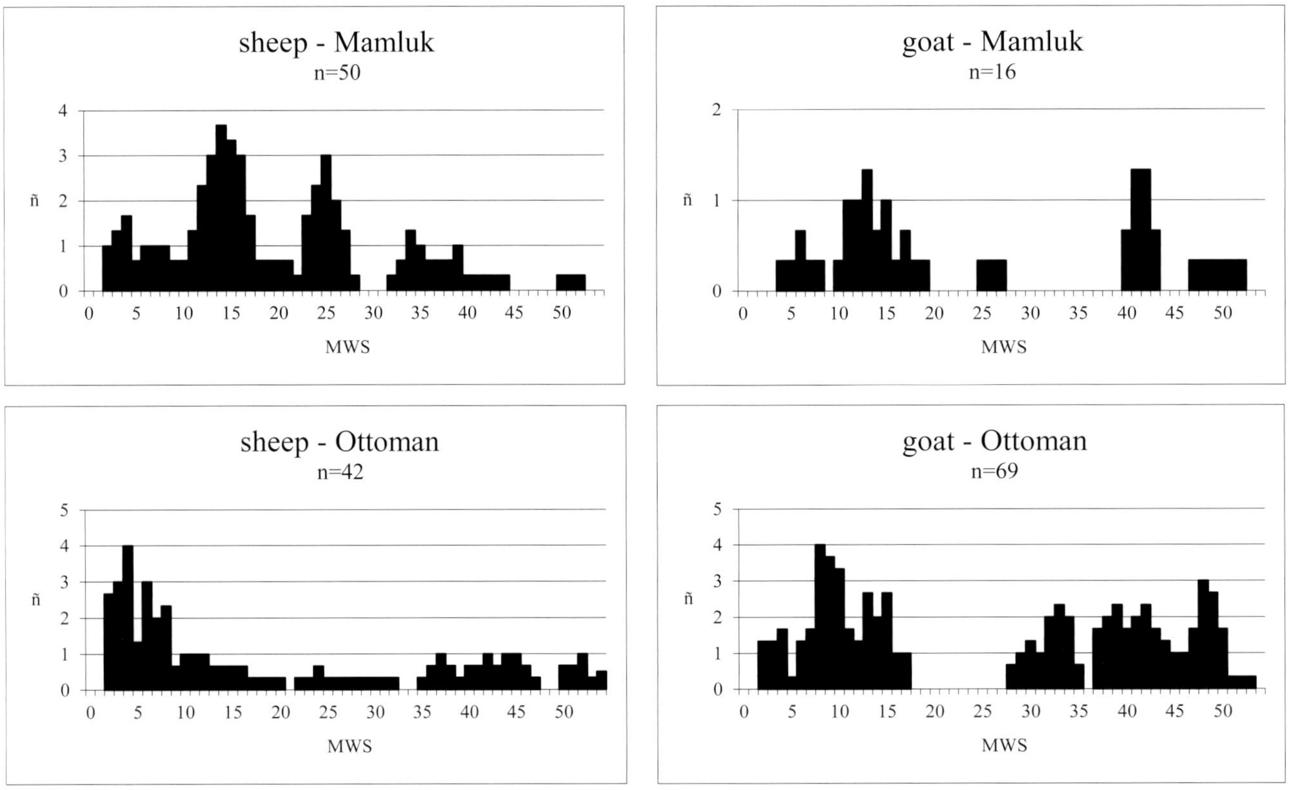

Fig. 31.13. Mandibular wear stages (MWS) of sheep and goat, according to Grant (1982); ñ: running mean.

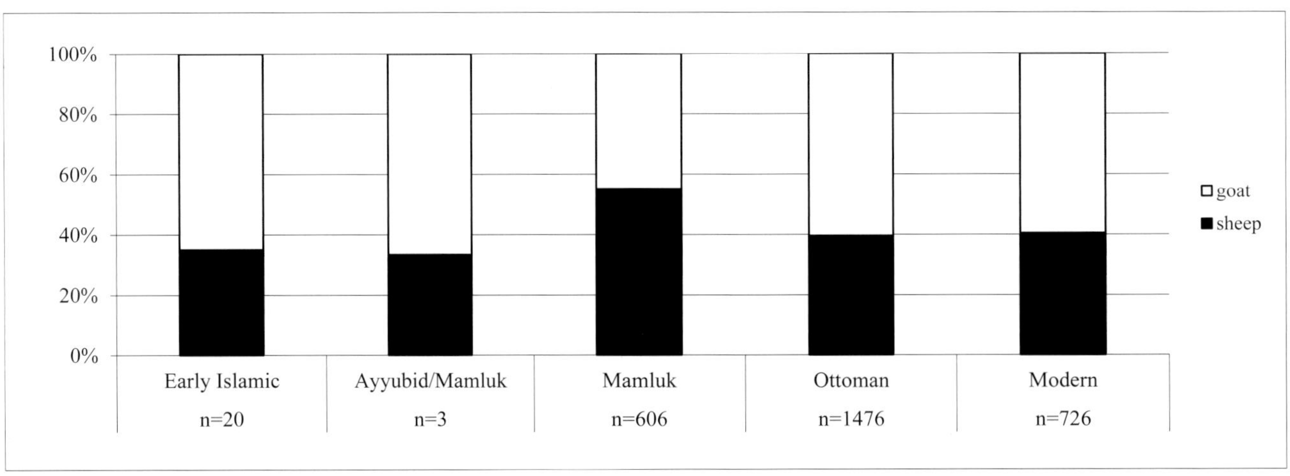

Fig. 31.14 Relative abundance of sheep and goat through time, based on NISP, excluding horn-cores and mandibulae.

of small gastropods, minute fragments of mammal bones, and some small fish bones. Another sample, from one of the older refuse layers that was rich in charcoal, contained a similar spectrum of finds but with a higher density. In this case, fish scales and a large number of plates of mail shells (Chitonidae) were recovered. A sample from the topsoil, the context that yielded most of the hand-collected bones and displayed the best preservation conditions, contained virtually no small animal remains.

The fill of a ceramic container without the base (AQ 2005 Y11), which had been put into the ground in order to serve as a sinkhole, contained a large number of fish bones and scales, mainly from Lethrinidae, Carangidae and Siganidae (Table 31.4). They were derived from both medium-sized and small fish. The sinkhole could be dated to the Mamluk period. The fill of a small water reservoir did not contain meaningful numbers of small animal remains.

Food provisioning at the Castle of Aqaba

The relative abundance of the different animal categories is shown in Figure 31.15; the few turtle remains are

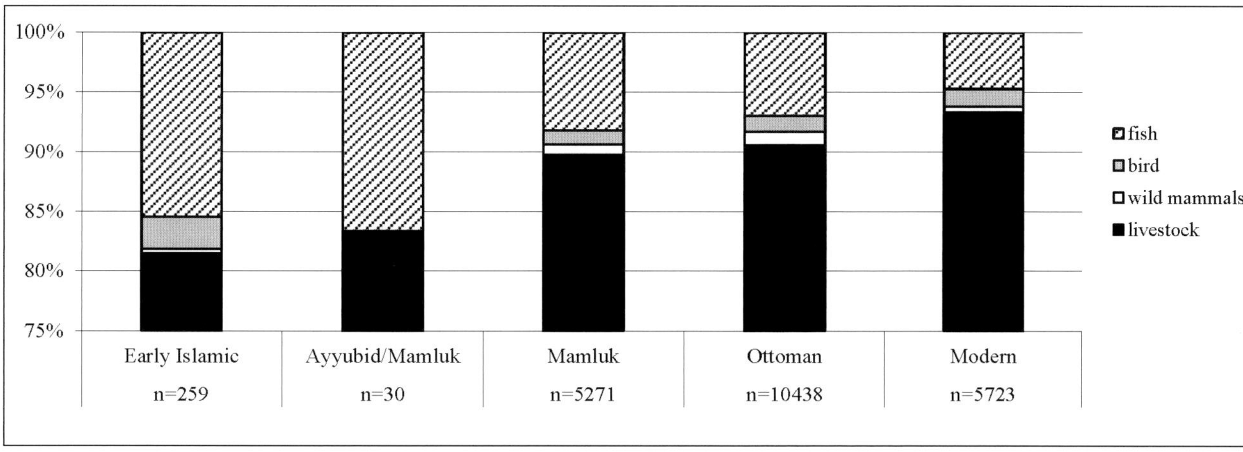

Fig. 31.15. Relative abundance of the main consumed animal groups through time, based on NISP.

not included here. The molluscs are not included either, but considering their large number within the faunal assemblage and the fact that the majority can be interpreted as consumption refuse, they indeed must have formed a substantial part of the diet of the inhabitants of the castle.

Fish consumption must also have been of importance during the occupation of the castle, especially in the Early Islamic period. Although fish remains were never very numerous in the excavated contexts from the later periods, the presence of large specimens corroborates this statement. It is difficult to evaluate the consumption of small fish, as sieved samples were limited and contained little useful material. Possibly, the preservation of small fish remains is not ideal at the site. The sinkhole produced remains mainly from Lethrinidae, followed by Siganidae and Carangidae, thus compared to the hand-collected material a somewhat different composition becomes apparent.

When the fish fauna from the Late Roman site of ez-Zantur (Petra) is considered (sample size: n=433), it is striking that no jacks at all have been found at this site; the groupers made up 35% of the identified fish remains, followed by the parrotfishes (21%), the emperors (19%) and the scombrids (17%) (Studer 2008). At Byzantine Jabal Harûn, located at about 5km southwest of Petra, parrotfish dominated the fish assemblage at 67% (sample size: n=2924) and medieval Wadi Farasa (Petra) almost exclusively yielded remains of parrotfish (sample size: n=218; Studer 2008). A high percentage of parrotfish was also observed at other inland sites, such as Deir ʻAin Abata and En Boqeq, both near the Dead Sea (Studer 2008). Although these sites are very distinct in nature, and certainly not comparable to the Castle of Aqaba, the high proportion of parrotfish in the Roman to medieval period indicates that these fishes must have been ideal for trading. The proportion of parrotfish in the Mamluk assemblage of Aqaba is 40%, but its amount declines during the Ottoman period to less than 10%. At the same time, a larger proportion of emperors are observed, although from specimens of a smaller size. Through time, fishermen ventured further offshore, towards the open sea, as indicated by the larger quantities of jacks. The causes of these changes (overfishing, possible destruction of habitats, changing fishing techniques) are not well understood and need to be investigated using material from more sites and by more refined analyses (e.g. of the growth rate of the fishes). The huge number of fish remains collected from Roman Aila is still awaiting analysis. They may add valuable information to the current picture.

Hunting of wild birds and mammals provided only a small portion of the diet, an observation which was also made at Roman Aila (Parker 1998). Most of the food provisioning of the castle relied on domestic mammals (Fig. 31.15). When the relative abundance of livestock (Fig. 31.16) is considered, it is obvious that the bulk is made up of bones of sheep and goat. Their relative number increases through time, while the relative number of camelid bones decreases. Excavations of Roman Aila have also yielded camelid bones, although without any further identification to species level (Parker 1996). At Aqaba the ratio of sheep and goat varies through time (Fig. 31.14), illustrating a growing dominance of goat. This is in accordance with other observations in the country. Indeed, in the Late Roman assemblage of ez-Zantur sheep remains are – in a ratio of 2:1 – more numerous than the bones of goat, while today the Bedouins have herds mainly consisting of goats (Studer 1996).

The environment

The presence of pig has not been attested at the Castle of Aqaba. The complete absence of this animal is not surprising, considering the religious context of the inhabitants. But pig remains are also very scarce at Roman Aila (Parker 1996) and Nabatean Petra (Studer 2007). The arid nature of the environment certainly played a role in the paucity of the pig remains. Similarly, the low frequency of cattle remains can be explained by the environmental conditions, which excluded large-scale cattle breeding. The presence of pig and cattle at Roman Aila was interpreted

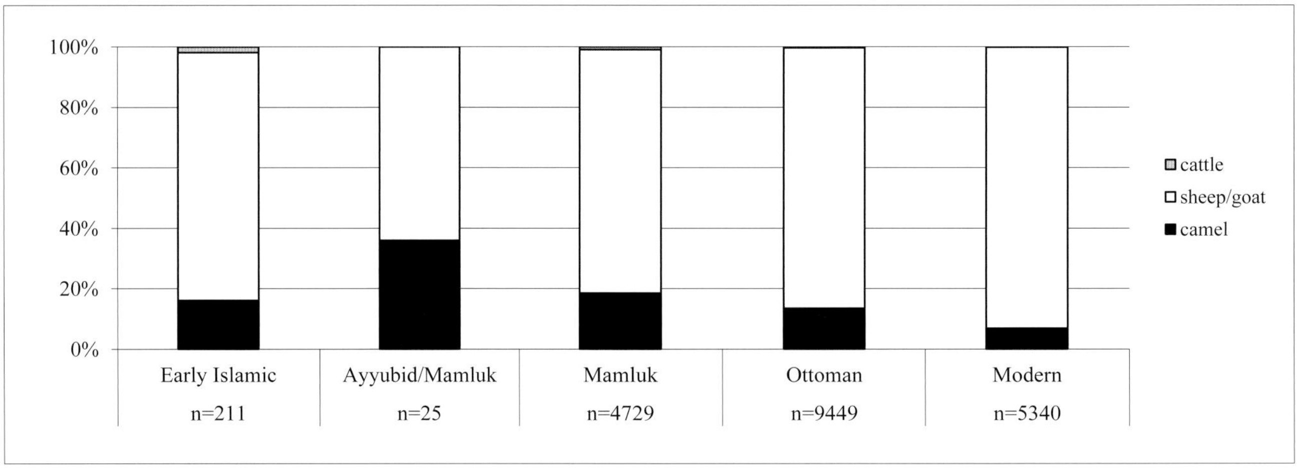

Fig. 31.16. Relative abundance of the consumed domestic mammals through time, based on NISP.

as imports from external sources, rather than being an exploitation of locally raised herds (Parker 1996; 1998). As far as small livestock is concerned, the shift from herds with more sheep than goats in Mamluk times towards herds with an emphasis on goat breeding in Modern times may be the consequence of either overexploitation of the environment (i.e. overgrazing) and/or changing climatic conditions. Humidity changes have indeed been described for the past centuries in Jordan (Shehadeh 1985). Although the described fluctuations for the Mamluk and Ottoman periods were not lasting long enough to have had a permanent impact on the environment, they may quite well have triggered a change in herd management, and also have induced the observed size decrease in sheep and goat (see appendix). Furthermore, several bird species that were still present during the occupation period of the Castle of Aqaba have now completely disappeared from the region. This may be the result of many factors: hunting, a loss of habitat under growing human pressure, climate changes.

Conclusion

The analysis of the faunal remains yielded insight in the food provisioning of the castle inhabitants, especially for the Mamluk, Ottoman and Modern periods. Sea fishing and harvesting of shells contributed substantially to the diet while the amount of meat provided by hunting wild birds and mammals must have been rather small. The bulk of the animal food products came from sheep and goat and to some extent from camel. The consumption of beef was rather minute, and occurred mainly during the oldest occupation periods. Diachronic changes of the relative abundance of the different fish taxa indicate a shift towards more fishing in the open sea, simultaneous with a decrease in the contribution of fish in general. At the same time, a smaller proportion of parrotfish and a larger proportion of the emperors (but of smaller size) is observed. A shift in the management of sheep can be suggested from the age-of-death-profiles. While subadult animals were preferentially slaughtered in the Mamluk period, an emphasis on lambs is observed in the Ottoman period. Slaughtering ages of the goats remain similar over the centuries. Through time, a shift towards goat dominated herds is observed, possibly to be related with a deterioration of the environment.

Acknowledgements

The excavations at Aqaba Castle have been organised by the Ministère de la Région wallonne (Belgium) and Cardiff University (UK) (2000–2003), and by Ghent University (Belgium), the UMR 5648 du CNRS/Lyon II (France) and Archaeologia Mediaevalis (Belgium) (2005–2008). The archaeozoological analysis in this text mainly represents research results of the Belgian Programme on Interuniversity Poles of Attraction (IAP 6/22) initiated by the Belgian Federal Science Policy Office. The contribution by Anton Ervynck is part of a former cooperation agreement between the Flemish Heritage Institute (Belgium) and Ghent University (Belgium). Scientific responsibility is assumed by the authors. We express our thanks to Sheila Hamilton-Dyer (Southampton) for the linguistic correction of the text.

This paper is dedicated to John De Meulemeester, the driving force behind the Aqaba Castle excavations, who died earlier in 2009. His support for environmental archaeology within the very traditional domain of medieval Mediterranean and Near Eastern archaeology has been much appreciated.

References

al-Shqour, R., De Meulemeester, J. & Herremans, D. 2008, 'The Aqaba Islamic Project'. *Studies in the History and Archaeology of Jordan* 2008, 641–656.

Andrews, I. J. 1995, *The Birds of the Hashemite Kingdom of Jordan.* Burns & Harris, Dundee.

Arndt, A., Van Neer, W., Hellemans, B., Robben, J., Volckaert, F. & Waelkens, M. 2003, 'Roman trade relationships at Sagalassos (Turkey) elucidated by ancient DNA of fish remains'. *Journal of Archaeological Science* 30, 1095–1105.

Beech, M. 1992, 'Animal bones', in K. D. Politis, *Excavations at Deir 'Ain 'Abata 1991. Annual of the Department of Antiquities of Jordan* 36, 284.

Benkheira, M. H. 2000, *Islâm et interdits alimentaires. Juguler l'animalité*. Presses Universitaires de France, Paris.

Clutton-Brock, J. 1989, 'A dog and a donkey excavated at Tell Brak'. *Iraq* 51, 217–224.

De Cupere, B., Van Neer, W., Monchot, H., Rijmenants, E., Udrescu, M. & Waelkens, M. 2005, 'Ancient breeds of domestic fowl (*Gallus gallus* f. domestica) distinguished on the basis of traditional observations combined with mixture analysis'. *Journal of Archaeological Science* 32, 1587–1597.

De Meulemeester, J. & Pringle, D. 2001, 'The archaeological investigation of 'Aqaba Castle', in *Research Reports from Jordan, CBRL 2001. Newsletter of the Council for British Research in the Levant*, 19–21.

De Meulemeester, J. & Pringle, D. 2002, 'The 'Aqaba Castle Project 2002', in *Research Reports from Jordan, CBRL 2002. Newsletter of the Council for British Research in the Levant,* 19–20.

De Meulemeester, J. & Pringle, D. 2004, 'The 'Aqaba Castle Project 2003', in *Research Reports from Jordan, CBRL 2004, Newsletter of the Council for British Research in the Levant*, 26–27.

De Meulemeester, J. & Pringle, D. 2005, 'The 'Aqaba Castle Project 2004–5', in *Research Reports from Jordan, CBRL 2005, Newsletter of the Council for British Research in the Levant,* 42.

De Meulemeester, J. & al-Shqour, R. 2006, 'The ´Aqaba Castle Project 2006', in *Research Reports from Jordan, CBRL 2006, Newsletter of the Council for British Research in the Levant,* 27–28.

Desse, J. & Desse-Berset, N. 2000, 'Julfar (Ras al Khaimah, Emirats Arabes Unis), ville portuaire du golfe arabo-persique (VIIIe–XVIIe siècles): exploitation des mammifères et des poissons', in M. Mashkour, A. M. Choyke, H. Buitenhuis & F. Poplin (eds), *Archaeozoology of the Near East IV B. Proceedings of the fourth international symposium on the archaeozoology of southwestern Asia and adjacent areas*, 79–93. ARC, Groningen.

Desse, J. & Desse-Berset, N. 2003, 'Les premiers pêcheurs de Chypre', in J. Guilaine & A. Le Brun (eds), *Le Néolithique de Chypre, Bulletin de Correspondance Hellénique supplément* 43, 279–291. De Broccard, Paris.

Ervynck, A. 2002, 'Sedentism or urbanism ? On the origin of the commensal black rat (*Rattus rattus*)', in K. Dobney & T. O'Connor (eds), *Bones and the Man. Studies in Honour of Don Brothwell*, 95–109. Oxbow Books, Oxford.

Frazier, J. & Salas, S. 1984, 'The status of marine turtles in the Egyptian Red Sea'' *Biological Conservation* 30, 41–67.

Grant, A. 1982, 'The use of tooth wear as a guide to age of domestic ungulates', in B. Wilson, C. Grigson & S. Payne (eds), *Ageing and Sexing Animal Bones from Archaeological Sites*. British Archaeological Report 109, 91–108. Britihs Archaeological Reports, Oxford.

Gruber, A. 1990, 'Vergleichend morphologische Untersuchungen an Einzelknochen in Ägypten vorkommender Ciconiidae'. Dissertation, Münich.

Halstead, P., Collins, P. & Isaakidou, V. 2002, 'Sorting the sheep from the goats: morphological distinctions between the mandibles and mandibular teeth of adult *Ovis* and *Capra*'. *Journal of Archaeological Science* 29, 545–553.

Harrison, D. L. & Bates, P. J. J. 1991, *The Mammals of Arabia*. Harrison Zoological Museum Publication, Kent.

Heinzel, H., Fitter, R. & Parslow, J. 1987, *Elseviers gids van de Europese vogels* (5th edn). Elsevier, Amsterdam-Brussel.

Horwitz, L. K., Cope, C. & Tchernov, E. 1990, 'Sexing the bones of mountain-gazelle (*Gazella gazella*) from prehistoric sites in the southern Levant'. *Paléorient* 16(2), 1–12.

Ibn al-Baytar, *Traité des simples* (trans. L. Leclerc 1987), Institut du monde arabe, Paris.

Kennedy, D. L & Bewley, R. 2004, *Ancient Jordan from the Air*. Council for British Research in the Levant, London.

Köhler-Rollefson I., 1989, 'Zoological analysis of camel skeletons', in R. H. Smith & L. P. Day (eds), *Pella of the Decapolis 2. Final Report on the College of Wooster Excavations in Area IX, the Civic Complex, 1979–1985*, 142–164. College of Wooster, Wooster.

Lorch, R. 1992, 'Vergleichend morphologische Untersuchungen am Skelett von *Pelecanus onocrotalus*, *Pelecanus crispus* and *Pelecanus rufescens*', Dissertation, Munich.

Marquez, M 1984, 'Sea turtles', in W. Fisher & G. Bianchi (eds), *FAO Species Identification Sheets for Fishery Purposes, Western Indian Ocean (Fishing Area 51)*. FAO, Rome.

Mason, I. L. 1984, 'Camels', in I. L. Mason (ed.), *Evolution of Domesticated Animals*, 106–115. Longmann, London and New York.

Parker, S. T. 1996, 'The Roman 'Aqaba Project. The 1994 campaign'. *Annual of the Department of Antiquities of Jordan* 40, 231–257.

Parker, S. T. 1998, 'The Roman 'Aqaba Project. The 1996 campaign'. *Annual of the Department of Antiquities of Jordan* 42, 375–394.

Payne, S. 1973, 'Kill-off patterns in sheep and goats: the mandibles from Asvan Kale'. *Anatolian Studies* 23, 281–303.

Payne, S. 1985, 'Morphological distinctions between the mandibular teeth of young sheep, *Ovis*, and goats, *Capra*'. *Journal of Archaeological Science* 12, 139–147.

Peters, J. 2001, 'Aspekte der Domestikations- und Kulturgeschichte des Dromedars (*Camelus dromedarius*)', in H. Buitenhuis & W. Prummel (eds), *Animals in the Past, Essays in Honour of Dr A. T. Clason*, 330–341. ARC, Groningen.

Potts, D. T. 2004, 'Camel hybridisation and the role of *Camelus bactrianus* in the ancient Near East'. *Journal of Economic and Social History of the Orient* 47(2), 143–165.

Richardson, H. D. 1857, *Dogs: their origin and varieties*. C. M. Saxton and Company, Agricultural Book Publishers, New York.

Roa-Quiaoit, H. A. F. 2005, 'The ecology and culture of giant clams (Tridacnidae) in the Jordanian sector of the Gulf of Aqaba, Red Sea'. PhD thesis, Bremen.

Rusmore-Villaume, M. L. 2008, *Seashells of the Egyptian Red Sea. The Illustrated Handbook*. American University in Cairo Press, Cairo-New York.

Schmidt, S. G. & Studer, J. 2003, 'The International Wadi Farasa Project (IWFP) preliminary report on the 2002 Season'. *Annual of the Department of Antiquities of Jordan – Amman* 47, 473–488.

Sharabati, D. 1984, *Red Sea Shells*. KPI, London-Boston-Melbourne-Henley.

Shehadeh, N. 1985, 'The climate of Jordan in the past and present', in A. Hadidi (ed.), *Studies in the History and Archaeology of Jordan II*, 25–37. Department of Antiquities, Amman.

Silver, I. A. 1963, 'The ageing of the domestic animals', in D. Brothwell & E. Higgs (eds), *Science in Archaeology*, 250–268. Thames and Hudson, London.

Smith, C. H. 1840, *Natural History of Dogs (Sir William Jardine's Naturalist's Library vols 9 & 10)*. W. H. Lizars, Edinburgh.

Smith, W. R. 1907, *Kinship and Marriage in Early Arabia* (3rd edn). Adam and Charles Black, London.

Steiger, C. 1990, 'Vergleichend morphologische Untersuchungen an Einzelknochen des postkranialen Skeletts der Altweltkamele'. Dissertation, Munich.

Studer, J. 1994, 'Roman fish sauce in Petra, Jordan', in W. Van Neer (ed.), *Fish Exploitation in the Past, Proceedings of the 7th meeting of the ICAZ Fish Remains Working Group, Annales du Musée Royal de l'Afrique Centrale, Sciences Zoologiques* 274, 191–196.

Studer, J. 1996, 'La faune romaine tardive d'ez-Zantur, à Petra', in A. Bignasca, N. Desse-Berset, R. Fellman Brogli, R. Glutz, S. Karg, D. Keller, B. Kolb, Ch. Kramar, M. Peter, S. G. Schmid, Ch. Schneider, R. A. Stucky, J. Studer and I. Zanoni (eds), *Petra, ez-Zantur I. Ergebnisse der Schweiserisch-Liechtensteinischen Ausgrabungen 1988–1992*, 359–375. Terra Archaeologica II, Monographien der Schweiserisch-Liechtensteinischen Stiftung für Archäologische Forschungen im Ausland (SLSA/FSLA), von Zabern, Mainz.

Studer, J. 2001, 'Observations on animal bones from Jabal Harûn', in J. Frössen, Z. T. Fiema, K. Koistinen, C. Danielli, J. Studer, Y. Gerber, D. Keller, H. Jansson, J. Lindblom, T. Tenhunen and E. Hertell (eds), *The 1998–2000 Finnish Jabal Harûn Project: Specialised Reports, Annual of the Department of Antiquities of Jordan, Amman* 45, 384–386.

Studer, J. 2002a, 'Dietary differences at Ez Zantur Petra, Jordan (1st century BC–AD 5th century)', in H. Buitenhuis, A. M. Choyke, M. Mashkour and A. H. Al-Shiyab (eds), *Archaeozoology of the Near East V*, 273–281. ARC- Publicaties 62, Groningen.

Studer, J. 2002b, 'The archaeozoological research', in J Frössen, K. Koistinen, J. Studer, Ch. Danielli, R. Holmgren, Y. Gerber, N. Heishka and A. Lahelma (eds), *The 2001 Finnish Jabal Harûn Project: preliminary report, Annual of the Department of Antiquities of Jordan, Amman* 46, 400–402.

Studer, J. 2007, 'Animal exploitation in the Nabataean world', in K. D. Politis (ed.), *The World of the Nabataeans*, 251–272. Proceedings of the conference on the 'World of the Herods and Nabataeans', held at the British Museum in London 17–19 April 2001 Vo. 2. Steiner Verlag, Stuttgart

Studer, J. 2008, 'Like a fish out of water: fish in the diet of Classical and Medieval populations in the Petra region (Jordan)', in P. Béarez, S. Grouard & B. Clavel (eds), *Archéologie du poisson. 30 ans d'archéo-ichtyologie au CNRS. Hommage aux travaux de Jean Desse et Nathalie Desse-Berset. XXVIIIe rencontres internationales d'archéologie et d'histoire d'Antibes, XIVth ICAZ Fish remains working group meeting*, 281–293. Editions APDCA, Antibes.

Studer, J. & Schneider, A. 2008, 'Camel use in the Petra region: 1st century BC to 4th century AD', in E. Vila, L. Gourichon, A. M. Choyke & H. Buitenhuis (eds), *Archaeozoology of the Near East VIII, Proceedings of the eighth international symposium on the archaeozoology of southwestern Asia and adjacent areas*, 581–596. Maison de l'Orient et de la Méditerranée, TMO 49, Lyon.

Tomek, T. & Bocheński, Z. M. 2000, *The Comparative Osteology of European Corvids (Aves: Corvidae), with a Key to the Identification of their Skeletal Elements.* Wydawnictwa Instytutu Systematyki I Ewolucji Zwierząt, Kraków.

Toplyn, M. R. 1994, 'Meat for Mars: livestock, *limitanei*, and pastoral provisioning for the Roman army on the Arabian frontier, A.D. 284–551'. PhD thesis, Harvard.

Toplyn, M. R. 2006, 'Livestock and *Limitanei*: the zooarchaeological evidence', in S. T. Parker (ed.), *The Roman Frontier in Central Jordan. Final Report on the Limes Arabicus Project, 1980–1989*, 463–548. Dumberton Oaks Research Library and Collection, Washington DC.

Uerpmann, H.-P. 1999, 'Camel and horse skeletons from protohistoric graves at Mleiha in the Emirate of Sharjah (U.A.E.)'. *Arabian Archaeology and Epigraphy* 10, 102–118.

Van Neer, W., Lernau, O., Friedman, R., Mumford, G., Poblome, J. & Waelkens, M. 2004, 'Fish remains from archaeological sites as indicators of former trade connections in the Eastern Mediterranean'. *Paléorient* 30(1), 101–148.

Vila, E. 1991, 'Note sur un dépôt de gazelles dans une fosse Uruk d'el-Kown 2 – Caracol (Syrie)'. *Cahiers de l'Euphrate* 5–6, 55–59. Editions Recherche sur les Civilisations, Paris.

von den Driesch, A. 1976, *A Guide to the Measurement of Animal Bones from Archaeological Sites*. Peabody Museum Bulletin 1, Peabody Museum, Harvard University, Cambridge MA.

von den Driesch, A. & Boessneck, J. 1974, 'Kritische Anmerkungen zur Widerristhöhenberechnung aus Längenmassen vor- und frühgeschichtlicher Tierknochen'. *Säugetierkundliche Mitteilungen* 22(4), 325–348.

von den Driesch, A. & Dockner, A. 2002, 'Animal exploitation in medieval Siraf, Iran based on the faunal remains from the excavations at the Great Mosque (seasons 1966–1973)'. *Bonner Zoologischer Beiträge* 50(3), 227–247.

Wapnish, P. 1984, 'The dromedary and the Bactrian camel in Levantine historical settings: the evidence from Tell Jemmeh'. in J. Clutton-Brock & C. Grigson (eds), *Animals and Archaeology 3. Early Herders and their Flocks*, British Archaeological Report S202, 171–200. British Archaeological Reports, Oxford.

Wilson, R. T. 1998, *Camels.* Tropical Agriculturalist Series, Macmillan Education, London and Basingstoke.

Zohary, M. 1945, 'Outline of the vegetation in Wadi Araba'. *Journal of Ecology* 32(2), 204–213.

Appendix – Aqaba castle. Selected measurements of the faunal remains (abbreviations after A von den Driesch (1976), all measurements in mm)

A. Measurements of wild birds – Alec: *Alectoris chukar*; Ammo: *Ammoperdix heyi;* Crex : *Crex crex*; Col : *Columba livia/oenas*; Corv: *Corvus corax*; Mam: Mamluk; Ott : Ottoman; Mod: Modern; A/M : Ayyubid/Mamluk.

	Alec	**Crex**	**Col**	**Col**	**Col**	**Corv**
coracoid	Mam	Mam	Ott	Mod	Mam	Mod
GL	36.5	23.5	33.6	33.3	32.4	49.5
Lm	35.3	22.8	32.4	32.3	30.9	44.0
Bb					13.3	
BF	7.8	(6.2)	8.7	7.2	8.6	15.0

	Alec	**Corv**
scapula	Ott	Mod
GL	60.4	
Dic	15.3	14.2

	Alec	**Ammo**	**Ammo**	**Crex**	**Crex**	**Crex**	**Col**	**Col**	**Col**	**Col**	**Strep**	**Corv**	**Corv**
humerus	Ott	Ott	Ott	Ott	Ott	Mam	Ott	Mam	Ott	Ott	Ott	Ott	Ott
GL	49.6			41.9	41.5		47.0	44.1	42.8			81.6	81.6
Bp	13.8	10.0		7.7	8.6	8.8	18.5	17.3	16.0			21.9	
SC	4.6	3.5		2.8	2.8	3.1	5.2	4.9	5.1	4.9	4.9	7.4	7.3
Bd	10.0		8.0	6.0	6.1		11.1	10.4	10.5	10.3	8.1	19.2	18.0

	Corv	**Corv**	**Corv**	**Corv**
humerus	A/M	Ott	Ott	Ott
GL	81.2	78.8	77.0	
Bp	21.1	20.6	20.0	
SC	7.5	7.3	7.1	6.8
Bd	20.0	17.2	17.9	17.4

	Strep	**Corv**
radius	Ott	Mam
GL	37.1	92.7
SC	2.6	2.4
Bd	3.9	7.0

	Ammo	**Col**	**Strep**	**Strep**	**Corv**	**Corv**	**Corv**	**Corv**	**Corv**	**Corv**	**Corv**	**Corv**	**Corv**
ulna	Mam	Mam	Ott	Ott	Mod	Ott	Ott	Mam	Mod	Ott	Ott	Ott	Ott
GL	35.0		41.5		104.6	97.0	92.2						
Bp	4.4	6.6	5.6		12.1	12.0	11.1	12.2	11.3				
Dip	6.3	7.8	7.9		13.8	8.5							
SC	2.0		2.7		5.4	5.3	5.0		5.5				
Bd	4.6		5.4	5.1	11.3	10.8	11.4			11.0	10.6	10.4	10.4

	Alec	**Crex**	**Col**	**Col**	**Corv**	**Corv**	**Corv**	**Corv**	**Corv**	**Corv**
carpometac.	Mam	Ott	Mam	Mam	Ott	Mod	Mam	Ott	Ott	Mod
GL	29.5	25.9	32.0		61.7	58.8	55.2	54.2	53.7	53.5
Bp	8.4		8.9			12.7	14.2	12.7	12.1	12.6
Did	5.1			5.7		12.5		11.7		12.6

	Alec	**Alec**	**Crex**	**Crex**	**Crex**	**Strep**	**Corv**
femur	Ott	Ott	Mod	Ott	Ott	Mod	Mam
GL	53.7		45.2			(34.3)	
Bp	11.2		7.0	8.0			11.4
Dp	7.0		4.7	5.1			
SC	4.1	4.0	2.7				
Bd	9.7	9.5	5.9		6.2		
Dd	7.6	7.4	5.6		6.0		

	Alec	**Alec**	**Alec**
tibiotarsus	Mod	Ott	Ott
GL	66.2	73.2	73.3
Dip	10.8	12.1	11.9
SC	3.3	3.6	3.7
Bd	6.8	7.1	6.9
Dd	6.9	7.2	7.2

	Alec	**Alec**	**Alec**	**Crex**	**Crex**	**Crex**	**Crex**	**Crex**	**Corv**	**Corv**
tibiotarsus	Ott	Mam	Mam	Mod	Ott	Ott	Mod	Mam	Ott	Ott
GL				60.9					106.9	
Dip	14.0	11.9		7.8	8.1	8.1			15.6	13.3
SC			3.6	2.7				2.9	5.4	
Bd			7.2	5.2			5.1	5.0	10.7	
Dd			7.0	5.5			5.5	5.6	10.3	

	Alec	**Ammo**	**Crex**	**Crex**	**Crex**	**Col**	**Corv**	**Corv**
tarsometat.	Ott	Ott	Ott	Ott	Ott	Ott	Ott	Mam
GL	74.6	32.6	39.6	38.4	37.5	31.2	65,0	63.0
Bp	9.2	6.0	5.9	5.6	5.4	7.4	11.4	11.0
SC	4.0	2.8	3.0	2.5	2.6	3.6	4.0	4.5
Bd	10.1	6.2	6.1	5.5	6.3		8.0	8.0

B. Measurements of chicken – Mam: Mamluk; Ott : Ottoman; Mod: Modern

coracoid	Mod	Mod	Mod	Ott
GL	56.2	52.9	49.7	
Lm	54.3	50.5		
Bb	13.6			13.4
BF	11.8	10.6		10.6

scapula	Ott	Ott
GL	66.6	
Dic	12.0	11.8

humerus	Mam	Mam	Ott	Mod	Mod	Ott	Mod	Mam	Mod	Ott	Ott	Mod
GL	75.0	65.7	65.6	65.3	64.7	63.7	63.0	60.5	60.1			
Bp	20.0	18.8	17.5	17.5		16.8	18.1	15.5				
SC	7.8	6.4	5.9			5.8	6.0	5.8				
Bd	16.0	14.4	13.3	13.6		13.1	13.3	12.6		13.2	(16)	12.5

radius	Mod	Ott	Ott	Ott
GL	74.3	58.8	55.1	51.1
SC	3.5		2.9	
Bd	7.5		6.0	

carpometac.	Ott	Ott	Mam
GL	39.3	33.0	31.9
Bp	12.3	10.6	10.0
Did	7.4	6.1	

ulna	Ott	-	Ott	Mod	Ott	Mod	Mam	Mod	Ott	Mod	Ott	-	Mam
GL	81.0	78.5	78.0	74.8	73,0	72.1	69.0	68.4	65.4	62.8	61.1	58.8	58.3
Bp	10.0		8.6		8.3		9.5		7.7				10.6
Dip	14.9		11.9		12.9		13.1		11.3				8.0
SC	4.5		4.1		4.4		4.1		3.8				3.2
Bd	10.3		9.3		9.3		9.5		8.1				8.1

ulna	Ott	Mod	-
GL	55.0		
Bp	7.0	7.8	
Dip	10.0	11.0	
SC	3.4		
Bd	7.5		9.1

femur	Ott	Mam	Ott	Mod	-	Mod	Mod
GL	89.1	86.0	(77)	76.4	75,0	74.7	72.2
Bp	18.7	18.5	16.7				
Dp	12.4	11.9	11.2				
SC	7.8	8.0	7.0				
Bd	19.5	18.0	15.7				
Dd	15.1	15.3	13.2				

femur	Mod	Ott	Ott	Mod	Ott	Ott	Ott	Mod	-	Mod	Ott	Mod	Ott
GL	72.0	71.4	70.8	70.4	69.0	67.4	66.9	66.0	65.8				
Bp					15.0	63.0	14.7		13.6	19.2	14.7	12.7	
Dp			8.8		9.5	13.1	9.2		9.0			7.4	
SC			6.2		6.3	8.5	6.1		5.9				
Bd			14.0		13.4	5.3	13.7		12.9				14.7
Dd			11.7		11.2	12.9	11.9		10.6				

femur	Ott	Mod	Mod
GL			
Bp			
Dp			
SC	7.0	8.1	
Bd	17.5	17.1	13.7

tibiotarsus	Mod	Mod	Mam	Mod	Mod	Ott	Mod
GL	134.6	134.4	123.6	121.1	117.0	107.2	102.6
Dip	24.1	24.0	22.0			17.5	
SC	7.8	7.8	7.6			6.4	
Bd	14.4	14.4	12.3			10.8	
Dd	15.0	15.0	13.0			11.8	

tibiotarsus	Mam	Ott	Mam	Ott	Mam	Ott	Ott	Ott	Mod	Ott
GL	99.2	89.4								
Dip	17.2	16.4	19.8							
SC	5.2	5.2		7.5		6.4	6.4	5.8	5.7	5.3
Bd	10.8	10.2		14.3	12.2	11.9	11.5	11.1	10.6	10.5
Dd	11.2	10.3		14.2		12.4	12.1	11.7	10.9	10.1

tarsometat.	Ott	Ott	Mam	Ott	Ott	Ott	Mod	Mam	Mod	Mam	Ott
GL	82.3	81.2	81.0	80.9	80.5	79.6	71.5	70.7	68.0	66.0	
Bp	14.9	15.0		13.8	13.4		13.1			12.1	11.8
SC	7.7	7.4	7.0	6.9	6.7	6.8	6.9			5.5	6.0
Bd	14.2	14.7	14.0		14.2	14.2	12.7			11.9	

C. Measurements of wild and domestic mammals – Hyae : *Hyaena hyaena*; Can: *Canis lupus* f. familiaris; Fel: *Felis silvestris* f. catus; Gaz : *Gazella gazelle*

	Hyae	**Fel**	**Gaz**	**Gaz**	**Gaz**	**Gaz**	**Gaz**	**Gaz**	**Gaz**	**Gaz**	**Gaz**	**Gaz**	**Gaz**
scapula	Mam	Ott	Ott	Ott	Ott	-	Mam	Mam	Ott	Ott	EI	Ott	Ott
GLP	41.2	11.9	29.1	29.0	28.8	28.2	28.1	28,0	27.3	27.1	26.8	26.4	25.5
LG	33.9	11.0	23.7	22.1	22.9		21.7	22.5	(21.2)	23.7	22.3	21.2	21.1
BG	23.7	6.9	20.5	21.1	21.3		19.7	20.4	18.8	18.7	17.8	18.7	17.6
SLC	34,0	10.6	16.0	16.0				15.0	12.8	14.4	(14.6)	13.2	13.5
HS								114.9					
DHA			140.9					123.8					

	Hyae	**Hyae**	**Hyae**	**Can**	**Can**	**Fel**	**Fel**	**Fel**	**Fel**	**Fel**	**Fel**	**Gaz**
humerus	Ott	Mam	Ott	Mam	Mod	Ott	Ott	mod	mod	Ott	mod	Mod
GL						91,0	86.6					
Dp/Bp	57.5			35.8		20,0	18.1	18.8				32.4
SC	15.7					6.8	6.0			7.3		
Bd		41.4	39.4		39.5	18.1	15.4		17.9	18.0	17.5	
BT												

	Gaz	**Gaz**	**Gaz**	**Gaz**	**Gaz**	**Gaz**	**Gaz**
humerus	Ott	Ott	Ott	Ott	Ott	Ott	-
SC		12.6					
Bd	27.4	27.0	26.6	26.2	25.0	24.6	24.2
BT	22.5	23.6	23,0	22.5	23.2	(23)	

	Hyae
axis	Ott
BFcr	36.1
SBV	21.7

	Hyae	**Can**	**Fel**	**Gaz**	**Gaz**	**Gaz**	**Gaz**	**Gaz**	**Gaz**	**Gaz**	**Gaz**	**Gaz**	**Gaz**
radius	Mam	Mod	Ott	Mod	-	Ott	Ott	Mam	Mam	Ott	Mod	Ott	Ott
GL			83.5	175.6	165,0								
Bp			7.3	27.0	25.0	26.4	26.0	25.3	25.3	25.0	24.5	24.2	24.1
BFp						23.3	23.2	21.9	23.2	23.8	22.5		22.3
SD	15.3		5.2	16.2	14.9	15;0	16.5	15.5		15,0	13.5		13.5
Bd	30.	20.6	11.0	24.6	23.8								

	Gaz	**Gaz**	**Gaz**	**Gaz**	**Gaz**	**Gaz**	**Gaz**
radius	Ott	Mod	Ott	Ott	Mod	Mam	-
Bp	23.9	22.7					
BFp	22.3						
SD			14.2	13.8			
Bd			23.2	22.6	22.3	20.3	19.4

	Fel	**Fel**	**Fel**
ulna	Mod	Ott	Ott
GL	109.0	108.6	97.7

	Can	**Can**	**Can**	**Fel**	**Fel**	**Fel**	**Fel**
pelvis	Ott	Mam	Ott	Ott	Ott	Ott	Mod
LAR	22.9	19.8	16.7	11.4	10.4	10.1	9.8

	Gaz	**Gaz**	**Gaz**	**Gaz**	**Gaz**	**Gaz**	**Gaz**
mc	Mam	Mod	Ott	Mam	Ott	Mam	Ott
GL	183.0	183.0					
Bp	19.4	19.3	20.7	19.7			
SD	12.0	12.0					
Bd	19.9	19.9			20.6	20.2	18.6

	Can
mc V	Mod
GL	61.5

	Fel	**Fel**	**Fel**	**Fel**	**Fel**	**Fel**	**Fel**	**Fel**	**Gaz**	**Gaz**	**Gaz**	**Gaz**	**Gaz**
femur	Mod	Ott	Ott	Ott	Mam	Ott	Ott	Mod	Ott	Ott	Ott	Mam	Mam
GL	97.5	96.0	94.2										
Bp	118.0	17.2	18.5	20.0	18.3	17.5			44.1	40.0	34.7	34.0	
DC									17.7	16.0	15.0	14.4	
SD	7.0	7.2	7.3	9.1	8.2	7.7	8.1						
Bd	16.6	16.4	17.4				17.9	18.1					33.1

	Gaz	**Gaz**
femur	Mam	Ott
SD	14.9	
Bd	32.2	31.1

	Gaz	**Gaz**
talus	Ott	Ott
GLl	26.0	25.0
GLm	23.5	23.1
Dl	14.1	13.6
Bd	15.8	15.6

	Gaz
calcaneus	Mam
GL	55.5

	Fel	**Fel**	**Fel**	**Fel**	**Gaz**	**Gaz**	**Gaz**	**Gaz**	**Gaz**	**Gaz**	**Gaz**	**Gaz**
tibia	Ott	Ott	Mam	Ott	Mam	-	Ott	Ott	-	Mod	Ott	Ott
GL	106.3											
Bp	18.8	17.5	17.5		(36)	34.9						
SD	6.8	7.5		6.3			15.4			12.7		
Bd	14.2			12.4			24.1	22.2	21.5	21.4	21.1	21.1

	Gaz	**Gaz**	**Gaz**	**Gaz**	**Gaz**	**Gaz**	**Gaz**	**Gaz**	**Gaz**	**Gaz**
mt	Ott	Ott	Mam	-	Ott	Mod	Ott	Ott	Mod	Mod
GL	207.7	181.6								
Bp	21.9	18.4	20.2	18.3	17.1	16.7	16.1			
SD	12;0	9.8	11.0				8.8	12.3		10.4
Bd	23.3	19.8						19.7	18.3	18.5

	Gaz	**Gaz**	**Gaz**	**Gaz**	**Gaz**	**Gaz**
ph1	Mod	Mod	Ott	Ott	Ott	-
Glpe	44.3	41.7	41.0	39.5	39.0	34.8
Bp		9.1	8.4	10.2	10.4	
SD		7.0	6.8	7.2	7.7	
Bd		8.5	8.5	9.0	8.7	

	Gaz
ph2	Ott
GL	23.2
Bp	9.8
SD	7.1
Bd	8.3

E. Selected measurements of sheep (*Ovis ammon* f. aries) and goat (*Capra aegagrus* f. hircus)

goat	**radius**				
Mamluk	mean	SD	min	max	n
GL	141.5	6.4	137.0	146.0	2
Bp	30.1	3.5	26.9	39.5	17
Bd	27.4	2.1	25.3	31.4	7
Ottoman	mean	SD	min	max	n
GL	148.3	6.8	138.8	157.0	10
Bp	29.2	3.2	21.3	40.0	107
Bd	28.6	2.8	25.0	37.0	33
Modern	mean	SD	min	max	n
GL	143.3	4.9	140.0	149.0	3
Bp	30.0	3.6	25.2	43.0	51
Bd	28.1	1.9	25.6	33.3	16

sheep	**radius**				
Mamluk	mean	SD	min	max	n
GL	-				-
Bp	34.0	1.9	30.2	36.8	11
Bd	30.2	0.5	29.6	30.6	3
Ottoman	mean	SD	min	max	n
GL	159.0	12.1	143.5	170.1	5
Bp	33.2	2.4	26.8	37.2	30
Bd	31.9	3.5	26.9	38.2	14
Modern	mean	SD	min	max	n
GL	-				-
Bp	33.4	3.0	25.2	39.6	28
Bd	29.0	7.8	3.9	35.4	14

goat	**metacarpus**				
Mamluk	mean	SD	min	max	n
GL	103.7	14.7	95.8	148.0	12
Bp	22.8	2.0	20.0	27.7	20
Bd	25.5	2.0	24.0	30.4	10
Ottoman	mean	SD	min	max	n
GL	100.8	5.6	89.8	112.3	23
Bp	22.5	2.2	17.2	27.4	50
Bd	25.2	2.4	20.5	31.2	29
Modern	mean	SD	min	max	n
GL	98.1	4.4	89.0	107.3	23
Bp	21.6	1.3	18.7	24.7	25
Bd	24.3	1.3	20.5	25.8	20

sheep	**metacarpus**				
Mamluk	mean	SD	min	max	n
GL	133.3	2.8	130.0	135.0	3
Bp	25.9	2.0	22.2	29.6	23
Bd	28.2	2.7	23.9	31.8	6
Ottoman	mean	SD	min	max	n
GL	125.2	14.2	89.6	142.6	15
Bp	24.6	2.1	21.1	30.0	28
Bd	27.5	2.1	23.7	32.7	21
Modern	mean	SD	min	max	n
GL	129.0	1.4	127.0	130.0	4
Bp	26.5	1.7	24.8	29.3	6
Bd	26.8	2.6	22.1	29.3	6

goat	**metatarsus**				
Mamluk	mean	SD	min	max	n
GL	111.7	8.4	104.0	125.6	5
Bp	20.2	3.4	17.3	29.1	12
Bd	23.7	2.3	21.3	27.6	7
Ottoman	mean	SD	min	max	n
GL	108.4	7.0	100.3	131.6	20
Bp	18.4	1.7	13.4	25.2	40
Bd	22.4	1.6	19.9	27.0	27
Modern	mean	SD	min	max	n
GL	-				-
Bp	20.9	3.5	17.5	25.0	5
Bd	-				-

sheep	**metatarsus**				
Mamluk	mean	SD	min	max	n
GL	138.5	4.9	134.5	144.0	3
Bp	21.7	1.5	19.3	24.8	22
Bd	25.4	1.4	24.2	27.3	6
Ottoman	mean	SD	min	max	n
GL	135.8	11.4	115.4	154.5	12
Bp	21.4	1.4	17.7	24.1	36
Bd	25.2	1.6	23.0	28.8	16
Modern					
GL	-				-
Bp	-				-
Bd	-				-